ACOUSTIC SYSTEMS IN BIOLOGY

ACOUSTIC SYSTEMS IN BIOLOGY

Neville H. Fletcher
CSIRO Australia
and Australian National University

New York   Oxford
OXFORD UNIVERSITY PRESS
1992

Oxford University Press

Oxford   New York   Toronto
Delhi   Bombay   Calcutta   Madras   Karachi
Kuala Lumpur   Singapore   Hong Kong   Tokyo
Nairobi   Dar es Salaam   Cape Town
Melbourne   Auckland

and associated companies in
Berlin   Ibadan

Copyright © 1992 by Oxford University Press, Inc.

Published by Oxford University Press, Inc,
200 Madison Avenue, New York, New York 10016

Oxford is a registered trademark of Oxford University Press

All rights reserved. No part of this publication may be reproduced,
stored in a retrieval system, or transmitted, in any form or by any means,
electronic, mechanical, photocopying, recording, or otherwise,
without the prior permission of Oxford University Press.

Library of Congress Cataloging-in-Publication Data
Fletcher, Neville H. (Neville Horner)
Acoustic systems in biology / by Neville H. Fletcher.
p.   cm.   Includes bibliographical references and index.
ISBN 0-19-506940-4
1. Bioacoustics.   I. Title
QP461.F52   1992   591.19' 14—dc20   91-37413

2 4 6 8 9 7 5 3 1

Printed in the United States of America
on acid-free paper

PREFACE

I first became interested in the acoustics of biological systems when I had the opportunity of discussing with some biology colleagues the experiments they were performing on the auditory systems of insects. The neurobiological techniques, with which I was unfamiliar, were immensely impressive, but I found that they tended to discuss the external acoustics of the system only in terms of greatly simplified physical concepts such as the resonances of open or stopped pipes, or the behavior of a simple Helmholtz resonator. Such ideas can often, it is true, provide a useful qualitative guide, but rarely anything more. It seemed to me that physical acoustics should be able to do better for the subject than this, and to provide a description of the peripheral aspects of the auditory system at a level at least comparable with the sophistication of the neurophysiological investigation. A trial foray into the field showed me that, indeed, the physical input could be refined without too much effort, and some productive collaborations developed.

I do not mean to imply by this that all biologists are unsophisticated in the physical aspects of their studies, and indeed I came to admire the blend of careful physical analysis and expert biological experimentation that some workers are able to bring to their studies. In many cases, however, this stems in part from the fact that these individuals, or members of their teams, began life as physicists or engineers before turning to the more complex field of biology. What I have observed, however, is that biologists are singularly poorly served when they search in the library for acoustics texts that could aid them in their work. The selection is between rather elementary and descriptive books on sound that stop at the level of stretched strings and organ pipes, and the fully developed mathematical texts that have been written for graduate level courses in physics departments. The required results are certainly there somewhere, but it is very difficult to dissect out just what is needed and to apply it to understanding the biological problem.

It is with this situation as background that the present book has been written. I see a need among biologists studying auditory communication in animals at the physiological level for an exposition of acoustics directed explicitly towards their needs, covering the necessary ideas from acoustics and showing how these can be applied quantitatively to understand the acoustic periphery of auditory and sound-producing systems. It is only when this relatively simple mechanical part of the system is properly understood that attention can be focused on the underlying physiological processes.

The task immediately presents difficulties, since biologists are not traditionally well trained in mathematics, and a certain amount of mathematics is inevitable if the treatment is to be more than descriptive in a hand-waving sort of way. What I have tried to do, therefore, is to write the book simultaneously on three different levels. For those who want a brief general survey of the field, each chapter begins with a completely non-mathematical Synopsis which summarizes the content and refers to the figures, all of which are designed to be understood even apart from the main text. At the next level, the reader should follow the main text, but need not give close attention to anything but the general shape of the equations involved. This will give a fairly detailed understanding of all the concepts and techniques involved, and will probably suffice for most readers. At the third level, the mathematical arguments should be followed in detail, and the discussion questions at the end of each chapter attempted. Each question has a reasonably detailed solution provided, and serves not just as a formal exercise but also as further discussion of particular cases of biological relevance.

I hope the book will prove suitable for a one-semester course at beginning graduate level for biologists with a general interest in auditory and vocal systems, though the instructor will need to supplement the physical emphasis of the text with appropriately realistic biological examples. Turning its purpose in reverse, it should also be suitable for a similar course at advanced undergraduate level for physics or engineering students, if the instructor is seeking a new approach to a classical field of study.

Since the book is designed as a textbook rather than a research monograph, there is no detailed list of references, and I have not tried in any way to give a survey of the current literature. Instead I have simply given a short and rather general annotated bibliography that refers the reader to standard sources from which additional information or formal detail can be found on topics treated in the book. Acoustics of the sort used here is one of the classical branches of physics, and the treatments of 50 or even 100 years ago remain completely valid—Lord Rayleigh's classic treatise of 1894 is fortunately readily available in reprint—and can still provide guidance to the modern worker! This does not mean, however, that modern approaches have been ignored. Electric network analogs, nonlinearity, and information theory all have their appropriate place.

It is a pleasure to express my thanks to those colleagues here in Australia, in the United States, and in Europe, who have taken time and trouble to educate me in the facts and theories of sensory biology. The blame for any misconceptions expressed in the book must, however, be mine alone. I hope that this book, by making the techniques of physics applied to biological systems more accessible to their students, will help repay that debt. I am particularly grateful to Ken Hill and Jack Pettigrew for reading and commenting upon a draft of the manuscript, and to Suszanne Thwaites for her collaboration in the earlier stages of this work.

Canberra, Australia
December 1991

N. H. F.

CONTENTS

	Common symbols		xi
1	**PHYSICS, BIOLOGY, AND MATHEMATICS**		3
	1.1	Physics and Biology	3
	1.2	Building Blocks and Models	4
	1.3	Appropriately Complex Models	6
	1.4	Mathematics	6
2	**SIMPLE VIBRATORS**		8
	2.1	The One-Dimensional Simple Oscillator	9
	2.2	Choice of Units	11
	2.3	Complex Notation	11
	2.4	Damping	14
	2.5	The Sinusoidally Driven Oscillator	16
	2.6	Impedance and Admittance	20
	2.7	Linearity and Superposition	23
	2.8	Transient Response	24
	2.9	Nonlinearity	26
3	**VIBRATION OF STRINGS AND BARS**		30
	3.1	Extended Systems	31
	3.2	The Wave Equation	31
	3.3	Normal Modes of a String	33
	3.4	Normal Modes and Eigenfunctions	35
	3.5	Damping	37
	3.6	The Sinusoidally Driven String	38
	3.7	String Excitation by a Distributed Force	39
	3.8	Point Admittance and Transfer Admittance	39
	3.9	Bars and Rods	42
	3.10	Sinusoidally Driven Bars	46
	3.11	Imperfectly Clamped Bars—Sensory Hairs	46
	3.12	End-Loaded Bars—Otoliths	47
4	**SENSORY HAIRS AND OTOLITHS**		53
	4.1	Sensory Hairs	53
	4.2	Viscosity and Boundary Layers	54
	4.3	Viscous Force on a Hair	56
	4.4	Sensory Hairs	57
	4.5	Sensory Thresholds	61
	4.6	Otoliths	63
	4.7	Nonlinearity	67
5	**VIBRATION OF MEMBRANES, PLATES, AND SHELLS**		70
	5.1	Extended Surfaces	71
	5.2	Vibration of Rectangular Membranes	71

5.3	Vibration of Circular Membranes	73
5.4	Tapered Membranes	76
5.5	Sinusoidally Driven Membranes	78
5.6	Elastically Braced Membranes	79
5.7	Loaded Membranes	81
5.8	Slack Membranes	83
5.9	Vibration of Plates	83
5.10	Vibration of Shells	85
5.11	Buckling of Shells	86

6 ACOUSTIC WAVES — 89

6.1	Waves	90
6.2	Plane Waves	90
6.3	Sound Pressure and Intensity	94
6.4	Reflection and Transmission at a Boundary	97
6.5	Wave Attenuation	99
6.6	Spherical Waves	100
6.7	Surface Waves on Water	103
6.8	Surface Waves in Solids	104
6.9	Scattering by Solid Objects	105
6.10	Scattering by Bubbles	107

7 ACOUSTIC SOURCES AND RADIATION — 110

7.1	Sound Generation	111
7.2	Simple Spherical Source	111
7.3	Mechanical and Acoustic Impedance	112
7.4	Source near a Reflector	114
7.5	Radiation from a Vibrating Disc	116
7.6	Radiation from a Vibrating Panel	117
7.7	Radiation from an Open Pipe	119
7.8	The Near Field	122
7.9	The Reciprocity Theorem	124
7.10	Acoustic Sources and Power	125
7.11	Underwater Sources	125

8 LOW-FREQUENCY NETWORK ANALOGS — 128

8.1	Electric Analogs	129
8.2	Analogs for Mechanical Components	130
8.3	Levers	134
8.4	Analogs for Acoustic Components	137
8.5	End Correction for a Pipe and an Aperture	141
8.6	The Helmholtz Resonator	142
8.7	The Tympanum	143
8.8	Solution of Networks	147
8.9	Computer Solutions	149

9 LOW-FREQUENCY AUDITORY MODELS — 152

9.1	Constructing Models	153
9.2	The Incident Sound Field—Diffraction	154
9.3	Response and Sensory Threshold	156
9.4	Baffled Diaphragm	157

CONTENTS

9.5	Simple Cavity-Backed Ear	159
9.6	Neural Transducer Matching	162
9.7	Cavity-Coupled Ears	164
9.8	A More Elaborate Model	168
9.9	Laboratory Studies	171
9.10	An Aquatic Auditory System	174
9.11	Conclusions	177

10 PIPES AND HORNS — 178
10.1	Acoustic Elements	179
10.2	Pipes and Tubes	179
10.3	Wall Losses	182
10.4	Helmholtz Resonator	183
10.5	Simple Horns	186
10.6	Directionality of a Horn	194
10.7	Obliquely Truncated Horns	195
10.8	Higher Modes in Horns	196
10.9	Shallow Asymmetric Horns	198
10.10	Hybrid Reflector Horns	201

11 HIGH-FREQUENCY AUDITORY MODELS — 204
11.1	High Frequencies	204
11.2	Horn-Loaded Simple Ear	204
11.3	Pipe-Coupled Ears	208
11.4	Complex Systems	212

12 THE INNER EAR — 215
12.1	Neural Transducers	215
12.2	The Transduction Problem	216
12.3	A Frequency-Dispersive Membrane	217
12.4	Cochlear Mechanics	219
12.5	The Middle Ear	225
12.6	Active Cochlear Response	228

13 MECHANICALLY EXCITED SOUND GENERATORS — 230
13.1	Sound Production	231
13.2	Impulsive Excitation	232
13.3	Frictional Excitation	234
13.4	Harmonics, Mode Locking, and Chaos	238
13.5	Pick and File Excitation	240
13.6	Buckling Tymbals	242
13.7	Conclusions	243

14 PNEUMATICALLY EXCITED SOUND GENERATORS — 245
14.1	Air-Driven Systems	246
14.2	Larynx and Syrinx	247
14.3	Vocal Systems	254
14.4	An Avian Vocal System	258
14.5	Other Vocal Systems	264
14.6	Aerodynamic Systems	265

15 SIGNALS, NOISE, AND INFORMATION — 272
- 15.1 Communication — 273
- 15.2 Time and Frequency — 273
- 15.3 Signals and Noise — 278
- 15.4 Information and Coding — 284
- 15.5 Animal Sonar — 289
- 15.6 Neural Processing — 293
- 15.7 Conclusion — 297

APPENDIXES
- A. Mathematical Appendix — 301
- B. Pipe and Horn Impedance Coefficients — 309
- C. Physical Quantities and Units — 312
- Glossary — 315
- Bibliography — 319
- Index — 325

COMMON SYMBOLS

While the number of different situations discussed in the book requires that the same symbol must be used to represent different things in different places, an attempt has been made to unify the usage, and some symbols have unique meanings. This list specifies the usual meaning of symbols; exceptional usage is local and is explained in the appropriate section of the text.

A, B, C	unspecified constants, amplitudes, etc.
(A)	(as superscript) acoustic, as of impedances
a	radius of a pipe, horn, or diaphragm; with subscript, a mode amplitude
C	compliance; electrical capacitance
c	speed of sound, generally in air, subscripts for other materials
E	Young's modulus
$\mathcal{E}$	efficiency
e	exponential function; $e = 2.7182\ldots$
F, F	force
f	frequency in hertz; acceleration
g	gravitational acceleration
H_{ij}	Z_{ij} coefficients for a horn
I	intensity
i, i	electric current in a network
i, j, k	(as subscripts) integers $1, 2, 3, \ldots$
Im	imaginary part of a complex quantity
J	(with subscript) Bessel function
j	imaginary number identifier, defined by $j^2 = -1$
K	elastic bulk modulus
k	(angular) wave number; $k = \omega/c$
L	level in decibels, with argument I for intensity, p for sound pressure, v for velocity; inertance; electrical inductance
l	length
(M)	(as superscript) mechanical
m, n	(as subscript) integers $1, 2, 3, \ldots$
m	mass; flare constant of an exponential horn
N	(with subscript) Neumann function (Bessel function of the second kind); as subscript, neural transducer

n	integer 1, 2, 3, ...
P	power
P_{ij}	Z_{ij} coefficients for a pipe
$p, \mathbf{p}$	acoustic pressure; with subscript E analog source pressure
Q	quality factor of a resonance
q	acoustic source strength (volume flow)
R	resistance; real part of impedance; as subscript, radiation
r	distance to observing point
Re	real part of a complex quantity
S	area or cross-section area; as subscript, solid, generally biological solid
T	tension; as subscript, tympanum
t	time
$U, \mathbf{U}$	acoustic volume flow
u	vibration velocity of a surface; speed of an air jet
$V, \mathbf{V}$	volume; electric potential difference in a network
$v, \mathbf{v}$	velocity, usually acoustic particle velocity
W	width dimension; as subscript, water
X	reactance; imaginary part of impedance
x, y, z	coordinates or displacements
Y	(acoustic) admittance; superscript (A) for acoustic or (M) for mechanical if necessary
Z	(acoustic) impedance; superscript (A) for acoustic or (M) for mechanical if necessary; subscript R radiation; subscript N neural transducer; subscript T tympanum
Z_{ij}	two-port impedance coefficients for levers, pipes, or horns
z	wave impedance
α	(alpha) attenuation coefficient
β	(beta) spring constant; elastic coefficient
γ	(gamma) resistive loss coefficient; Poisson's ratio; nonlinear coefficient; ratio of specific heats of air
Δ	(capital delta) symbol for a small increment
δ	(delta) boundary layer thickness; symbol for a small increment
ε	(epsilon) nonlinear parameters
η	(eta) coefficient of viscosity
θ	(theta) angle

SYMBOLS

κ   (kappa) radius of gyration of the cross-section of a bar; transverse wave number in a pipe of horn

λ   (lambda) wavelength

μ   (mu) mechanical compliance of the root of a sensory hair; with subscript H, compliance of hair; with subscript B, compliance of bar; elastic shear modulus

ν   (nu) kinematic viscosity, $\nu = \eta/\rho$

ξ   (xi) acoustic displacement

π   (pi) $\pi = 3.14159\ldots$

ρ   (rho) density, generally density of air; ρ_s density of solid biological material

Σ   (capital sigma) summation

σ   (sigma) elastic stress; surface tension; with subscript, parameter ± 1

τ   (tau) time; time interval

φ   (phi) angle, generally phase angle

ψ   (psi) shape function for a normal mode, generally normalized

ω   (omega) angular frequency in radians per second, $\omega = 2\pi f$

ACOUSTIC SYSTEMS IN BIOLOGY

1 PHYSICS, BIOLOGY, AND MATHEMATICS

1.1 Physics and Biology

This book is about the application of physics to the understanding of biological systems, so that it is appropriate to think briefly about the difference between the approaches usually adopted in the physical and the biological sciences. This will serve to explain the background to the approach adopted, which shows physics not as an end in itself but as a tool to be used by biologists.

Biology is immensely complex, and it is in that systematic complexity that much of the interest of the subject lies. Leaving aside ecologists, most biologists work with individual animals (or, of course, plants) or with the subsystems that make up those individuals. A good understanding of the biochemical and biophysical systems that make up a functioning individual has been achieved, and the same is true of many of the features of the neurophysiological system. The biologist makes free use of many physical concepts down to the molecular level, but beyond that the subject ceases to be biology and becomes biochemistry or even physics.

Physics, on the other hand, deals essentially with simple laws and theories that describe the behavior of the building blocks (from the interactions of elementary particles to the structure of space-time) from which the universe is made. The fact that the universe is complicated reflects the complicated pattern in which the blocks are arranged. Philosophically, physics does not claim that it will ever truly understand the nature of the universe, but only that it may succeed in constructing a mathematical model, or hierarchy of models, that predict adequately the behavior of the systems to which they apply. Ultimately we might hope that the phenomenon of life—leaving aside consciousness—might also be predictable from physical principles within this philosophical context, but the time when that prediction will be possible is not yet near. Even if physics were to be successful in this attempt, it would still not usurp the role of biology, nor, on a more philosophical plane, would it necessarily imply a "clockwork" universe. Modern developments in the mathematics of nonlinearity and chaotic behavior preserve us from that unpalatable possibility!

The theories of physics have been developed as a hierarchy, and the choice of one level or another depends upon the degree of sophistication and accuracy that we want in our model. To model the interaction of elementary particles, quantum chromodynamics is essential, but it merges into ordinary quantum mechanics at the atomic and molecular level. This, in turn, gives results that are essentially identical with those of Newtonian mechanics for all ordinary macroscopic phenomena but, if we increase scale and speed, we must use the theories of special and then general relativity. The seams between these domains are not yet entirely smooth, but in principle we could put together a grand theory unifying them all. To use a model in that form, however, particularly for ordinary laboratory-scale applications, would not make sense, for many of the large-scale and small-scale features of the theory would simply turn out to have no effect. The whole art of investigation in physics is to use an appropriately complex theoretical model. If the complexity is too small, we may miss out on modeling important features of the real system, while if it is too great we may find the mathematics too difficult or too tedious to complete.

With these ideas as background, it makes sense to look at the areas of biology, particularly the areas of biological acoustics, to which physical theories and models can make a useful contribution. In doing so, we should again be guided by the hope of finding an appropriately complex model. In fact there are two areas to which physical ideas can immediately contribute. The first is at the behavioral level, where acoustic and vibratory signals often serve as a means of communication among individuals of the same or even of different species. Physical theories can tell us a great deal about the propagation of such signals in the environment, about the information content they can convey, and about appropriate coding schemes to optimize performance. The second area is at the mechanical level in an individual. The way in which sound signals are captured and led to an appropriate neural transducer is simply physical, and we might hope for a physical understanding of the important anatomical and physiological features that have developed under evolutionary pressures. The same is true of the sound-production mechanism, though the analysis here may be more complicated since the system is active, rather than simply passive. All this can be done within the domain of ordinary classical Newtonian mechanics. The next level in the biological organism involves the transduction of mechanical signals to neural form, and is much less straightforward. The physics involved is at the level of quantum mechanics, which is no great difficulty, but the biological mechanisms and membrane processes are so complex that only parts of the system have yet been elucidated in detail.

Our discussion will therefore stop after brief consideration of the mechanical behavior of the transducer organ. An understanding of the mechanical behavior of the auditory system to this level is a good guide to many features of total system behavior, and it does provide for the neurobiologist a specification of the mechanical input to the neural transducer cells. At a more preliminary level, even a partial understanding of the behavior of the external part of the system allows the design of crucial experiments to extend the model. Similar remarks can be made in relation to vocal systems.

1.2 Building Blocks and Models

Physics proceeds by studying the simplest possible building blocks, because for these the mathematical analysis is simplest and the physical behavior is most easily defined. When the behavior of the simplest cases has been understood, then we can go on to more complex cases until we have enough building blocks to build a conceptual approximation to the system we wish to study. The physical behavior of this conceptual model can then be evaluated quantitatively by solving the equations describing its components and their interactions.

That is the plan we shall follow in this book. The very simplest component is a massive particle, constrained to move only along a straight line and bound to a fixed point on that line by a force that is proportional to the displacement of the particle from that point—the so-called simple-harmonic oscillator. The analysis, we shall find, is very simple, and indeed it is probably familiar to all readers. Nevertheless the behavior of this oscillator when acted upon by external forces shows many of the features that turn out to be important in more complex systems. Even the mathematics has a strong family resemblance, so that the more complex formulae met later have an element of familiarity.

This one-dimensional point oscillator is a reasonable model for only one biological component—the mass-loaded hair cell; in all other cases the biological system has a significant extension in space. The first move towards this is to discuss the behavior of a stretched string. Once again, this is the simplest such system to analyze, and it shows up many of the features that are important in more complex and realistic system elements. A string stiffened to the extent that it becomes a rod is closer to a biological system component, so this is considered next. From here we move on to the behavior of taut membranes—certainly important biological components—and stiffened membranes, which behave more like plates or elastic shells. This provides an adequate stock of mechanical components from which to construct our model system.

We then focus attention on the medium with which the vibrating element interacts and through which the sound or vibration is transmitted. The media of greatest biological importance are air and water, and we consider waves in both these fluids, as well as waves that travel along the surface of water. The other major components of biological acoustic systems are air-filled tubes, horns or cavities, generally connecting in some way with the outside environment. Understanding and modeling the way in which sound propagates through these components is an important part of understanding the whole system.

This list provides an adequate stock of components to model most of the systems in which we are interested. Our understanding of each component must, of course, be quantitative, since we now wish to assemble a selection of the components into a simplified model representing the real biological system, and to calculate its acoustic behavior. The rules for this model building are mostly intuitive, but a few precautions must be observed and, of course, we need to have available some simple way of carrying out the necessary final calculations to determine important elements of the acoustic behavior such as sensitivity, frequency response, and directionality. We consider this for a variety of model systems, closely related to those found in the peripheral auditory apparatus of insects and vertebrates. We also give a brief discussion of the neural transducer organ from a mechanical point of view, but stop short of considering transduction at the cellular level.

In the case of active sound-producing systems, the model building proceeds in much the same way, but we supply a source of steady energy, often in the form of a supply of air under a small excess pressure. An important extra feature, the necessity for nonlinearity in at least one of the system components, enters when we begin to analyze such an active system to determine its acoustic output. We will have met nonlinearity in the earlier discussion, but now we will put our knowledge to use.

Finally, we examine the external environment in which acoustic communication takes place: the propagation of sound in the atmosphere and the ocean, the competing sources of interfering noise, and the coding strategies that animals use to transmit information. This is a large subject—the whole basis of modern communications theory—and a rather brief survey must necessarily suffice.

1.3 Appropriately Complex Models

The whole art of applying physical ideas to the analysis of biological systems rests on constructing an appropriately complex model for the system under study. Experience acts as a guide here, but the general philosophy should be to construct first the simplest model that appears to be a possible representation of the system. This will generally have a relatively small number of components connected together quite simply, the physical dimensions, densities and elastic properties of all these elements being reasonably well known. The behavior of such a simple model will be easy to calculate and to compare with such experimental data as are available. When this has been done, refinements can be added to improve the agreement, again constrained by anatomical and physical information, and their effect evaluated. Any refinements that make negligible difference to the system behavior can be discarded.

In the course of the book we shall try to give some guidance, by example, on the way this should be done. A model must often be based, however, on some sort of physical or biological intuition, and the process of analysis allows this intuition to be critically tested through additional measurements. In this way, we hope that the procedures outlined will be of value to active researchers in the field, as well as to those simply seeking a general understanding.

1.4 Mathematics

The object of this book is to show how we can construct simplified models that will help us to understand the behavior of biological systems. It would be possible to build these models in physical reality and then to measure them. Though this would be easier than making measurements on biological systems, it would tell us little more and, every time we wished to vary a physical parameter to see its effect on the behavior of the system, we would have to build a new model. For this reason all our models are mathematical—they are quick to build, trivially simple to change, and calculation provides predicted behavior over very wide ranges of frequency or other parameters.

This means, however, that we must be prepared to use mathematics at every stage of the development, and this may not come naturally to many people trained in the biological sciences. In writing the book, therefore, I have assumed very little in the way of prior mathematical knowledge or technique apart from elementary algebra and the ability to differentiate and integrate, and I have introduced only those mathematical ideas that are essential for analyzing the acoustic models. When a mathematical technique is needed, it is developed at that point, so that its use is clear, and a few practice examples are given at the end of the chapter. For convenience of reference, some relevant mathematical results are collected in Appendix A.

The same comments apply to computers. In the present context they are simply machines for quickly doing arithmetic, and we should recognize that there are powerful numerical procedures that can help us in the calculation. The final stage of evaluating a model will almost always be the calculation and graphical display of its performance, giving, for example, membrane vibration amplitude as a function of frequency or of the direction of sound incidence. Most of these calculations are quite straightforward for anyone with a modest background in computer programming in a language such as Basic, and require only a small desk-top microcomputer.

Finally, a word of caution. With modern computers, and even hand-held calculators, it is very easy to arrive at results to many decimal places, and it is tempting to believe that this precision is significant. For the simple entities of physics—electrons, atoms, electromagnetic waves, and so on—this is often true, for our mathematical models take proper account of all the features of the system. For the complex systems of biology, however, the models with which we are working are only first approximations to the real world, and many subtle factors have been omitted in the interests of simplicity. Even anatomical shapes, and basic physical quantities such as density and elastic moduli for many biological materials, are not well defined in many cases. While, therefore, we hope that our models are able to mimic and explain the behavior of the real system, we should not expect very close numerical agreement. In general, agreement to within 10% would be extremely good, while agreement to within a factor 3 in either direction (or ±5 dB) might be completely acceptable, provided the qualitative behavior of the system is predicted over a wide frequency range. With these limitations in mind, it is rarely useful to give numerical results to more than two significant figures, and often one figure will suffice. This convention has been adhered to in the text.

2 SIMPLE VIBRATORS

SYNOPSIS. The simple vibrator, or simple harmonic oscillator, is perhaps the most important system in all of physics. In particular, it is vital for the understanding of acoustic phenomena, and for this reason we treat it in some detail. Essentially it consists of a small mass, able to move only along a line and bound to a fixed point on that line by a spring that exerts a restoring force proportional to the displacement. Such a system vibrates about the fixed point with a natural frequency ω_0 which is determined by the mass of the particle and the strength of the spring. This behavior is illustrated in Fig. 2.1.

In all real oscillators there is dissipation of energy, the most usual mechanism in simple systems being losses caused by the viscosity of the medium in which the mass moves. Even air has appreciable viscosity. The viscous force is proportional to the speed of the motion, and we can define the constant of proportionality, divided by twice the mass, to be the damping constant α of the oscillator. The oscillations of a damped vibrator decay with time as shown in Fig. 2.2. We define the quality factor, or Q value, of an oscillator to be $\omega_0/2\alpha$.

The response of an oscillator to an external force of frequency ω depends on its natural frequency and Q value as shown in Fig. 2.3 for displacement and in Fig. 2.4 for velocity. The response is a maximum when $\omega = \omega_0$, a phenomenon called resonance. The resonance is sharply defined in frequency if the Q value is high, but becomes very broad for Q values near 1 or less. Indeed the full-width of the resonance at points where the amplitude has dropped to $1/2^{1/2} = 0.707$ of its peak value is just ω_0/Q. The Q values for biological oscillators typically range from about 1 to 10. The phase angle for the response is also shown in these two figures. If the phase angle for displacement is zero, then the displacement exactly follows the applied force. If the phase angle is negative, then the displacement lags behind the force. Similar remarks apply to the phase of the velocity. The velocity is always 90° ahead of the displacement in phase. It is often useful to plot the displacement or velocity response on logarithmic scales, as shown in Fig. 2.5.

In discussing vibrating systems it is very helpful to introduce complex numbers, written here in ***bold italic*** type. The imaginary part of a complex number is labeled with the symbol j, which is defined to have the property $j^2 = -1$. A complex quantity has real and imaginary parts, or equivalently a phase and an amplitude, so that its connection with vibrations is formally very close. In interpreting complex quantities representing physical variables such as velocity or position, we always take the real part of the quantity.

We define the admittance $\boldsymbol{Y}^{(M)}$ of a mechanical system to be the ratio of the velocity response to the exciting force. The admittance of a simple oscillator thus has the form shown in Fig. 2.4, with a sharp maximum at the resonance frequency. Because the velocity differs in phase from the exciting force, except at the resonance frequency, the admittance is a complex quantity with both a magnitude and a phase angle. Another way to express this is to split the admittance into the sum of two parts, one exactly in phase with the force and one 90° ahead of the force in phase. We refer to these as the real and imaginary parts of the admittance. Fig. 2.6 shows how they behave. The real part is always positive, but the imaginary part may change sign. The mechanical impedance $\boldsymbol{Z}^{(M)}$ is defined to be the ratio of the force to the velocity, and is thus the reciprocal of the mechanical admittance. Though the magnitude of the impedance is just the reciprocal of the magnitude of the admittance, the same is not true individually for the real and imaginary parts, because of complications caused by phase changes. The real and imaginary parts of both admittance and impedance, for a simple oscillator, are shown in Fig. 2.6.

Often in biological systems we are concerned not with steady signals but rather with transients. The response of an oscillator to a transient force can be calculated. The simple case of the behavior of an oscillator with natural frequency ω_0 and quality factor $Q = 10$ in response to an abruptly applied sinusoidal force of frequency ω is shown in Fig. 2.7. It takes a time equal to about 10 cycles of the natural oscillation before a steady-state response is reached. In general the duration of the transient is about Q cycles of the natural frequency. A similar transient occurs when the force is switched off.

In quite a general way we conclude that oscillators with low damping and therefore high Q values give a large response at resonance but have a narrow frequency bandwidth for that response. They also have transients of long duration. In contrast, oscillators with low Q values have good transient response and relatively wide bandwidth, but are handicapped by low sensitivity.

2.1 The One-Dimensional Simple Oscillator

Much of the theory of acoustics, and indeed of many other branches of physics, is based upon the behavior of a mass connected by a simple linear spring to a fixed point and constrained to move only along a line. This is often called a one-dimensional simple-harmonic oscillator. Most of this chapter will be devoted to examining the behavior of this simple system, not because anything exactly like it occurs in biological systems, but because it is often a good first approximation, and the mathematics of more complex models is usually quite similar.

In the simplest case, let us suppose that the mass of the oscillating particle is m and that its displacement from the fixed point $x = 0$ along a straight line is measured by the coordinate x. Let the elastic force in the x direction when the particle is at position x be $F_E = -\beta x$, where β is a constant identified with the stiffness of the spring, so that the force is always directed back towards the origin $x = 0$. Then the motion of the particle is described by the law that the force is equal to the mass times the acceleration or, in calculus notation,

$$m\frac{d^2x}{dt^2} = -\beta x. \tag{2.1}$$

This differential equation tells us all about the behavior of the system if we know the position and velocity of the particle at some initial time, usually taken as $t = 0$. Let us see how this comes about.

We can guess a formal solution to (2.1) based on our knowledge of the properties of elementary functions. If we try

$$x = a \cos(\omega_0 t + \theta) \tag{2.2}$$

then this satisfies the equation provided that

$$\omega_0 = (\beta/m)^{1/2}. \tag{2.3}$$

No condition is placed upon the other parameters a and θ. This gives us a formal solution for the motion, but we must look to see what it means. (An equally good

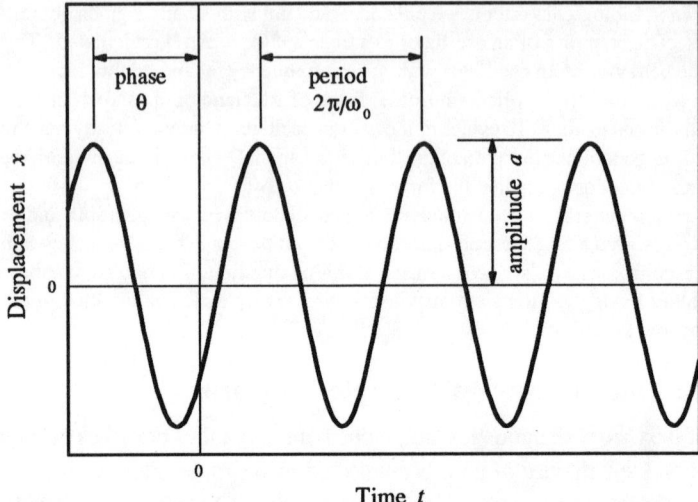

Figure 2.1 The sinusoidal displacement of a simple oscillator as a function of time. Because we have chosen a cosine representation, the phase angle θ is defined to be such that the displacement is $a \cos \theta$ at $t = 0$.

solution could have been obtained with a sine rather than a cosine function, as we discuss at the end of this section.)

Figure 2.1 shows a plot of the behavior of the particle displacement x with time t. It undergoes a regular repetitive oscillatory motion with period $2\pi/\omega_0$ and amplitude a. The number of periods executed in each unit time is

$$f = \omega_0/2\pi \qquad (2.4)$$

and f is called the frequency of the vibration and is measured in hertz (previously called "cycles per second") after the German physicist Heinrich Hertz (1857–1894), celebrated for his experimental discovery of radio waves. The quantity ω_0 is called the angular frequency of the natural oscillation, and is measured in radians per second. Clearly, from (2.3), the frequency of the motion increases if the particle is made lighter or the spring stiffer—a physical insight that we need to keep in mind. Note that, if we know the mass m and measure the free oscillation frequency ω_0, then (2.3) gives us the spring stiffness β. Finally, θ is called the phase of the oscillation and indicates in just what part of the cycle the system is at $t = 0$. The phase is measured in radians, but can be referred to in degrees provided we remember to convert back to radians for calculations.

Since there are still two parameters a and θ to be determined, we need to know two independent pieces of information about the system. These might be

SIMPLE VIBRATORS

the displacement x_0 and the velocity v_0 at time $t = 0$. Substituting these values in the general solution (2.2), we easily find that

$$a^2 = x_0^2 + \left(\frac{v_0}{\omega_0}\right)^2 \quad \text{and} \quad \tan\theta = -\frac{v_0}{\omega_0 x_0}. \tag{2.5}$$

This then specifies the solution completely.

We remarked above that we could have carried out this discussion using in (2.1) the trial function $a \sin(\omega_0 t + \theta')$ rather than a cosine function, and indeed the interpretation of the solution would have been just the same. The only difference would have been in the phase constant or, equivalently, in the choice of the zero of time. The two solutions are identical if we choose $\theta' = \theta + \pi/2$. It is simply a matter of convenience which form we use and, in this book, we shall keep consistently to the form (2.1).

2.2 Choice of Units

It is possible to carry out calculations in acoustics in a variety of different systems of units, provided we are absolutely consistent about it. Older books often use the c.g.s. system and, indeed, we may feel that its fundamental units are of more nearly appropriate size for problems in sensory biology than are other systems. There are, however, considerable advantages to be gained from consistent use of the SI system, now standard in almost all of science. Indeed, when dimensions are given in microns (micrometers) and masses in milligrams, it is just as easy to convert to meters and kilograms as to anything else. We have therefore used the SI system exclusively in this book, and we must warn against the random use of other units if the results of calculations are to be correct. Details of these units are given in Appendix C.

In the case of the simple oscillator it is easy to think of the mass and convert it to kilograms. The force constant β for the spring is less familiar, however. Its value must be given in the appropriate SI units, which are newtons per meter, since the force F_E is in newtons and the displacement x in meters. The examples at the end of the chapter give some practice in performing the necessary calculations.

2.3 Complex Notation

We now discuss a mathematical device that may seem unnecessarily complicated at this stage, but that leads to immense simplification in all our later discussion. It is therefore worthwhile to spend a little effort to master the ideas. It is as well to remember, from the outset, that the development is a piece of pure mathematics in which we explore the consequences of particular formal assumptions, so that we should not be worried by the assumption of "imaginary" numbers with particular unusual properties—they exist only within the framework of the mathematical theory that defines them. More detail is given in Appendix A.

We know that the exponential function $\exp nt$ or e^{nt} can be differentiated with respect to t to give

$$\frac{d}{dt}e^{nt} = ne^{nt}. \qquad (2.6)$$

Let us try $x = ae^{nt}$ as a solution to equation (2.1). Substitution and use of (2.6) twice shows that this is a satisfactory solution provided that

$$n^2 = -\beta/m = -\omega^2. \qquad (2.7)$$

The problem is that the right-hand side of this equation is negative, while if n is an ordinary number then its square is necessarily positive.

We can get over this difficulty, and in doing so extend the domain of mathematics, by introducing a symbol j defined by the relation

$$j^2 = -1. \qquad (2.8)$$

Clearly j cannot be an ordinary number, but is a new sort of number that has been traditionally called an imaginary number. There is a related imaginary number i defined by

$$i = -j \qquad (2.9)$$

which similarly satisfies the equation $i^2 = -1$. We could develop our discussion in terms of either i or j. Traditionally physicists and mathematicians have used i, while engineers have used j, and books on acoustics may use either. In this book we shall use j, because some of our methods will turn out to be closely related to those of electrical engineering, but it is possible to convert any result to the other convention by simply writing $-i$ for j wherever it occurs.

Equation (2.8) contains nearly all we need to know about imaginary numbers. Mathematics is a formalism that follows from basic definitions such as this equation, and all we need to do is follow the rules. Since imaginary numbers such as j or $3.7j$ are different from ordinary numbers such as 1 or 3.7, they must be kept separate from them. A general number, traditionally called a complex number, has a real part and a complex part and is written $a + jb$, for example. In this book, to avoid confusion, we shall print the symbols for all complex numbers in ***bold italic typeface***. Other books may not use this convention, relying upon the context to indicate whether the number is real or complex, or may use a different convention. Complex numbers can be added or multiplied, simply remembering to apply the definition (2.8). Complex numbers can also be divided one into the other, but the details are a little more complicated and are set out in Appendix A.

SIMPLE VIBRATORS

It turns out, as again we show in more detail in Appendix A, that there is a close relationship between the ordinary trigonometric sine and cosine functions and the exponential function with an imaginary argument. The connection is expressed by the result

$$e^{j\theta} = \cos\theta + j\sin\theta \qquad (2.10)$$

where we take θ to be real. This leads to the inverse relations

$$\cos\theta = \frac{e^{j\theta} + e^{-j\theta}}{2} \quad \text{and} \quad \sin\theta = \frac{e^{j\theta} - e^{-j\theta}}{2j}. \qquad (2.11)$$

Finally we note a further definition, that of the magnitude or absolute value of a complex number. If $z = a + jb$, then the magnitude of z, written as $|z|$ or simply as z, is defined to be

$$|z| = z = |a + jb| = (a^2 + b^2)^{1/2}. \qquad (2.12)$$

This is almost as though real and imaginary numbers are geometrically at right angles to one another, and indeed a diagram in which complex numbers are drawn on a plane with the x axis representing their real component and the y axis their imaginary component is often useful. This too is discussed in Appendix A. From this definition (2.12) and (2.10) it follows that

$$|e^{j\theta}| = (\cos^2\theta + \sin^2\theta)^{1/2} = 1 \qquad (2.13)$$

for any real θ.

This digression into complex number notation finally allows us to write formal solutions for the displacement x and velocity $v = dx/dt$ in the simple oscillator problem, which is defined by the equation (2.1), as

$$x = ae^{j(\omega_0 t + \theta)} \qquad v = j\omega_0 x = a\omega_0 e^{j(\omega_0 t + \phi)} \qquad \phi = \theta + \pi/2. \qquad (2.14)$$

The second form for v comes about from (2.10), from which we see that $e^{j\pi/2} = j$. These complex expressions, because they contain j, are mathematical constructs that require interpretation when applied to physical systems. Indeed they have no physical meaning at all until we have decided how this interpretation is to be made.

Comparing the first of (2.14) with (2.10), we see that the complex expression contains two versions of the simple oscillation (2.2) that we identified as the physical solution to the simple oscillator problem. The real part gives the cosine version and the imaginary part the sine version, though with the same value of θ so that the two are always 90° out of phase. We need only one of these versions, and we adopt the convention of always choosing the real part to represent the physical quantity. We could just as easily have chosen the imaginary part, with the symbol j removed, but the convention is to take the real part, just as we chose a cosine rather than a sine in (2.2). This interpretation now allows us to identify a as the amplitude, ω_0 as the angular frequency, θ as the phase angle of the displacement, and ϕ as the phase angle of the velocity. We shall usually write all oscillatory quantities in forms similar to (2.14), with the time variation $e^{j\omega t}$ included. Sometimes, however, it is convenient to omit this factor and refer to what remains, for example $a = ae^{j\theta}$, as a complex amplitude.

The reason for adopting what looks like a great deal of mathematical sophistication only to arrive back at our original simple result is that it saves us a great deal of difficulty when we must treat more complicated physical situations. If we represent physical quantities as complex numbers, as in (2.14), then the single symbol x or v includes information about amplitude, frequency, and phase, and these are all dealt with correctly and automatically by the procedures of complex algebra. It is easy, for example, to multiply together two or more exponential functions, since we simply add their arguments, whereas the multiplication of two or more cosine functions generates sines and cosines of sum and difference angles, which are very difficult to keep track of.

In the sections that follow, and indeed throughout this whole book, we shall come to appreciate the power and simplicity of this complex notation applied to problems involving waves and vibrations. It is very far from being just a demonstration of mathematical sophistication.

2.4 Damping

To bring the simple oscillator a little closer to physical reality, we must allow for some imperfections in its behavior. The most important of these is the introduction of energy losses, either through imperfection in the spring or because of viscous losses in the medium through which the mass is moving. For our present purposes it does not matter which of these forms of loss we consider, for in each of them the force opposing the motion is proportional to the velocity of the oscillating particle. This is quite different from the loss forces arising from sliding friction, which are nearly independent of all but the sign of the velocity until the particle is brought to rest.

We can represent the viscous drag forces by a term $F_R = -\gamma \, dx/dt$ added to the right-hand side of the equation of motion (2.1) so that, after rearrangement, it can be written

$$m\frac{d^2x}{dt^2} + \gamma\frac{dx}{dt} + \beta x = 0. \tag{2.15}$$

To solve this equation we assume a solution in the form of an exponential, $x = ae^{nt}$ where a is real and n may be complex. If we substitute this into (2.15), then we

SIMPLE VIBRATORS

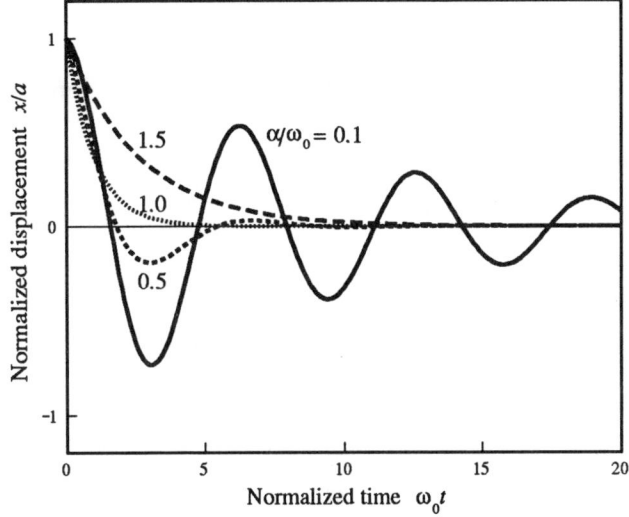

Figure 2.2 The decay behavior of a simple damped oscillator, with natural frequency ω_0 and damping constant α, as a function of the normalized damping α/ω_0. For small damping ($\alpha/\omega_0 < 1$), the decay is oscillatory; if $\alpha/\omega_0 = 1$ the damping is critical and the displacement tends to zero in the shortest possible time; if $\alpha/\omega_0 > 1$ the return to zero displacement is prolonged.

can simply divide out the amplitude a and we are left with a quadratic equation for n, which is simply solved to give $n = j\omega' - \alpha$ where

$$\omega' = \omega_0 \left[1 - \frac{\alpha^2}{\omega_0^2} \right]^{1/2} \qquad \alpha = \frac{\gamma}{2m} \qquad (2.16)$$

and $\omega_0 = (\beta/m)^{1/2}$ as before. The motion is then given by

$$x \approx a e^{-\alpha t} e^{j\omega' t} \qquad (2.17)$$

and α is called the damping coefficient of the oscillator. If the damping is small, in the sense that $\alpha/\omega_0 \ll 1$, then $\omega' \approx \omega_0$, but if the damping is large the frequency may be significantly lowered. If $\alpha/\omega_0 = 1$, the damping is termed critical, and the mass returns to rest without any oscillation at all. For damping greater than the critical value, the return to rest is slower still. This behavior is illustrated in Fig. 2.2.

All real oscillators are damped to some extent, and the damping of springs made of biological material is generally very much higher than that of the metal springs found in mechanical systems. We defer more detailed comparisons until we have a model that is more directly related to a real system.

2.5 The Sinusoidally Driven Oscillator

A damped oscillator simply settles down to its stationary equilibrium position after a time long compared with the reciprocal of the damping coefficient α. The situation is then completely uninteresting unless something happens to excite the oscillator into motion again. In the next few sections we examine how this happens, starting with the simplest case in which the oscillator settles down into a steady-state vibration under the influence of a sinusoidal external force.

Suppose the oscillator has mass m, restoring force constant β and damping γ, so that its natural frequency is $\omega_0 = (\beta/m)^{1/2}$ and its damping constant $\alpha = \gamma/2m$. Let the external force have magnitude F and frequency ω so that, using complex notation, we can write it as $F e^{j\omega t}$ with the zero of time chosen so that the phase of the force is zero. The equation of motion is

$$m\frac{d^2x}{dt^2} + \gamma\frac{dx}{dt} + \beta x = F e^{j\omega t} \qquad (2.18)$$

or, dividing by m,

$$\frac{d^2x}{dt^2} + 2\alpha\frac{dx}{dt} + \omega_0^2 x = (F/m)e^{j\omega t}. \qquad (2.19)$$

In the steady state, it is obvious that the displacement x must vary sinusoidally with frequency ω, and we need to find the amplitude of that motion and its phase relative to the exciting force. To do this, we assume that $x = a e^{j(\omega t + \theta)}$ as usual and substitute this in (2.19). Dividing out the common factor $e^{j\omega t}$ then gives the result

$$a e^{j\theta} = \frac{(F/m)}{(\omega_0^2 - \omega^2) + 2j\omega\alpha}. \qquad (2.20)$$

To find the amplitude a we take the absolute value of each side of this equation, giving

$$a = \frac{(F/m)}{\left[(\omega_0^2 - \omega^2)^2 + 4\omega^2\alpha^2\right]^{1/2}} \qquad (2.21)$$

To find the phase θ involves a little more algebra, but simplifies to

$$\tan\theta = \frac{-2\omega\alpha}{\omega_0^2 - \omega^2}. \qquad (2.22)$$

Here θ is the phase angle of the displacement x relative to the exciting force F.

This somewhat involved bit of algebra has given us the results we need in the form of equations (2.21) and (2.22). We now need to see what they mean in physical

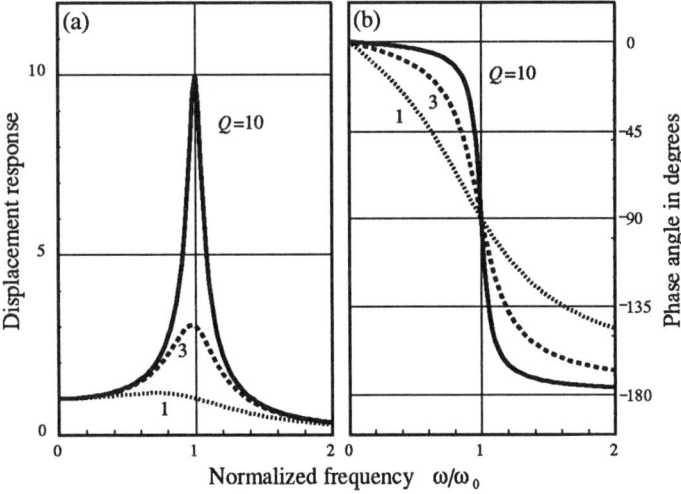

Figure 2.3 Response of a simple oscillator, with natural frequency ω_0 and Q value as shown, to an exciting force of constant amplitude as the frequency ω is varied. The first panel shows the displacement amplitude response a, and the second panel the phase response θ of the displacement.

terms. The most obvious question to ask of the present system is: what is the amplitude of the motion as the frequency is varied, keeping the magnitude of the external force constant? The answer is given by equation (2.21), and typical response curves are shown in Fig. 2.3(a). As is clear from the equation, the amplitude is a maximum when $\omega = \omega_0$, a condition known as resonance. The amplitude response at low frequencies tends to a constant value, while at high frequencies it declines steeply as $1/\omega^2$. Two curves are plotted in the figure, one for $\alpha = 0.1$ and one for $\alpha = 0.3$.

Another way in which to display these results, and one that is more often used, is to plot not the displacement amplitude a but rather the velocity amplitude $v = \omega a$. The change to the formula (2.21) is obvious, simply amounting to multiplying by ω, and the effect on the response curves is shown in Fig. 2.4(a). The response maximum is still very close to ω_0 but the curve has now a nearly symmetrical bell shape, going to zero in the limit of both low frequencies, where the response is proportional to ω, and of high frequencies, where the response varies as $1/\omega$.

To describe these curves, and particularly their variation as the damping is changed, it is usual to define a quality factor, denoted by Q, for the resonance. The definition relates to the amplitude or velocity response curves and, provided the damping is not too large, the Q value is defined to be the ratio of the frequency of the peak response divided by the full-width $\Delta\omega$ of the curve at the two points where its amplitude is $1/\sqrt{2} \approx 0.707$ times the peak value (also called the half-power full

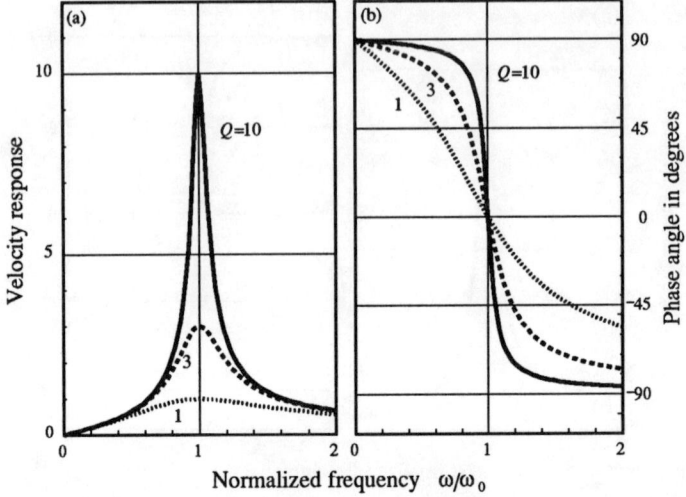

Figure 2.4 Response of a simple oscillator, with natural frequency ω_0 and Q value as shown, to an exciting force of constant amplitude as the frequency ω is varied. The first panel shows the velocity amplitude response v, and the second panel the phase response ϕ of the velocity.

width, since power is proportional to the square of the amplitude). This leads immediately to the result

$$Q = \frac{\omega_0}{\Delta\omega} = \frac{\omega_0}{2\alpha}. \qquad (2.23)$$

In the case of very large damping $\Delta\omega$ becomes uncertain, and so Q is defined quite generally as $\omega_0/2\alpha$. The situation of critical damping, referred to in Fig. 2.2 in relation to the decay of free vibration, corresponds to $\alpha/\omega_0 = 1$ and thus to $Q = 0.5$. The Q value for a vibrator is found by measuring the frequency response curve and dividing the resonance frequency ω_0 at the response peak by the peak full-width $\Delta\omega$.

If we plot the displacement and velocity response curves using logarithmic scales for both frequency and response axes as shown in Fig. 2.5, then the behavior is seen to be particularly simple. The "skeleton" of the response is the two straight lines, shown dotted in the figure. Superimposed upon their intersection is a peak of height Q times the value of the response at the intersection, the width of the peak being inversely proportional to Q. We shall often use this sort of logarithmic plot in the later development.

In biological systems, the oscillators we meet have relatively small Q values, in the range from 1 to about 30. Metallic vibrating systems such as tuning forks

SIMPLE VIBRATORS

or gongs may have Q values of several hundreds, while specially prepared systems such as vibrating quartz crystals can have Q values of many thousands. Systems with high Q values have sharp response peaks, giving large response at the peak frequency, but small bandwidth. Conversely, systems with low Q values have modest peak response but broad bandwidth. In fact there is always a trade-off as expressed by the relation

$$(\text{peak response}) \times (\text{bandwidth}) = \text{constant}. \tag{2.24}$$

We shall later find that a law of this kind is quite general.

We can look at the phase response $\theta(\omega)$ or $\phi(\omega)$ of the oscillator in much the same way, using equation (2.22). As far as the displacement response is concerned, $\theta \to 0$ at low frequencies, so that the displacement is exactly in phase with the force. At the resonance $\omega = \omega_0$, $\theta = -\pi/2$ or $-90°$, and at very high frequencies $\theta = -\pi$ or $-180°$. The phase change takes place essentially over the width of the response peak near ω_0, as shown in Fig. 2.3(b). The phase $\phi(\omega)$ of the velocity response has a similar variation but, because $v = j\omega x$ and $j = e^{j\pi/2}$, we find that $\phi = \theta + \pi/2$. This means that the velocity is exactly in phase with the external force at the resonance frequency ω_0. The variation in phase for the velocity response is shown in Fig. 2.4(b).

Figure 2.5 Logarithmic plot of (a) the displacement response and (b) the velocity response, for a simple oscillator, with natural frequency ω_0 and Q value as shown, to an exciting force of frequency ω. Note the straight dotted lines of the skeleton that show the asymptotic response for ω well away from ω_0. The peak response at resonance is Q times the skeleton response.

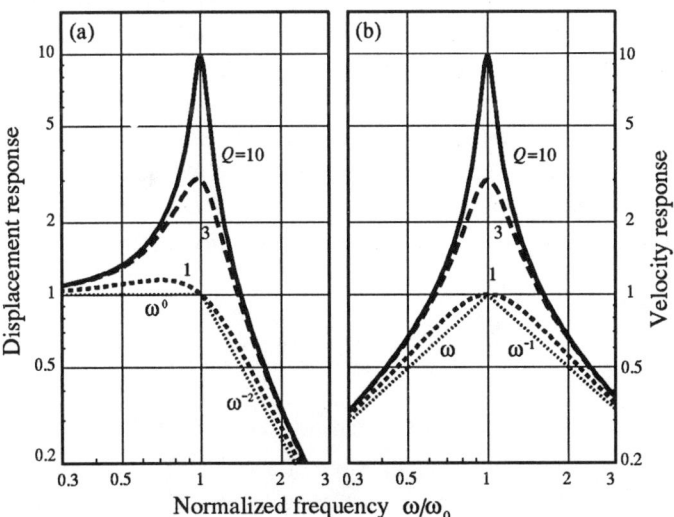

2.6 Impedance and Admittance

We now define two quantities which, in various guises, will be immensely important in our development of the subject. It is clearly necessary to be able to describe the response of a system to an applied force, and this is most usefully specified for oscillatory systems by supposing that the force has oscillatory behavior with frequency ω and amplitude F. It turns out to be best to deal with the velocity response v of the system, rather than the displacement response, so that altogether we can write

$$F = Fe^{j\omega t} \quad \text{and} \quad v = ve^{j\phi}e^{j\omega t}. \tag{2.25}$$

We define the mechanical admittance $Y^{(M)}(\omega)$ at frequency ω to be the complex quantity

$$Y^{(M)}(\omega) = \frac{v}{F} = \frac{v}{F}e^{j\phi} \tag{2.26}$$

with the superscript (M) implying "mechanical" to differentiate this admittance from acoustic admittance, which we shall meet later. From (2.20), an explicit expression for the admittance is

$$Y^{(M)}(\omega) = \frac{j\omega}{m[(\omega_0^2 - \omega^2) + 2j\omega\alpha]}. \tag{2.27}$$

The magnitude $|Y^{(M)}|$ of the mechanical admittance is, except for a constant factor, the quantity illustrated in Figures 2.4 and 2.5(b). It measures how easy it is to cause the mass of the oscillator to move at a given frequency. The admittance itself is, however, a complex quantity. It does not have a physical existence like a pressure or a displacement, but is rather an operator that converts one such physical quantity (force) into another (velocity). Because of its complex nature it generally introduces a phase shift into this conversion, so that this phase shift appears between the two physical quantities that it connects. Note that at resonance, when $\omega = \omega_0$, the admittance as given by (2.27) is a real quantity and, by (2.26), $\phi = 0$ and the velocity is in phase with the force.

The mechanical impedance is a related quantity, just the reciprocal of the admittance,

$$Z^{(M)}(\omega) = [Y^{(M)}(\omega)]^{-1} = \frac{F}{v}. \tag{2.28}$$

and is often used for convenience in some contexts. $Z^{(M)}(\omega)$ is given explicitly by taking the reciprocal of the complex expression in (2.27). The impedance, too, is real at the resonance frequency ω_0 but has a minimum rather than a maximum

SIMPLE VIBRATORS

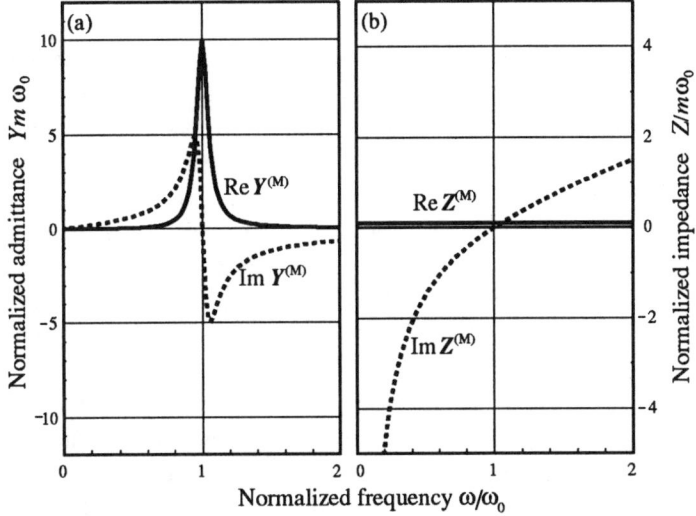

Figure 2.6 Real and imaginary parts of (a) the admittance $Y^{(M)}(\omega)$ and (b) the impedance $Z^{(M)}(\omega)$ for a simple oscillator of mass m and natural frequency ω_0 for the case $Q = 10$.

value there. We shall have occasion to use both impedance and admittance at different times, and their conversion involves no difficulty. Remember, however, that one cannot simply take the reciprocal of the real or imaginary part of the admittance to obtain the real or imaginary part of the corresponding impedance; the simple formula for dividing by a complex quantity, as set out in Appendix A, must be used.

If we plot the absolute magnitude of the admittance using logarithmic scales for both magnitude and frequency, then we have a curve that is identical to that in Fig. 2.5(b), except for the scale associated with the magnitude axis. It is also important to note that, since $|Z^{(M)}| = |Y^{(M)}|^{-1}$, we can just invert the figure if we wish to plot the magnitude of the impedance.

It is often convenient to think of admittance and impedance in terms of their real and imaginary parts, rather than as an amplitude and a phase. This simply involves using the relation $e^{j\phi} = \cos\phi + j\sin\phi$ and plotting the two parts separately. This has been done for the admittance in Fig. 2.6(a) and for the impedance in Fig. 2.6(b). In the case of physical quantities such as force or velocity, our convention is that only the real parts of the complex quantities F and v have physical meaning. The admittance $Y^{(M)}$ is not a physical quantity in the same sense, however, since it is derived by dividing one physical quantity v by another, F. To see the implication of the complex nature of $Y^{(M)}$, let us write it as $Y^{(M)} = G + jB$, where G and B are both real. We can therefore write the relation between force and velocity in the form

$$v = Y^{(M)}F = (G + jB)F(\cos \omega t + j \sin \omega t)$$
$$= (G \cos \omega t - B \sin \omega t)F + j(B \cos \omega t + G \sin \omega t)F. \quad (2.29)$$

Now we take the physically meaningful part of each side to be the real part, so we can write

$$v \cos(\omega t + \phi) = GF \cos \omega t + BF \cos(\omega t + \pi/2). \quad (2.30)$$

Thus the velocity can be considered to be made up of two parts, one of amplitude GF, proportional to the real part of the admittance, which is in phase with the exciting force, and another part with amplitude BF, proportional to the imaginary part of the admittance, which is $\pi/2$ in advance of the phase of the exciting force (also referred to as being "in quadrature with it"). These add together to give a simple cosine vibration of intermediate phase. A similar interpretation applies to the impedance, the real and imaginary parts of which give the components of the force that are respectively in phase and in quadrature relative to the velocity.

The important thing about this decomposition of velocity into components that are in phase and in quadrature relative to the exciting force is that only the in-phase component takes energy from the force, so maintaining or increasing the amplitude of the oscillation. This comes about because the work done by the force is the product of the instantaneous magnitudes of the force and the velocity, integrated over a whole cycle of the motion. For the in-phase component this product involves $(\cos \omega t)^2$, which is always positive and so integrates to a positive quantity, while for the in-quadrature component the product is $\cos \omega t \sin \omega t = \frac{1}{2} \sin 2\omega t$ which integrates to zero. The real part of the admittance or impedance of a system with no internal energy source is always positive, since it cannot supply energy.

If we let the mass and damping of the oscillator go to zero, then the system becomes a simple spring. From (2.3), $m\omega_0^2 = \beta$, and so, from (2.27) and (2.28),

$$Y^{(M)}_{\text{spring}} = \frac{v}{F} = j\frac{\omega}{\beta} \quad \text{and} \quad Z^{(M)}_{\text{spring}} = \frac{F}{v} = -j\frac{\beta}{\omega}. \quad (2.31)$$

We sometimes say that the spring is a compliant impedance, or a compliance for short, and the quantity $1/\beta$ is the magnitude of the compliance.

In a similar way, if we let the spring stiffness β and the damping α go to zero, then the system becomes a simple mass. From (2.3), $\omega_0 \to 0$, and so, from (2.27) and (2.28),

$$Y^{(M)}_{\text{mass}} = -\frac{j}{\omega m} \quad \text{and} \quad Z^{(M)}_{\text{mass}} = j\omega m. \quad (2.32)$$

We refer to a mass as being an inertive impedance, or as possessing an inertance, measured by the mass m.

SIMPLE VIBRATORS

Finally, if we let the mass m and the spring stiffness β both go to zero, then we have a purely viscous load $2m\alpha = \gamma$, for which

$$Y_{\text{visc}}^{(M)} = 1/\gamma \quad \text{and} \quad Z_{\text{visc}}^{(M)} = \gamma. \tag{2.33}$$

Both of these are purely real quantities.

2.7 Linearity and Superposition

The simple oscillator we have been discussing has a property called linearity. It derives in part from the linear relation between force and displacement for the spring, but has a deeper significance, for it is necessary for all terms in the differential equation describing the system to have a similar linearity. Equation (2.18) is the most general equation we have met so far, since it describes the response of the system to an arbitrary external force. In this equation the dependent variable is x, and we see that, since the force is simply an explicit function of time and does not involve x, the highest power of the dependent variable in the equation is unity. The equation is said to be linear in x.

The important thing about this is that different solutions to linear equations can be simply added together. Suppose the responses of the system to two external forces F_1 and F_2 are denoted by x_1 and x_2 respectively, then we can simply add together the two equations analogous to (2.18) and see that the response of the system to the combined force $F_1 + F_2$ is just $x_1 + x_2$.

A useful general case arises when the individual forces F_n all vary sinusoidally with different frequencies ω_n and different phases. If, as is usually the case, we want to know the velocity response of the system when it is acted upon by a number of such forces simultaneously, then we can simply write

$$v(t) = \sum_n Y^{(M)}(\omega_n) F_n(\omega_n) e^{j\omega t}. \tag{2.34}$$

Even more generally, if the system is acted upon by a continuous spectrum of external forces with force $F(\omega)d\omega$ in the frequency range $d\omega$ at ω, then the sum becomes an integral and

$$v(t) = \int Y^{(M)}(\omega) F(\omega) e^{j\omega t} d\omega. \tag{2.35}$$

This sort of situation might arise when considering the response of a vibrator to the random noise forces generated by ocean turbulence or breaking waves. Similar cases arise in acoustics. We shall also find later on, making use of a mathematical technique called the Fourier transform, discussed in Chapter 15, that a completely general force can be considered as made up of sinusoidal components distributed over a wide frequency range, so that the result (2.35) is really very general indeed. The fact that the integrand of (2.35) is complex need not cause concern about its evaluation. We can evaluate the product $\mathbf{y}_F e^{j\omega t}$ that forms the integrand and, by the ordinary rules of complex algebra, convert it into a single complex expression $A + jB$ with real and imaginary parts. These can then be evaluated as two separate integrals, one of which is accompanied by a factor j. Actually doing this in a general case can, of course, be quite laborious, but this is to be expected—the result is the total behavior of a resonant system under a completely general time-varying force. Usually we shall be concerned with much simpler problems.

In (2.34) and (2.35) it should be noted that the phases of the forces may vary, generally randomly, from one frequency to another. In the formal relations given, we must therefore remember that each F is a complex quantity and include the appropriate phase factor $e^{j\theta'}$ along with each amplitude F.

2.8 Transient Response

It is not appropriate to consider transient response in detail at this stage. A general method for calculating transients is given in equation (2.35), once the exciting force has been transformed into a frequency-domain representation using the Fourier techniques discussed in Chapter 15. However it would not be wise to leave the present discussion without giving at least the flavor of the transient phenomena experienced with even a simple oscillator. We therefore examine a particular case that is quite representative of transient phenomena, and that can be worked out without any complicated mathematics.

This special case is the response to what is called in acoustics a tone-burst—a sinusoidal excitation starting suddenly and ceasing equally suddenly after a short period of time. For convenience we consider the onset stage first, and suppose it to occur at $t = t_0$. We know that for $t < t_0$ the oscillator is stationary, so that $x(t_0) = 0$, $v(t_0) = 0$. For $t > t_0$, however, the motion is described by the equation

$$\frac{d^2x}{dt^2} + 2\alpha\frac{dx}{dt} + \omega_0^2 x = \frac{F}{m} e^{j\omega t} \qquad (2.36)$$

where F is the complex amplitude of the force and ω its frequency. We can consider the solution to this equation in two parts, the first being the steady response to the force and the second the natural response to the shock at the beginning of the excitation. We adjust the natural response to satisfy the condition that the system is stationary at $t = t_0$.

Mathematically, we add to the ordinary solution for the steady motion (which satisfies (2.36) at all times) another function of arbitrary amplitude representing the free vibrations of the system with $F = 0$ at the frequency $\omega' \approx \omega_0$ given by (2.16). The first part of the solution is clearly inapplicable for $t < t_0$, but this is outside the domain of the equation and we know that the system is stationary for $t < t_0$. The second part of the solution satisfies (2.36) with the right-hand side set to zero, and so can be added to the steady solution of (2.36) without affecting it, since the equation is linear. Thus we take a solution of the form

$$x = \frac{(F/m)e^{j\omega t}}{(\omega_0^2 - \omega^2) + 2j\alpha\omega} + Ce^{-\alpha t}e^{j\omega' t} \qquad (2.37)$$

and determine the complex constant C, which gives the amplitude and phase of the natural vibration, by the requirement that $x(t) = 0$, $v(t) = 0$ at $t = t_0$, these conditions referring to the real parts of (2.37) and of its time derivative since these represent the physical quantities. The result depends upon the phase of the force F at time t_0, and is algebraically complicated for the general case. A specific case is treated in the discussion examples at the end of the chapter. From the form of (2.37), however, it is clear that the response to the onset of the tone burst consists of the steady response with, superposed upon it, an oscillation at the free damped frequency that is fixed in magnitude and phase so that it exactly cancels the displacement and velocity of the steady response at time $t = 0$. Since the steady response velocity is a maximum when the steady response displacement is zero, there is always a significant transient. The time necessary for the system response to reach a steady state is the same as the time for the damped natural vibration to die exponentially away to negligible amplitude.

Figure 2.7 shows examples of this onset transient for several values of the exciting frequency ω relative to the resonance frequency ω_0, and the initial condition discussed in the examples. The Q value used to calculate these figures

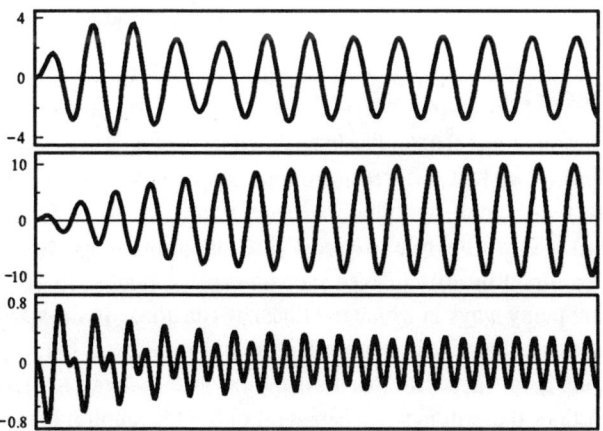

Figure 2.7 Response of a simple oscillator with natural frequency ω_0 and $Q = 10$ to a tone burst of frequency ω beginning at left edge of the frame. The three cases shown are for $\omega/\omega_0 = 0.8$, 1.0, and 2.0 respectively. Note the change of scale between the three plots.

was 10, corresponding to $\alpha = 0.05\omega_0$, and we note that the transient takes about 10 cycles of the natural frequency to die away. This is a quite general rule, the decay of the transient always taking about Q cycles of the natural frequency.

The transient at the end of the tone burst is just an exponentially decaying damped oscillation of the form given in (2.17). We simply have to match the amplitude and phase of this oscillation to that of the initial response given by (2.37) at the time of the end of the tone burst $t = t_0 + \tau$. The general result is again algebraically complicated, and it is not worthwhile to present it here, though a particular case is discussed in the examples at the end of the chapter. The initial amplitude of the decay transient can vary, depending on the phase of the exciting signal when the tone burst terminates. If the force is turned off when the displacement of the mass is a maximum, then the decay transient begins with this same initial amplitude and decays smoothly with its natural frequency ω' and decay constant α. However if the force is turned off just as the mass passes through zero displacement, the initial amplitude of the decay transient is a factor ω/ω_0 times the quasi-steady amplitude during the tone burst itself, so that there is overshoot behavior for high-frequency excitation. The final transient again takes about Q cycles of the natural frequency of the oscillator to die away, as for the initial transient.

We need not be concerned here with the details of this behavior. The important general conclusion is that resonant vibrators do not respond very faithfully to transient signals unless the duration of the signal is a good deal greater than $1/\alpha$ seconds. For shorter signals the response is dominated by the transient behavior of the resonator at its own natural frequency. On the other hand, as we saw in our discussion of (2.21) and in Fig. 2.5, the magnitude of the resonant response is proportional to Q and thus to $1/\alpha$, so that we have another general relation similar to (2.24) in the form

$$\text{(peak response amplitude)} \times \text{(time resolution)} = \text{constant.} \tag{2.38}$$

This is just the same relation expressed in a rather different way. We shall make use of this in our more general discussion of auditory systems.

2.9 Nonlinearity

All the discussion in this chapter has been based on the fact that the simple vibrator is a linear system, so that different solutions can be simply added to treat more complex cases. We shall find this to be true of most of the systems we meet in this book, but it is worthwhile to give a little attention at an early stage to the consequences of nonlinearity in even such a simple system.

There are many ways in which nonlinearity can arise. In the simple vibrator, the possibilities are that the simple linear relation $F_E = -\beta x$ for the behavior of the restoring spring fails to apply, or that the damping force γv tends to a more complex behavior. Both of these things can happen if either the amplitude or the velocity becomes large. We concentrate, for simplicity, on the force behavior, but we should note that, for high velocities, the viscous damping behavior tends to become proportional to v^2 rather than to v because of turbulence, and this nonlinearity can also be significant.

It is common experience that, if we stretch a spring, it unwinds and often becomes stiffer. A similar thing happens if it is compressed until the turns begin to touch. Biological springs often rely upon similar mechanisms on the molecular level, and we might expect the relation between force and extension to be not a straight line, but rather some sort of curve. If the extensions away from the natural length of the spring are small, then the small section of curve involved is close enough to a straight line that a linear relation can be assumed, but this is no longer valid for large extensions. We must then write the force relation as

$$F_E = \beta x + \varepsilon_2 x^2 + \varepsilon_3 x^3 + \ldots \qquad (2.39)$$

where β, ε_2, ε_3 are constants. If the displacement x is small then x^2 is very small, or rather $\varepsilon_2 x^2 \ll \beta x$, so that the second and later terms can be ignored. For large x we must keep these terms in the equation. If the behavior of the spring is symmetrical under tension and compression, then the term $\varepsilon_2 x^2$ vanishes, and the equation of motion becomes

$$\frac{d^2x}{dt^2} + 2\alpha \frac{dx}{dt} + \omega_0^2 \left(x + \frac{\varepsilon_3}{\beta} x^3 \right) = F/m . \qquad (2.40)$$

This equation, known as Duffing's equation, has been very much studied and is still leading to surprising conclusions relating to chaotic behavior. We do not need this sophistication, and it is enough to simply consider the response amplitude as a function of frequency as we did for the linear vibrator.

The behavior depends upon the amplitude, because of the cubic term in (2.40). For amplitudes small enough that $(\varepsilon_3/\beta)x^3 \ll x$, the response is nearly identical with that of the corresponding linear oscillator. For larger amplitudes, however, the effective stiffness and thus the resonance frequency rises, if $\varepsilon_3 > 0$ as is usually the case. Other interesting hysteresis and chaotic phenomena occur for still larger amplitudes, but we shall not need to discuss these here. For our present purpose it is enough to point out that we can no longer simply add the responses to two forces acting simultaneously, because of the cross terms introduced when their displacements are cubed.

To see in simple terms what happens, suppose that we have a general nonlinear elastic force response of the form (2.39) and that the vibration is, to a first approximation, of the simple form $x = a \cos \omega t$. When we substitute this in (2.39) we get terms $\varepsilon_2 a^2 \cos^2 \omega t$ and $\varepsilon_3 a^3 \cos^3 \omega t$. Since $\cos^2 \omega t = (1 - \cos 2\omega t)/2$, the second-order term introduces an extra force $(\varepsilon_2 a^2/2) \cos 2\omega t$, which then produces a second harmonic of the motion with frequency 2ω and amplitude proportional to a^2. Similarly, the third-order nonlinear term produces a third harmonic with amplitude proportional to a^3. This behavior is known as harmonic distortion, and is common to all real systems driven to very high amplitudes. Clearly it becomes progressively more severe as the amplitude increases but can be neglected for very small amplitudes.

More generally, if a nonlinear system is made to vibrate at two unrelated frequencies ω_1 and ω_2, for example by two external forces, then its response at large amplitude will also contain motions with frequencies $\omega_1 \pm \omega_2$ or, more generally, $m\omega_1 \pm n\omega_2$, where m and n are integers.

Nonlinearity introduces many new features and requires new techniques for their analysis. Fortunately, passive systems, such as the auditory system, usually operate at such low amplitudes that their behavior is linear to an adequate approximation in nearly all cases. Active systems however, such as the vocal system, are inherently nonlinear and rely upon that nonlinearity to produce their characteristic acoustic output. We discuss such systems in Chapters 12 and 13.

References

The references given below refer by number to the books and papers listed in the Bibliography. Only a selection of such references is given.

Simple oscillators: [1] Ch 2; [3]; [7] Ch 2; [8] Ch 1; [10] Ch 1
Complex numbers: [1] Ch 1
Transients: [1] Ch 2 Sect 6; [10] Ch 1
Internal losses: [3] Ch 3

Discussion Examples

1. The newton may not be familiar—how large a force is 1 newton? A small apple has a mass of 100 grams. Taking the gravitational acceleration to be 10 m s^{-2}, calculate the weight of the apple in newtons.
2. If $A = 3 + 4j$ and $B = 1 - j$, calculate (a) $|A| \equiv A$, (b) $|B|$, (c) $A + B$, (d) $A \times B$ (e) A/B. Use the methods described in Appendix A.
3. Express $1 + j$ in the form $re^{j\theta}$.
4. Calculate $\sqrt{4j}$.
5. A grain of sand of mass 2 mg attached to the end of a sensory hair causes it to deflect through 0.5 mm under the influence of gravity. Calculate the spring constant β for the hair considered as a simple spring.
6. A grain of sand of unknown mass attached to the end of a sensory hair causes it to deflect 0.1 mm under the influence of gravity. Calculate the oscillation frequency of the loaded hair.
7. A system consisting of a grain of mass 1 mg on the end of a sensory hair has a resonance frequency of 200 Hz. The damping is such that the Q value of the system is 3. Treating it as a simple vibrator, calculate the amplitude of the vibration when a force of fre-

SIMPLE VIBRATORS

quency 200 Hz and amplitude 2 μN acts on the mass. What is the amplitude when a force of the same magnitude but with a frequency of 150 Hz acts on the grain?

8. An oscillator has a resonance frequency of 300 Hz and a Q value of 8. If it is set into free vibration with an initial amplitude of 100 μm, how long is it before the amplitude has decayed to 10 μm?

9. A vibrator, having mass m, natural resonance frequency ω_0, and damping constant α, is acted upon by a tone-burst force of amplitude F and frequency ω. The onset phase of the burst is timed so that it would correspond to a moment of zero displacement in the forced motion of the mass. Calculate the form of the onset transient.

10. After the oscillator of question 10 has settled into steady motion, the exciting force is turned off just as the particle displacement passes through zero. Calculate the form of the final transient.

Solutions

1. $F = mg \approx 0.1 \times 10 = 1$. So the weight of an apple is about 1 newton!

2. (a) 5, (b) $\sqrt{2}$, (c) $4+3j$, (d) $7+j$, (e) $(-1+7j)/2$.

3. $\sqrt{2}e^{j\pi/4}$.

4. $4j = 4e^{j\pi/2}$ so $\sqrt{4j} = 2e^{j\pi/4} = 2\cos(\pi/4) + 2j\sin(\pi/4) = 2/\sqrt{2} + 2j/\sqrt{2} = \sqrt{2} + j\sqrt{2}$.

5. $F_E = mg = 2 \times 10^{-6} \times 10 = 2 \times 10^{-5}$ N and $x = 5 \times 10^{-4}$ m. So $\beta = F_E/x = 0.04$ N m^{-1}.

6. Let the mass be m, then the spring constant is $\beta = mg/x = m \times 10/0.0001 = 10^5 m$. By (2.3) $\omega_0 = \sqrt{\beta/m} = \sqrt{10^5} \approx 316$ so $f_0 = 316/2\pi \approx 50$ Hz.

7. $\omega_0 = 400\pi \approx 1257$ s^{-1}. By (2.23) $\alpha = \omega_0/2Q \approx 210$ s^{-1}. From (2.21) the amplitude at resonance is 3.8 μm. At 150 Hz, from (2.21), the amplitude is about 0.8 μm.

8. $Q = 8$ and $\omega_0 = 2\pi \times 300 \approx 1890$ s^{-1} means $\alpha \approx 118$ s^{-1}. A decay from 100 μm to 10 μm means $e^{-\alpha t} = 0.1$ from (2.37). Therefore $t = \alpha^{-1} \ln 10 \approx 2.3026/118 \approx 20$ ms.

9. Use (2.20)–(2.22) to rewrite (2.37) in simpler form. Choose the onset time t_0 so that $\omega t_0 = -\theta + \pi/2$. Write $C = A + jB$. Set the real part of $x(t_0)$ equal to 0, giving $A = 0$. Differentiate the new form of (2.37) to get velocity and set the real part of $v(t_0)$ equal to 0, giving $B = -a\omega/\omega'$. Take the real part of the final expression for $x(t)$, giving

$$x(t) = -a\sin[\omega(t-t_0)] + a\frac{\omega}{\omega_0}e^{-\alpha(t-t_0)}\sin[\omega'(t-t_0)], \qquad t > t_0.$$

This is plotted in Fig. 2.7 for several values of ω.

10. Proceed just as in example 9, choosing the cut-off time to be t_1.

$$x(t) = \pm a(\omega/\omega')e^{-\alpha(t-t_1)}\sin[\omega'(t-t_1)], \qquad t > t_1.$$

The result has an ambiguity in sign referring to two possible cut-off phases. Note that, in this case, the initial amplitude of the decay transient is ω/ω' times that of the steady response.

3 VIBRATION OF STRINGS AND BARS

SYNOPSIS. A perfectly flexible string stretched between two fixed points is the simplest example of an extended system, and by studying it we can understand many of the properties of more complex systems. The behavior of the string is described by a wave equation according to which any disturbance of the position of the string moves with a characteristic velocity c_S as shown in Fig. 3.2. We defer detailed discussion of such wave-like behavior to a later chapter. The stretched string can vibrate only in patterns that are combinations of certain characteristic vibrations or normal modes, sometimes referred to pictorially as standing waves, and known mathematically as eigenfunctions. The first four normal modes for a stretched string are shown in Fig. 3.3. The points on the string at which the amplitude of a normal mode is zero remain fixed when the string vibrates in that mode, and are known as nodes of the motion. The frequencies of the normal modes of a string are integer multiples $n = 2, 3, 4, ...$ of the frequency of the first mode or fundamental. Such a sequence is called a harmonic series, though this term properly refers to the wavelengths on the string, which vary as $1/n$.

It is important to understand the way in which the string responds when excited at some point along its length by a force that varies sinusoidally with time. Generally the response motion will go through maxima when the frequency of the exciting force coincides with that of one of the normal modes, but if the excitation point is a node for a particular mode then this mode will not be excited. This is illustrated in Fig. 3.4, which shows the magnitude of the point admittance, or ratio of velocity to force, for a string excited at a point 0.2 of its length from one end. Note that mode 5 is not excited. If the measurement point is different from the excitation point, the ratio of velocity to force is called the transfer admittance between the two points. Fig. 3.5 shows this same information plotted on a logarithmic scale. Note the fairly sharp minima between each pair of maxima. The magnitude of the point impedance is simply the reciprocal of the magnitude of the point admittance, and can be obtained by turning this diagram upside down, as shown on the right in the figure.

A stiff rod or bar, clamped rigidly at one end and free at the other, is another simple extended system that has biological application in discussing sensory hairs. The stiffness of the bar depends on both the area and shape of its cross-section, as shown in Fig. 3.6, where the stiffness is proportional to κ^2 for a given area. The first four normal modes of a bar clamped at one end are shown in Fig. 3.7. They are similar in appearance to the normal modes of a string, except that the free end is a maximum in the motion, rather than being fixed like the two ends of the string. The frequencies of the bar modes n behave approximately as $\beta_n^2 = (n - \frac{1}{2})^2$, so that the upper modes are not harmonics of the fundamental.

It is possible to calculate the mode frequencies if the end of the bar is not rigidly clamped but instead has some angular compliance μ. The behavior of the mode frequencies as a function of the parameter μ/μ_B, where μ_B is the bending compliance of the whole bar, is shown in Fig. 3.8. The frequency of the first mode $n = 1$ declines steadily towards zero as the compliance of the clamp is increased, while the β_n values for higher modes decrease only towards $n - \frac{3}{4}$. The mode of biological interest is the fundamental, $n = 1$.

We can also calculate the behavior of a bar rigidly clamped at one end and loaded with a concentrated mass m at the other. If m_B is the mass of the bar itself, then Fig. 3.8

also shows the variation of the frequency parameters β_n as m/m_B is varied. Again the mode of biological interest is the fundamental, $n = 1$. It is possible to combine these two cases and calculate the behavior of a bar set in a compliant mount and loaded at its free end.

3.1 Extended Systems

With this chapter we come a little closer to physical reality, but we are still in the preliminary stages of our discussion of acoustic systems. Strings certainly occur in biological systems in the form of flexible sinews, and bars provide a simple model for the various bones in an animal, but neither of these has a very close connection with acoustic systems, which generally involve membranes, plates and tubes. As with the simple oscillator discussed in Chapter 2, our purpose is to set out the physics and mathematics first for the simplest possible system and, in this way, to provide an easier path to the understanding of more complex acoustic components.

Strings and rods are an extension of the simple oscillator in the sense that they have one major physical dimension—their length—whereas the simple oscillator was a point mass. The next step is on to membranes and plates, which have extension in two dimensions, but we reserve this for Chapter 5.

3.2 The Wave Equation

If we think of a thin flexible string stretched between two fixed points, then it is clear that we can consider two types of displacement of an element of the string. In the first case we can suppose the element to be translated along the direction of the string itself, so that the string remains a straight line. Such a displacement is called longitudinal. In the second case we can suppose the element to be moved sideways, normal to the line of the string. This type of displacement is termed transverse. A thorough discussion of the motion of a string requires that both types of displacement be considered, but fortunately it turns out that they are almost independent of each other so that we can discuss them one at a time. It also turns out that we can forego discussion of longitudinal displacements here, for the transverse case is the one that is important at this stage of our exposition.

Figure 3.1 shows a section of string that has been displaced transversely from its normal position, shown as a broken line. What we want to do is to work out the equation describing its motion. To do this we note that the string is held in place between its two fixed ends by a tension force T, determined by how much it has been stretched when mounted. If a is the radius of the string, taken to be very small compared with its length, and σ the elastic stress in the string, then

$$T = \pi a^2 \sigma = \pi a^2 E \left(\frac{l - l_0}{l_0} \right) \tag{3.1}$$

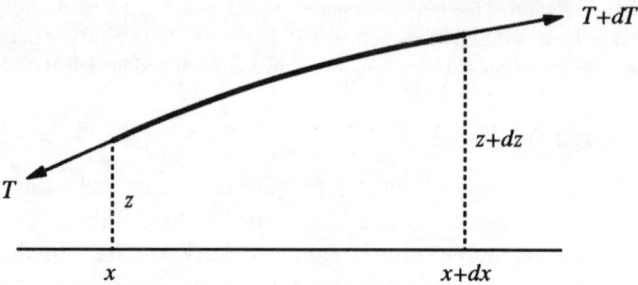

Figure 3.1 Tension forces on a short element of string displaced from its rest position.

where E is the Young's modulus of the string material, l_0 is its unstretched length, and l is its length when stretched between the end clamps.

This force is constant along the length of the string, and the transverse displacement, provided it is very small, does not appreciably increase the length of the string, so that the tension of the displaced string is still T. If we take x to be the coordinate along the string and z the coordinate in the transverse direction, then the tension force acts everywhere in the direction of the tangent to the curve of the string, as shown in Fig. 3.1, and thus in the direction of its slope, which is very nearly equal to dz/dx provided the slope is small.

Consider the small element of string of length dx shown. Then the net force acting on it in the z direction is

$$T\left[\frac{dz}{dx}\right]_{x+dx} - T\left[\frac{dz}{dx}\right]_{x} = T\left\{\left[\frac{dz}{dx}\right]_{x} + dx\left[\frac{d^2z}{dx^2}\right]_{x}\right\} - T\left[\frac{dz}{dx}\right]_{x} = T\frac{d^2z}{dx^2}dx. \quad (3.2)$$

The mass of the element of string is $\pi a^2 \rho_s dx$, where ρ_s is the density of the string material, so that the equation of motion is

$$\pi a^2 \rho_s \frac{\partial^2 z}{\partial t^2} = T\frac{\partial^2 z}{\partial x^2} \quad (3.3)$$

where we have cancelled out the common factor dx. The derivatives have been written in partial form, $\partial z/\partial x$ and so on, since z is a function of both x and t and we wish to differentiate with respect to one variable while keeping the other constant. We can use (3.1) to write this in terms of elastic stress σ rather than tension, which gives, after a little rearrangement,

$$\frac{\partial^2 z}{\partial x^2} = \frac{\rho_s}{\sigma}\frac{\partial^2 z}{\partial t^2}. \quad (3.4)$$

VIBRATION OF STRINGS AND BARS

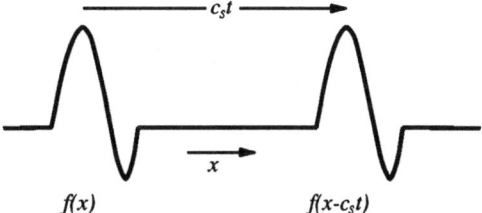

Figure 3.2 An arbitrary displacement $f(x)$ on a taut string moves with velocity c_s, so that at time t it is displaced to position $c_s t$ unchanged in shape. Its general form is represented by $f(x - c_s t)$.

Equation (3.4) is a particular form of the wave equation, which we shall need to discuss in great detail later. For the moment it is enough to note that, if we assume a form

$$z(x,t) = f(x - c_s t) \tag{3.5}$$

for the displacement z, where f is any real function of its argument as shown in Fig. 3.2, then this will satisfy (3.4) provided

$$c_s = \sqrt{\sigma/\rho_s}. \tag{3.6}$$

We can see what this means by looking at the behavior of the suggested solution (3.5). Whatever the shape of the function f, it is unchanged if we increase t by δt and increase x by $c_s \delta t$. The arbitrary disturbance f thus moves along in the $+x$ direction with a speed c_s. We call c_s the wave speed on the string. From (3.6), the wave speed is proportional to the square root of the elastic stress in the string and inversely proportional to the square root of the string density.

3.3 Normal Modes of a String

For the moment we are not particularly concerned with wave propagation along a string, but rather with the way in which the string vibrates under the influence of external forces, and so we concentrate on this aspect. To do so we rewrite (3.3) to include an external force $F(x,t)$ which acts, in general, at several points along the string, or is perhaps distributed continuously along its length, as would be the case for viscous forces from the medium in which the string might be immersed. Using (3.6) to introduce the wave speed, the result is

$$c_s^2 \frac{\partial^2 z}{\partial x^2} - \frac{\partial^2 z}{\partial t^2} = \frac{1}{\pi a^2 \rho_s} F(x,t). \tag{3.7}$$

The simplest case is one in which the external force F is zero. We then facilitate matters by seeing whether a solution of the form

$$z(x,t) = a \sin(kx + \xi)e^{j(\omega t + \theta)} \tag{3.8}$$

might exist, where we have used the complex notation introduced in Chapter 2 for the time variation. Substituting (3.8) in (3.7) shows that

$$\omega = kc_S. \tag{3.9}$$

The phase constant θ need not cause us any problem, for it is familiar from Chapter 2, but we have still to do something about the other two constants k and ξ. If we know k then we immediately know the free vibration frequency ω from (3.9). We can find values for k and ξ by looking at the boundary conditions at the two ends of the string, which we have assumed to be clamped so that $z = 0$ at $x = 0$ and l, where l is the length of the string. The first of these conditions requires that $\xi = 0$, while the second requires that $\sin kl = 0$, which in turn requires that

$$k \equiv k_n = \frac{n\pi}{l} \qquad n = 1, 2, 3, \ldots \tag{3.10}$$

Thus there is a set of possible values of k, which we have called k_n, for which the equation is satisfied. These values are called characteristic values or eigenvalues (from the German) for the particular string problem. We look at their significance further in a moment, but first we see that the general solution is one of the functions

$$z_n(x,t) = \psi_n(x)e^{j(\omega_n t + \theta_n)} \tag{3.11}$$

where

$$\psi_n(x) = \left(\frac{2}{l}\right)^{1/2} \sin\left(\frac{n\pi x}{l}\right). \tag{3.12}$$

The factor $(2/l)^{1/2}$ is included in order to simplify later calculation. The vibrations that these functions $z_n(x,t)$ represent are called the normal modes of the string system, and the spatial parts $\psi_n(x)$ of the functions themselves are called mode functions, characteristic functions, or eigenfunctions for the string problem. In this book we shall always use the Greek letter ψ (psi) to denote eigenfunctions for the problem we are discussing, though the explicit form will vary from problem to problem. The first few eigenfunctions for the string problem are shown in Fig. 3.3. As is clear from this figure, the normal modes are simply stationary sine waves along the string, the amplitudes of which vary sinusoidally in time so that the displacement passes simultaneously through zero all along the string. The envelope swept out by any mode has a sequence of maxima along the string, separated by zeros or nodes. The amplitude maxima are sometimes referred to as antinodes.

We are, of course, very interested in the frequencies of the normal modes of the system. These are given by (3.9) and (3.10) to be

VIBRATION OF STRINGS AND BARS

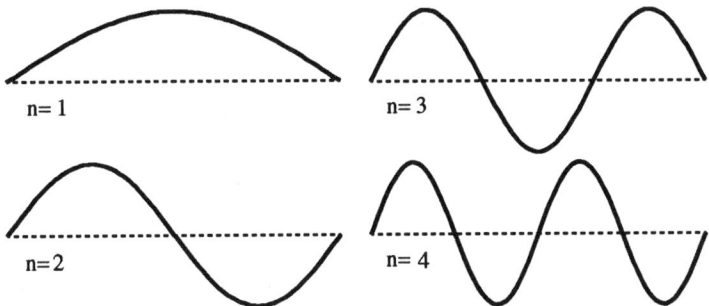

Figure 3.3 The first four normal modes or eigenfunctions for a string fixed at its two ends.

$$\omega_n = n\left(\frac{\pi c_s}{l}\right) \qquad n = 1, 2, 3, \ldots \qquad (3.13)$$

so that they form a series of integer multiples of the lowest or fundamental frequency ω_1. Such a series is called a harmonic series, and the mode frequencies are said to be harmonics of the fundamental. (Strictly speaking, a harmonic series has terms $1/n$, and so refers to the wavelengths involved.)

3.4 Normal Modes and Eigenfunctions

Almost the whole treatment of the vibration of extended bodies relies upon the concept of normal modes, so that it is important to understand it thoroughly. The string is a good first example, as was remarked above, because it has extension in only one direction and its eigenfunctions are mathematically simple.

Suppose a string, clamped at both ends, is set into some arbitrary vibration. This might be done, for example, by striking it with a hammer or by plucking. Once the exciting force has completed its action and returned to zero, the motion of the string is described by the equation (3.4), or by (3.7) with $F = 0$, and possible motions are limited to the normal modes (3.12). Because the system is linear, these normal modes can be added together with any desired amplitude, however, so that the complete motion has the form

$$z(x,t) = \sum_{n=1}^{\infty} a_n \psi_n(x) e^{j(\omega_n t + \theta_n)} = \sum_{n=1}^{\infty} \mathbf{a}_n \psi_n(x) e^{j\omega_n t} \qquad (3.14)$$

where the a_n are real constants and $\mathbf{a}_n = a_n e^{j\theta_n}$. The motion is completely described once we have evaluated the amplitudes a_n and phases θ_n, or equivalently the complex amplitudes $\mathbf{a}_n$.

To show how this is done, we first need to discover one general mathematical property of eigenfunctions. From the definition (3.12), a simple integration shows us that

$$\int_0^l \psi_n(x)\psi_m(x)dx = 0 \quad \text{if} \quad m \neq n$$

$$= 1 \quad \text{if} \quad m = n. \qquad (3.15)$$

This is a property called orthonormality—the eigenfunctions for different modes are orthogonal in the sense that the integral of their product over the whole length of the string vanishes, while if we integrate the square of the value of any one of the eigenfunctions over this same domain we get unity. We shall see later that this is true of eigenfunctions for any problem, even in cases much more complex than the flexible string, and this is the reason we have adopted a rather general notation.

Now, to find the constants a_n, what we have to do is to match the real part of the general solution (3.14) to the given displacement $z(x,0)$ and velocity $v(x,0)$ of each point of the string at the initial time $t = 0$. We can write an equation for the displacement matching quite directly, replacing n by m for convenience, as

$$\text{Re}\left[\sum_m a_m \psi_m(x)\right] = z(x,0). \qquad (3.16)$$

Multiply both sides of this equation by $\psi_n(x)$ and integrate over the length l of the string. Then, using the orthonormality relations (3.15), this picks out just the term with $m = n$ from the left-hand side of the equation and gives us the result

$$\text{Re}(a_n) = \int_0^l \psi_n(x)z(x,0)dx. \qquad (3.17)$$

If we differentiate $z(x,t)$ in (3.14) to find the velocity, then this gives, writing m for n,

$$\left[\frac{dz}{dt}\right]_{t=0} = j\sum_m \omega_m a_m \psi_m(x) \qquad (3.18)$$

We then proceed just as for the displacement. The factor j on the right-hand side, however, means that, when we take the real part, we get the result

$$\text{Im}(a_n) = -\frac{1}{\omega_n}\int_0^l \psi_n(x)v(x,0)dx. \qquad (3.19)$$

These two equations (3.17) and (3.19) completely determine the expansion coefficients a_n, so that we then know the subsequent motion of the string from (3.14). In cases where the initial velocity is zero, the a_n are entirely real (or in other words the phases θ_n are all 0 or π), while if the initial displacement is zero the a_n are entirely imaginary (and all θ_n are $\pm\pi/2$).

This somewhat complicated piece of mathematical analysis is extremely important for calculating the response of the string system to external forces. It is comforting to know that the analysis for more complicated systems is formally identical, so that we shall not need to go through the mathematics again. All that is necessary is to use the eigenfunctions appropriate to the particular problem with which we are dealing, and integrate over the whole extent of the vibrating object.

We shall not be specially interested in calculating transient response for the string system, or indeed for most of our biological systems, but it is worthwhile to point out that, because each mode of the system vibrates with a different frequency, the time evolution is very complicated. For an ideal string, since the mode frequencies are in the ratio of simple integers as given by (3.13), the undamped motion repeats once every cycle of the fundamental, but this is not true for more complex systems, or indeed for real strings, the mode frequencies of which are not in exact harmonic relation, as we note later.

3.5 Damping

The motion of a real string is damped by two mechanisms, one internal and one external. Internal damping generally arises from two basic causes. In strings of polymeric materials, including biological materials, there is damping caused by molecular interactions, and also damping caused by deformation of cells and of the liquids contained in them. In metallic wires the same role is played by the motion of dislocations, and there is an additional loss from thermal conduction, since compression generally heats the material and elongation cools it. These forces are roughly proportional to rate of deformation, and thus to the transverse velocity of the string at the point in question, but we can realistically allow a more general dependence on frequency. The second mechanism is external viscous damping in the fluid, even if it is only air, in which the string is vibrating. This viscous force is proportional to the local transverse velocity of the string, but also has a dependence on frequency as we shall discuss further in Chapter 4. One might reasonably ask about losses by sound radiation, but these turn out to be small compared to the viscous losses.

These losses, whatever their origin, simply contribute a term $-\gamma(\omega) \partial z/\partial t$ to the right-hand side of the equation of motion (3.3). This ends up modifying the equation for the time dependence of $z(x,t)$ to the form

$$\frac{d^2z}{dt^2} + 2\alpha(\omega)\frac{dz}{dt} + \omega_n^2 z = 0 \qquad (3.20)$$

which is of just the same form as equation (2.15) for the damped simple oscillator, in the absence of an external force. All our discussion in Sections 2.4 and 2.5 can therefore be applied quite directly to the damped vibrating string; in particular, the string modes decay as $e^{-\alpha(\omega)t}$. Generally the internal damping increases with increasing frequency, so that the decay of higher modes is more rapid than that of the fundamental.

3.6 The Sinusoidally Driven String

Rather than working out the decay behavior, as sketched above, in more detail, let us immediately attack the problem of the response of a sinusoidally driven string, for this is of much more practical relevance. The equation of motion is obtained by simply adding the damping term and the external force $F(x)e^{j\omega t}$ to the simple equation of motion (3.3) to give

$$\pi a^2 \rho_s \frac{\partial^2 z}{\partial t^2} + \gamma(\omega)\frac{\partial z}{\partial t} - \pi a^2 \sigma \frac{\partial^2 z}{\partial x^2} = F(x)e^{j\omega t} . \tag{3.21}$$

The force is taken to be distributed along the string as described by the function $F(x)$, but of course it could be taken to be concentrated at a point, as a special case. If we divide by $\pi a^2 \rho_s$ and introduce the wave velocity c_S as defined by (3.6), then this equation can be written in the simpler form

$$\frac{\partial^2 z}{\partial t^2} + 2\alpha(\omega)\frac{\partial z}{\partial t} - c_S^2 \frac{\partial^2 z}{\partial x^2} = \frac{F(x)}{\pi a^2 \rho_s} e^{j\omega t} \tag{3.22}$$

where $\alpha = \gamma/4\pi a^2 \rho_s$ has been introduced as before to simplify the notation. We know that in the steady state the displacement z must oscillate sinusoidally with the same frequency ω as the exciting force, and we also know that $z(x)$ can always be represented as a sum of the eigenfunctions of the string with appropriate coefficients. We therefore try a solution of the form

$$z(x,t) = \sum_n a_n \psi_n(x) e^{j\omega t} \tag{3.23}$$

where $\psi_n(x)$ are the normalized eigenfunctions for the string, as given by (3.12). Substituting (3.23), with n replaced by m, in (3.21), multiplying by one of the eigenfunctions $\psi_n(x)$, integrating and using the orthonormality properties (3.15) of the eigenfunctions, we arrive at the result

$$a_n = \left(\frac{1}{\pi a^2 \rho_s}\right)\left[\frac{\int_0^l F(x')\psi_n(x')dx'}{(\omega_n^2 - \omega^2) + 2j\omega\alpha(\omega)}\right] \tag{3.24}$$

where x' is a dummy integration variable, introduced to avoid later confusion. This expression tells us the extent to which mode n is excited by the applied force.

This result, together with the expansion formula (3.23), is one of the most important in the whole theory of vibrations, and it can be applied with very little modification to other systems such as bars, membranes, plates, and even solid bodies, simply by using the appropriate form for the eigenfunctions ψ_n and integrating over the whole of the body. In the following sections we shall look at some particular applications of expression (3.24), but initially it is helpful to note its close resemblance to expression (2.20) for the response of a simple point vibrator to an external sinusoidal force. All that is different is the mass factor at the head of the result, and the necessity to somehow integrate the effect of the force over the whole length of the string. This integral is different for each mode because of its distinctive shape.

3.7 String Excitation by a Distributed Force

A practically important example of the application of these results is the case in which the excitation force is evenly spread along the whole length of the string. This might occur, for example, if the string were immersed in a liquid which flowed back and forth across it with frequency ω. A detection system of this kind could be used by aquatic animals, though in fact they use sensory rods or hairs rather than taut strings. To see how this works out, we simply take $F(x) = F = $ constant in (3.24), which, when inserted in (3.23), gives

$$z(x,t) = \left(\frac{F}{\pi a^2 \rho_s}\right) \sum_n \left[\frac{\int_0^l \psi_n(x')dx'}{(\omega_n^2 - \omega^2) + 2j\omega\alpha(\omega)}\right] \psi_n(x) e^{j\omega t}$$

$$= \left(\frac{F}{\pi a^2 \rho_s}\right) \sum_n \frac{(2l)^{1/2}}{n\pi} \left[\frac{1 - \cos n\pi}{(\omega_n^2 - \omega^2) + 2j\omega\alpha(\omega)}\right] \psi_n(x) e^{j\omega t} \quad (3.25)$$

the last expression coming from use of the explicit form (3.12) for $\psi_n(x)$. This rather complicated expression simply tells us the amplitude and phase for each of the normal modes $\psi_n(x)$ involved. All the even-numbered modes are eliminated because, for them, the numerator vanishes, while the fundamental $n = 1$ is most efficiently coupled to the excitation, because of the factor $1/n$ in the response. The reason for this is plain. In the fundamental mode, all parts of the string move in the same direction, while for the higher modes parts of the string are moving against the direction of the excitation and therefore lose energy rather than gaining it. For even-numbered modes, the sections moving against the flow exactly balance those moving with the flow, giving zero excitation.

3.8 Point Admittance and Transfer Admittance

When we are performing a measurement on a vibrating system, one method is to excite it with a sinusoidal force at a particular point, and to measure the velocity response, either at the same point or at a different point. This is what we did, in effect, for the simple vibrator, and it allowed us to define the mechanical admittance $Y^{(M)}(\omega) = v/F$. For an extended body such as a string, membrane, or plate, we need a more general result, which is called the mechanical transfer admittance $Y_{xx'}^{(M)}(\omega)$ between two points x and x'. By differentiating (3.23) to find the velocity, we immediately have, since $F(x)$ is concentrated at the point $x = x'$, the result

$$Y_{xx'}^{(M)}(\omega) = \left(\frac{j\omega}{\pi a^2 \rho_s}\right) \sum_n \left[\frac{\psi_n(x')\psi_n(x)}{(\omega_n^2 - \omega^2) + 2j\omega\alpha(\omega)}\right]. \tag{3.26}$$

We note that we get just the same result if we interchange x and x', a consequence called reciprocity, which can be helpful in measurements. If we take both measurements at the same place, so that $x = x'$, then the quantity $Y_{xx}^{(M)}(\omega) \equiv Y^{(M)}(\omega)$ is called the point admittance, or sometimes the input admittance, at x.

Both the point admittance and the various transfer admittances have a series of maxima at the frequencies of the normal modes, as shown by the example in Fig. 3.4. The envelope function determining the prominence of the peaks depends on the positions of the exciting and measuring points relative to the nodes, or stationary points, of the various modes. For any of these admittances, expression (3.26) shows that we are simply adding up the admittances $Y_{xx',n}^{(M)}(\omega)$ of the individual modes n:

$$Y_{xx'}^{(M)}(\omega) = \sum_n Y_{xx',n}^{(M)}(\omega). \tag{3.27}$$

This too is illustrated in Fig. 3.4. We can sometimes make use of this by concentrating on just one mode with frequency approximately matching that of the external excitation and ignoring all others. The result will not be accurate, but it may be good enough for the purpose in hand. Note that, except for integration over the string length, the individual mode admittances $Y_{xx,n}^{(M)}(\omega)$ are essentially the same as the admittance of a simple vibrator, as given by (2.27).

It is an interesting feature of the admittance function of an extended vibrator that between each pair of maxima there is a quite deep minimum. This is clearly illustrated in Fig. 3.5, which shows the point admittance for a string, excited as in Fig. 3.4, plotted on a logarithmic scale. The point impedance is obtained simply by inverting the scale, as on the right in the figure. The maxima in the admittance are located at frequencies corresponding to the natural frequencies of the normal modes, and represent frequencies for which the overall motion of the system is large. They are often called the resonances of the system. The minima, on the other hand, occur at frequencies at which the velocity excitation of the next higher mode (which will be nearly +90° in phase relative to the force) is almost the same as that of the next lower mode (which will be nearly −90° in phase relative to the force) so that they nearly cancel at the measurement point. These minima in the admittance, which are, of course, maxima in the impedance, are sometimes called

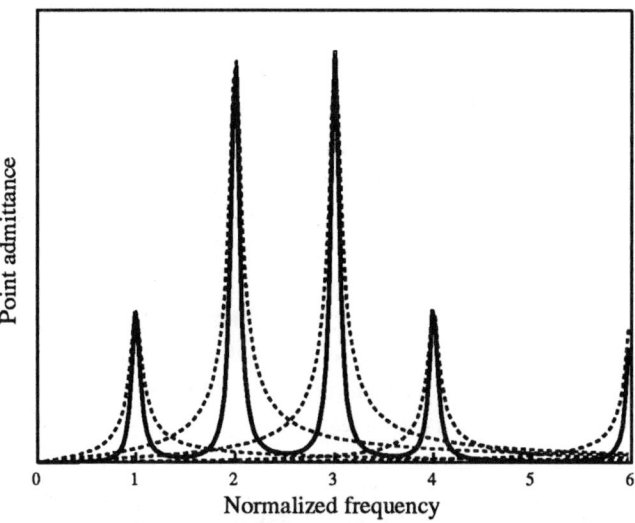

Figure 3.4 The magnitude of the point admittance of a taut string, measured at a point one fifth of its length from one end (full curve). The damping is assumed to give a Q value of 10. The admittance maxima associated with excitation of the normal modes occur at integral values of the normalized frequency ω/ω_1. The envelope of these maxima depends on the measurement point, and shows a peak near normalized frequency 2.5 and a zero at normalized frequency 5, so that mode 5 is completely unexcited. The broken curves show the contributions of individual modes, which tend to subtract rather than add because of phase differences.

Figure 3.5 The magnitude of the point admittance and point impedance for the string of Fig. 3.4, plotted on a logarithmic scale. Note that the impedance scale is simply the reciprocal of the admittance scale.

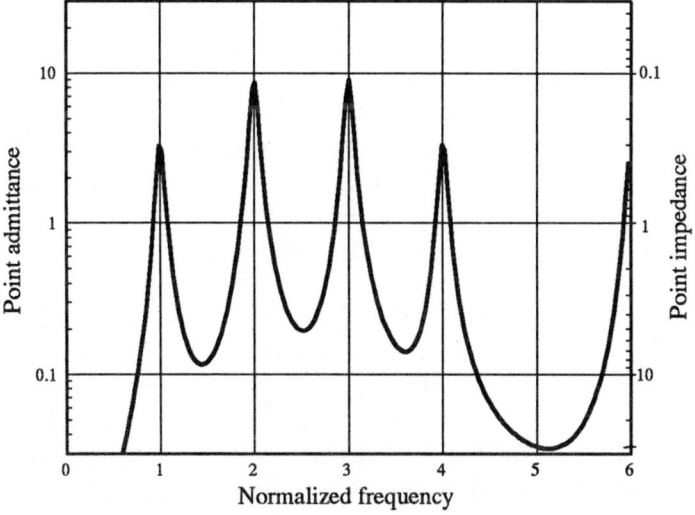

anti-resonances, but this is misleading, since their frequencies are not a property of the system itself but depend also upon the points of excitation and measurement. We shall therefore refer to them simply as admittance minima or impedance maxima.

It is worthwhile to mention at this stage a particular case of transfer admittance or transfer impedance that we shall take up later. This is the lever, which we treat in detail in Section 8.3. A lever is an important link in the auditory chain of many animals, and clearly it has two arms that are connected respectively to a driving point and a receiving point. It could be treated directly using the ideas discussed above, but it turns out that a rather simpler and more practically oriented approach is more useful for our present purposes.

3.9 Bars and Rods

Bars and rods are another step nearer to real biological objects. An object is properly called a strut when its length is great compared with its transverse dimensions and it is stiff enough to deflect only a small amount under its own weight. For simplicity we suppose that the cross-section is uniform along the whole length of the strut. A rod is simply a strut with more-or-less circular cross-section, while a bar has a generally rectangular cross-section. Let us refer to all these variants simply as bars in what follows. The important thing for our discussion is the stiffness of the bar, and experience tells us that this increases greatly as the cross-sectional area increases, and that it also depends upon the elastic modulus of the bar material. Derivation of an expression for the stiffness of a bar would take us too far from our subject, so we simply quote the result. A detailed derivation can be found in the references listed in the Bibliography.

Suppose the Young's modulus for the material of the bar is E, and that its cross-section has one of the forms shown in Fig. 3.6. We can define a quantity κ called the radius of gyration for the cross-section, which is given by one of the expressions in the figure, and we also know the area S of the cross-section. For a solid circular section of radius a, $\kappa = a/2$ and $S = \pi a^2$. The quantity that we shall call the stiffness of the bar is defined by $ES\kappa^2$. If the bar is bent, then the force tending to return the bent section to its original position is $-F_E$ where

$$F_E = -ES\kappa^2 \frac{d^4 z}{dx^4}. \tag{3.28}$$

The equation of motion of an undamped bar therefore has the form

$$S\rho_s \frac{\partial^2 z}{\partial t^2} = -ES\kappa^2 \frac{\partial^4 z}{\partial x^4} \tag{3.29}$$

where ρ_s is the density of the solid material of the bar. If we assume that the time variation can be written $e^{j\omega t}$ as usual, then (3.29) becomes

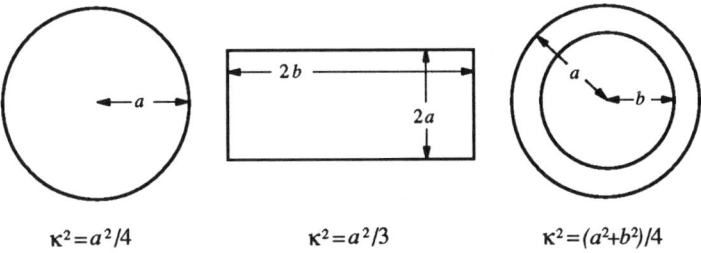

$\kappa^2 = a^2/4$ $\kappa^2 = a^2/3$ $\kappa^2 = (a^2+b^2)/4$

Figure 3.6 The cross-section radius of gyration κ for a circular rod, a rectangular bar bending in the a direction, and a circular pipe.

$$\frac{\partial^4 z}{\partial x^4} = \frac{\rho_S}{E\kappa^2}\omega^2 z \equiv k^4 z \tag{3.30}$$

where k is a constant, not quite the same as the k used for the string. Guided by knowledge of the differentiation of particular functions, we can try a solution of the general form

$$z(x) = A\cos kx + B\sin kx + C\cosh kx + D\sinh kx \tag{3.31}$$

where A, B, C, D are constants. The hyperbolic functions cosh and sinh are discussed in Appendix A, and it is enough here to note that they are analogous to the cosine and sine functions and are defined by

$$\cosh x = (e^x + e^{-x})/2 \qquad \sinh x = (e^x - e^{-x})/2 . \tag{3.32}$$

In the case of an infinitely long bar, both C and D must vanish, since the hyperbolic functions tend to infinity for large arguments. In this case we have simple sinusoidal propagating waves and $k = \omega/c_B$ where c_B is the wave speed on the bar. If we substitute this expression for k in (3.30), we find immediately that c_B is proportional to $\omega^{1/2}$ so that high-frequency waves travel faster than do those of low frequency. This behavior is called dispersion, or dispersive propagation, and contrasts with the case of a string for which all waves travel at the same velocity. Dispersion leads to distortion of the shape of a complex propagating wave.

Returning to a finite bar, the fact that there are four constants in (3.31) reflects the fact that the differential equation (3.29) is of fourth order. This also means that we require four boundary conditions to evaluate the constants and to find the eigenvalues k_n corresponding to allowed solutions. There are three simple ways in which we can treat each end of the bar, since it does not lose its shape if left free as a string would do. Thus we can either leave an end free (which corresponds to conditions of no bending moment, $d^2z/dx^2 = 0$, and no transverse force, $d^3z/dx^3 = 0$), we can attach it to a hinge (which corresponds to $z = 0$, $d^2z/dx^2 = 0$)

or we can clamp it rigidly (corresponding to $z = 0$, $dz/dx = 0$). Each combination of these sets of boundary conditions gives a different set of mode frequencies and a different set of mode shapes. References to more detailed discussion are given in the books listed in the Bibliography. We recognize that only one of these cases is really of interest in the biological situation, that of a bar that is clamped at one end where it grows out of the biological substrate and free at the remote end. We shall consider later the possibility of incomplete clamping.

If we apply the conditions

$$z(0) = 0 \quad \left[\frac{dz}{dx}\right]_{x=0} = 0 \quad \left[\frac{d^2z}{dx^2}\right]_{x=l} = 0 \quad \left[\frac{d^3z}{dx^3}\right]_{x=l} = 0 \quad (3.33)$$

for a bar of length l clamped at $x = 0$ and free at $x = l$, then a certain amount of algebra leads to the normal mode functions (eigenfunctions) in the form

$$\psi_n(x) = N_n \left(\frac{\cosh k_n x - \cos k_n x}{\cosh k_n l + \cos k_n l} - \frac{\sinh k_n x - \sin k_n x}{\sinh k_n l + \sin k_n l} \right) \quad (3.34)$$

where N_n is a normalizing factor and the eigenvalues k_n are the solutions of the equation

$$\cos kl \cosh kl = -1. \quad (3.35)$$

The function $\cosh kl$ is always greater than unity, and increases as kl increases. The function $\cos kl$, on the other hand, goes through a complete cycle from 1 to -1 whenever kl increases by π, so that there must be one solution to the equation in each range π of kl. Detailed solution requires numerical methods, but the mode frequencies ω_n, which from (3.30) vary as k_n^2, can be expressed in a convenient general form if we define $\beta_n = k_n l/\pi$. With this definition,

$$\omega_n = \frac{\pi^2}{l^2} \left(\frac{E\kappa^2}{\rho s} \right)^{1/2} \beta_n^2 \quad (3.36)$$

and

$$\beta_1 \approx 0.597 \quad \beta_2 \approx 1.494 \quad \beta_3 \approx 2.500 \ldots, \quad \beta_n \approx \left(n - \frac{1}{2}\right). \quad (3.37)$$

The sequence of mode frequencies thus has the approximate ratios 1.0, 6.3, 17.5, 34.4, ..., which are very widely stretched compared with the harmonic series 1, 2, 3, 4, ... of a string. Although we will have no call to examine the properties of bars supported in other ways, it is interesting to note that for free–free and clamped–clamped bars, the frequencies are still given by an expression of the form (3.36) but with $\beta \approx n + 1/2$ instead of $\beta \approx n - 1/2$. Rather surprisingly, a

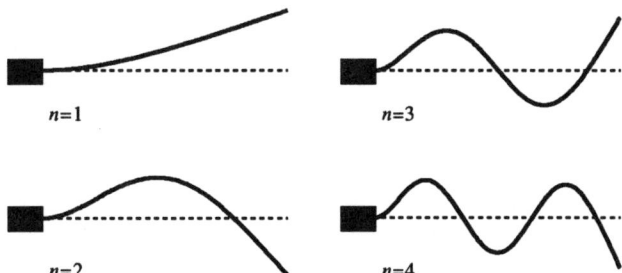

Figure 3.7 The first four normal modes or eigenfunctions for a bar clamped at one end and free at the other.

clamped-clamped bar has the same set of frequencies as the same bar in a free-free configuration.

Returning to (3.36), which we shall later use to represent the behavior of sensory hairs, we note that the mode frequencies all vary inversely with the square of the length of the bar, rather than linearly as for a string, and that they also vary linearly as the radius of gyration κ, and thus linearly with the diameter of the bar. These general observations will be of use when we come to applications of the theory.

Finally in this section we look at the shape of the mode function $\psi_n(x)$ for the first few modes of a bar in the clamped-free configuration. These are given by (3.34) and shown in Fig. 3.7. They show the same sinuous form as the mode functions for a string, but the nodes are no longer quite evenly spaced, those near the clamped end being wider apart than those near the free end. While we need to recognize the existence of higher modes, we shall find that the fundamental mode $n = 1$ is the only one of importance in most biological applications.

As an addendum, it is interesting to note that this discussion allows us to make somewhat more realistic our treatment of strings. A real string differs from the idealized model considered earlier in this chapter in that it has some stiffness, rather than being ideally flexible. This feature can be included in the equation of motion (3.3) in the form of an extra fourth-order differential term in the restoring force, derived as for the bar equation (3.28). Provided the stiffness term is small relative to the tension term, the equation can be solved approximately with no difficulty and leads to the expression

$$\omega_n \approx \frac{n}{l}\left(\frac{\sigma}{\rho_s}\right)^{1/2}(1+\varepsilon_1+\varepsilon_2 n^2) \qquad (3.38)$$

where ε_1 is a small correction to the fundamental frequency, proportional to $(E\kappa^2/\sigma)^{1/2}$, the square root of the ratio of the stiffness to the tension, and ε_2 is an inharmonicity parameter proportional to $E\kappa^2/\sigma$. The effect of the stiffness, as we might have expected from our discussion of the mode frequencies of bars, is to spread the mode frequencies of the real string rather farther apart than those of a perfect string.

3.10 Sinusoidally Driven Bars

It is straightforward to generalize the bar equations to include the effect of dissipative forces, whether from losses internal to the bar itself or from external viscous forces. The analysis goes ahead just as for the case of a string, and there is no need to repeat it. The physical consequences, too, are similar, the main practical differences being that, because the mode frequencies of a bar are quite widely spread, there is not so much overlap between the responses of the different modes. This makes the approximation of neglecting all modes except the single mode closest to the excitation frequency more nearly valid than for the string case.

The response of the system to a single sinusoidally varying force is similarly easily derived from our earlier discussion of the string. The only amendments required to the string-related result (3.24) are first to replace the string eigenfunctions $\psi_n(x)$ by the bar eigenfunctions, given by (3.34), and then to replace the mass factor $\pi a^2 \rho_S$ by the corresponding factor $S \rho_S$ for the bar. We defer a discussion of this to Chapter 4, where we shall also examine the damping by viscous forces in more realistic detail.

3.11 Imperfectly Clamped Bars—Sensory Hairs

Even in a macroscopic laboratory system it is impossible to clamp the end of a bar, or even a string, with perfect rigidity. This feature is even more apparent in biological systems, where the clamping tissue is often relatively soft and where, indeed, it is necessary to have motion communicated to the substrate tissue in order to activate some sort of neural transducer. Leaving aside this second aspect for the present, let us examine the effect of non-rigid clamping on the mechanical behavior. The simplest case is naturally that of the string, but it is better to examine the more biologically relevant case of a sensory hair, and thus of a bar imperfectly clamped at one end.

At the clamped end of a bar we have to satisfy two boundary conditions, which were taken for ideal clamping to be $z = 0$, $dz/dx = 0$. We could relax either or both of these conditions, but the more realistic one biologically is the constraint on the tilt dz/dx where the hair follicle is attached to the substrate cells. The attachment will itself have some sort of mechanical admittance, perhaps with a frequency dependence, but it is adequate for our present discussion to suppose it to be simply compliant, so that the angular deflection dz/dx is proportional to the bending moment. Thus

$$\frac{dz}{dx} = \mu E S \kappa^2 \frac{d^2 z}{dx^2} \tag{3.39}$$

where μ could be called the angular compliance of the mounting and the remainder of the right-hand side is the bending moment exerted on the mounting by the bent bar. We now need to choose the four constants in the general solution (3.31) for the bar so that this condition is satisfied instead of $dz/dx = 0$. The algebra is straightforward, but is not worth pursuing in detail. The results can, however, be presented in quite general form and are particularly illuminating.

If we define the angular compliance μ_B of the bar as a whole to be the angle through which its tip is deflected by unit force in a static situation, then $\mu_B = l/ES\kappa^2$ where l is the length of the bar. A useful general parameter to describe the behavior of any such compliantly mounted bar is then the ratio of the mount compliance to the bar compliance, $\mu/\mu_B = \mu ES\kappa^2/l$. The general behavior is now deduced from solution of an equation rather like (3.35) but having extra terms that are proportional to the compliance ratio μ/μ_B. We look at this in the discussion questions at the end of the chapter. In Fig. 3.8 we show the behavior of the factors identified as β_n in (3.36) for the first two modes of vibration of the bar, as functions of the parameter $2\mu/\mu_B$, the reason for the extra factor 2 becoming clear later. We recall that the mode frequencies are proportional to β_n^2, and that $\beta_n \approx n - 1/2$ for $n > 1$ for a bar clamped rigidly at one end. If $2\mu/\mu_B \ll 1$, implying that the mounting is much stiffer than the bar, then the values of all the β_n are essentially the same as those given in (3.37) for a rigid clamp. As the compliance ratio of the mounting increases towards 1, the values of the β_n all fall, implying a fall in the mode frequencies. For the fundamental mode $n = 1$, this decrease in β_1 continues as the parameter value rises above 1, and indeed $\beta_1 \to 0$ in the limit of large values of the parameter. This is clearly physically correct, for the attachment then becomes a simple hinged support, and the bar can rotate freely about it. For the higher modes however, as exemplified by the behavior of β_2, β_n simply tends to a value of about $n - 3/4$, the value appropriate for a bar hinged at one end and free at the other, which is to be compared with the value $n - 1/2$ for the bar when rigidly clamped at one end. The relative change in the upper mode frequencies is thus not large.

It is the behavior of the fundamental, however, that concerns us in the biological context, as we see in the next chapter. If the compliance ratio of the support is much less than 1, then the eigenfunction $\psi_1(x)$ for this first mode is still given approximately by (3.34) with $n = 1$. If the compliance ratio is much greater than 1, however, $k_1 \to 0$ and the first eigenfunction is approximately the linear deflection function

$$\psi_1(x) \approx (3/l^3)^{1/2} x. \qquad (3.40)$$

3.12 End-Loaded Bars—Otoliths

As another case of interest we consider the behavior of a bar, clamped rigidly at one end and loaded at its free end with a concentrated mass m. The free motion of the bar is once again described by the equations (3.29) and (3.30), and the general solution for a single mode still has the form (3.31). All that is different from the initial case is that, instead of the boundary condition $d^3z/dx^3 = 0$ at the free end, expressing the fact that the shearing force normal to the bar vanishes, we must substitute the condition that this force is related to the acceleration of the attached mass. The appropriate relation is

$$-ES\kappa^2 \left[\frac{\partial^3 z}{\partial x^3}\right]_{x=l} = m\left[\frac{\partial^2 z}{\partial t^2}\right]_{x=l} = -m\omega^2 z(l) \qquad (3.41)$$

which, using (3.30), can be transformed to

$$\left[\frac{d^3 z}{dx^3}\right]_{x=l} = -\frac{m}{\rho_s S} k^4 z(l). \qquad (3.42)$$

Again these boundary conditions lead to values for the coefficients in (3.31), which define the shape of the modes, and to an equation related to (3.35), the solutions to which give the normal mode frequencies. Again this equation is given in the discussion questions. If we define $m_B = \rho_s lS$ to be the mass of the bar, then a general parameter to describe the behavior of any bar is the ratio of the added mass to the mass of the bar, m/m_B. If we derive an equation, analogous to (3.35), from which to determine the mode frequencies, then it turns out that it has exactly the same form, in terms of the parameter m/m_B, as the analogous equation determining the mode frequencies of a compliantly mounted bar in terms of the parameter $2\mu/\mu_B$. This means that the behavior is once again as illustrated in Fig. 3.8. If the added mass is small compared with the mass of the bar, so that $m/m_B \ll 1$, then it has very little effect. As m increases, the value of β_1, and hence the frequency of the first mode, tends towards zero for $m/m_B \gg 1$. The frequencies of all the higher modes also fall as the mass load is increased, but their frequencies reach new limiting values given approximately by $\beta_n \to n - 3/4$ for $n > 1$. These are the mode frequencies for a bar clamped at one end and hinged at the other.

As in the previous example, it turns out that the important mode in biological sensory systems is the first mode, so that weighting a hair at its free end has a large effect on frequency response if the added mass is comparable to or larger than the mass of the hair. The form of this mode in the biologically important limit of large mass loading can be found by neglecting the terms in ω in the equation of motion (3.30) for the bar. The shape $z(x)$ of the bar then satisfies the simple equation $d^4z/dx^4 = 0$, and so has the form

$$z = A + Bx + Cx^2 + Dx^3 \qquad (3.43)$$

where A, B, C, D are new constants. When we satisfy the boundary conditions at $x = 0$ corresponding to a mounting with compliance μ and allow for a mass load m at the tip, which introduces a frequency term again, we find that a reasonable approximation to the frequency is given by

VIBRATION OF STRINGS AND BARS

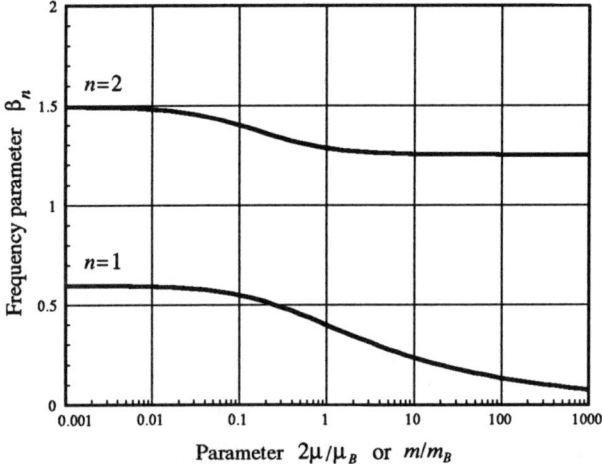

Figure 3.8 The frequency of the nth mode of a rod or bar is proportional to the quantity β_n^2. The figure shows how β_n varies for the first two modes as the ratio of the angular compliance μ of the clamp at the fixed end to the compliance μ_b of the bar itself is varied. The same curves describe the behavior of a bar loaded with a mass m at its free end as the ratio of m to the mass m_b of the bar itself is varied.

$$\omega^2 \approx \frac{3E\kappa^2 S}{ml^3}\left[1+\frac{3E\kappa^2 S\mu}{l}\right]^{-1}. \tag{3.44}$$

If the compliance μ of the hair root is very large, this dominates the effect of bending of the hair itself, and

$$\omega \approx (\mu m l^2)^{-1/2}. \tag{3.45}$$

We can treat the case of a bar, compliantly mounted at one end and attached to a mass load at the other, in just the same way. This is, indeed, the precise case applicable to an otolith detector. If we define the compliance ratio μ/μ_B and the mass ratio m/m_B just as before, then the equation analogous to (3.35) from which the mode frequencies can be determined is

$$\cos kl \cosh kl + 1 + \left(\frac{2\mu}{\mu_B}+\frac{m}{m_B}\right)kl(\sinh kl \cos kl - \cosh kl \sin kl)$$

$$-\frac{2\mu m}{\mu_B m_B}(kl)^2 \sinh kl \sin kl = 0. \tag{3.46}$$

The individual equations for the cases treated in the previous two sections can be recovered by setting either $m=0$ or $\mu=0$. The effects are not simply additive, because of the final term which involves both parameters. This cross term is very small, however, for the fundamental mode when both parameters are large, because the decreasing value of k more than outweighs the increasing values of m and μ.

The shape of the fundamental mode is approximately

$$\psi_1(x) \approx N\left[x + \frac{(3lx^2 - x^3)}{6lES\kappa^2\mu}\right] \tag{3.47}$$

where N is a new normalization constant. Note that if the bar is moderately stiff and the mounting moderately compliant, then the first term dominates this expression, and the mode displacement is nearly a linear function of x. Evaluation of the normalization constant N requires some thought to allow for the mass load at $x = l$. To correct for this we imagine a further section of bar, of sufficient length $l' = m/S\rho_s$ for its mass to equal that of the added load, compressed to a point at the end of the bar $x = l$, and moving with it. We then describe the eigenfunction by a sort of internal coordinate ξ running from 0 to $l + l'$ with $\psi_1(\xi) = \psi_1(x)$ for $\xi < l$ and $\psi_1(\xi) = \psi_1(l)$ for $l < \xi < l + l'$. The normalization constant N is then determined by squaring $\psi_1(\xi)$, integrating with respect to ξ from 0 to $l + l'$, and setting the result equal to unity as demanded by the normalization equation (3.15). The result is complicated in the general case but, for a bar stiff enough that a linear approximation is adequate in (3.46), we find

$$N \approx \left[\frac{3}{l^3(1 + 3m/\rho_s lS)}\right]^{1/2} \tag{3.48}$$

which can be compared directly with the approximation (3.40) for the unloaded bar. We can use the expression (3.46) with (3.47) directly in formulas such as (3.24) to calculate the response of the loaded bar to an external force.

The procedure outlined for calculating the eigenfunctions for a loaded bar can be refined and generalized, using appropriate notation, but it is not worthwhile to elaborate this here since we meet a similar problem only once again in this book.

References

Vibrating strings: [1] Ch 3; [7] Ch 5; [8] Ch 2; [10] Ch 2
Vibrating bars: [1] Ch 4; [7] Ch 6; [8] Ch 3; [10] Ch 2

Discussion Examples

1. By matching the boundary conditions for the general bar solution (3.31) for the case of a bar ideally clamped at one end and free at the other, derive the equation (3.35) that determines the possible values of the parameter k_n and thus, by (3.30), the mode frequencies ω_n.
2. Repeat this procedure for the case of a bar mounted in a clamp of angular compliance μ and show that the equation determining the values of k_n is now

    $$\cos kl \cosh kl + 1 + 2(\mu E\kappa^2 S/l)kl(\sinh kl \cos kl - \cosh kl \sin kl).$$

VIBRATION OF STRINGS AND BARS

3. Repeat this procedure for the case of a bar rigidly clamped at one end and loaded at the other end with a mass m and show that the equation determining the values of k_n is now

$$\cos kl \cosh kl + 1 + (m/\rho_s lS)kl(\sinh kl \cos kl - \cosh kl \sin kl).$$

4. Calculate the frequency of the fundamental mode for a rod of solid bone material ($E \approx 1 \times 10^{11}$ N m^{-2}, $\rho \approx 3000$ kg m^{-3}) of length 10 mm and diameter 1 mm, assuming one end to be clamped and the other free.

5. If the internal structure of the bone is cellular rather than solid, thus reducing both the effective density and effective Young's modulus by a factor 2, how does this affect the stiffness and resonance frequency? Compare these results with simply shrinking the cross-sectional area by a factor 2.

6. A sinew is made of biological material with density 1000 kg m^{-3} and breaking strength 2×10^9 N m^{-2}. If a 10 cm length of this sinew is stretched between rigid supports, calculate the maximum fundamental transverse mode frequency that can be attained before it breaks.

7. A bar has cross-sectional dimensions 5 mm × 1 mm and length 10 cm and is clamped at one end. It can be set into vibration either in the direction of its smaller or its larger cross-sectional dimension. Calculate the ratio of the fundamental mode frequencies in the two cases.

8. A string is stretched between two rigid supports 10 cm apart. A vibration exciter is attached to the string a distance of 2 cm from one end and a vibration detector a distance of 2.5 cm from the other, and vibration modes are detected by varying the frequency of the exciter and looking for maxima in the response of the detector. Are there any modes that would not be detected by this arrangement?

9. A hair has length 2 mm and diameter 20 µm. If the density of keratin is 1000 kg m^{-3} and its Young's modulus is 2×10^9 N m^{-2}, calculate the first resonance frequency of the hair when one end is clamped. About how large must a sand grain (density 3000 kgm^{-3}) be in order for its mass to greatly influence this resonance frequency?

10. If the sand grain in example 9 has diameter of 500 µm, what is the approximate resonance frequency for the loaded hair?

Solutions

1. Use the general solution (3.31). The boundary conditions (3.33) can then be used to eliminate A, B, C, D, and the result simplifies to the form (3.35). This takes about a page of algebra.
2. Proceed as for Example 1, but use the boundary condition (3.39).
3. Proceed as for Example 1, but use the boundary condition (3.41).
4. Frequency ≈ 16 kHz.
5. Subject to assumptions about the nature of the porosity, the stiffness is simply reduced by a factor 2. Since the density is similarly reduced, there is no change in resonance frequency. Shrinking the cross-section by a factor 2 reduces the radius by $1/\sqrt{2}$ and reduces the stiffness by a factor 4. The mass per unit length is reduced by a factor 2, so the resonance frequency is reduced by a factor $1/\sqrt{2} = 0.707$.

6. Maximum frequency ≈ 2250 Hz.
7. The ratio of κ^2 values for the two cases is 1:25. The frequency varies as κ, since the cross-section area is unchanged, so the frequency ratio is 1:5.
8. We are essentially measuring the transfer impedance between the two points x, x', as defined by (3.26). This has maxima at all mode frequencies unless $\psi_n(x)$ or $\psi_n(x')$ vanishes. In fact $\psi_n(x) = 0$ for $n = 5, 10, \ldots$ and $\psi_n(x') = 0$ for $n = 4, 8, \ldots$ so none of these modes will be detected.
9. Fundamental frequency ≈ 1.0 kHz. Hair mass = 0.6 µg so sand mass must be greater than about 1 µg and diameter greater than about 80 µm.
10. From (3.44) with $\mu = 0$, frequency ≈ 270 Hz.

4 SENSORY HAIRS AND OTOLITHS

SYNOPSIS. We discuss two different types of sensory hairs. The first is a simple hair in a compliant mounting, the deflection of which at its root activates neural transducer cells. Such a sensor responds to the velocity of relative movement between the animal to which it is attached and the surrounding fluid medium. It may be used by an animal either to sense the velocity disturbance created by motion of the wings or fins of a predator, or to sense the animal's own motion through the fluid. The second sensor is an otolith, consisting of a sensory hair loaded at its tip with a relatively large mass. Such a sensor responds to acceleration or to gravitational force, and can be used to detect the orientation of the animal relative to the vertical, the acceleration of its motion, or, in the case of aquatic animals, the acceleration component of disturbances in the medium.

A crucial determinant of performance is the viscous interaction between the medium and the hair when the two are in relative motion. When there is an oscillatory disturbance in the medium, viscous effects create a boundary layer near the stationary surface of the animal, across which the velocity increases from zero to its free-field value in the manner shown in Fig. 4.1(a). The thickness of the boundary layer is typically about 0.1 mm in biologically interesting situations, but it varies as the inverse square root of frequency. If the animal is moving through the fluid, then there is again a boundary layer in which the fluid tends to be carried along with the animal. The thickness of this boundary layer increases as the square root of distance from the leading edge of the animal as shown in Fig. 4.1(b), and is inversely proportional to the square root of velocity. For small animals at typical velocities the boundary layer may be a few millimeters in thickness.

Viscous forces cause considerable damping for thin hairs in air, so that their Q value is ordinarily low. This means that the hair follows the fluid motion closely, with some decrease in response away from the resonance frequency as shown in Fig. 4.2. It is customary in sensory biology to plot the stimulus threshold for detectable neural response. These threshold curves are normally plotted logarithmically, as in Fig. 4.3, and are given in decibels (dB) relative to some standard stimulus value. An increase in threshold of 10 dB represents an increase in the stimulus power necessary for detectable response by a factor 10. Since power is proportional to the square of amplitude, 10 dB also corresponds to an increase by a factor 3.16 in amplitude. Apart from the definition of the reference value corresponding to 0 dB, threshold curves are simply response curves plotted logarithmically and turned upside down.

Otoliths have the general appearance shown in Fig. 4.4. The otolith is normally surrounded by watery liquid and can occur either internally in land-based animals or connected to the environment in aquatic animals. The mass loading generally gives the otolith a low resonance frequency and the viscosity of the water generally means that it is nearly critically damped. Its angular displacement therefore corresponds closely to the acceleration of the animal, if the otolith is protected from outside influences. If exposed, it functions as a combination acceleration detector and low-resonance velocity detector.

4.1 Sensory Hairs

We now have nearly enough background to discuss the behavior of sensory hairs and similar systems. Our aim is not so much to attempt a detailed description of any real system as to set out the principles by which such a system might be analyzed. There are two basic types of hair receptors we shall examine. The first is a simple hair, mounted in a sensory cell that responds to angular displacement of the hair, or rather to the velocity of angular displacement. This has the function of detecting fluid movement, usually the vibrations generated by the wings or fins of a predator, but also the slower movements caused by air or water currents or by the motion of the animal itself. The second is a similar hair loaded at its remote end by a grain of sand or similar mass, constituting the otolith (literally "ear stone") detector used by some aquatic animals. This generally has the function of detecting either the static force of gravity, thus giving an indication of orientation, or the changing force caused by acceleration of the animal during its motion. We consider these in turn, but first we need to discuss the viscous forces acting on the hair in the surrounding medium.

4.2 Viscosity and Boundary Layers

To simplify matters as far as possible, let us suppose that the sensory hair in which we are interested is located on a surface of the animal that is flat over a distance large compared to the length of the hair. We then idealize matters to look at the behavior of a single hair on an infinite plane surface. This will give a good first approximation to the real behavior, and is sufficiently simple that we have the hope of an explicit solution.

The forces exerted by the fluid on the hair when they are moving relative to one another result predominantly from viscous forces in the fluid. These forces are specified by a quantity called the coefficient of viscosity η, which measures the resistance of the fluid to shearing forces. In some applications it is more useful to specify the kinematic viscosity $\nu = \eta/\rho_F$, where ρ_F is the density of the fluid. (The Greek letter nu (ν) is very like an italic v, but the context should allow the distinction to be made.) For air and water at an ordinary temperature around 20°C,

$$\eta_{air} \approx 1.8 \times 10^{-5} \text{ Pa s} \qquad \eta_{water} \approx 1.0 \times 10^{-3} \text{ Pa s}$$

$$\nu_{air} \approx 1.5 \times 10^{-5} \text{ m}^2 \text{ s}^{-1} \qquad \nu_{water} \approx 1.0 \times 10^{-6} \text{ m}^2 \text{ s}^{-1} \qquad (4.1)$$

The viscosity of air and other gases, which arises from molecular motion, increases with increasing temperature as $T^{1/2}$, where T is the absolute temperature, while the viscosity of water, which arises from molecular bonding, decreases with increasing temperature by about 2 percent per degree.

Let us suppose that the disturbance in which we are interested consists of a sinusoidally varying displacement of the fluid parallel to the surface and with frequency ω. We have already mentioned that any arbitrary signal can be decomposed into such simple sinusoidal frequency components by Fourier analysis, and we look at the details in Chapter 14. The signals generated by beating wings or fins are, in any case, more or less sinusoidal. Suppose that the fluid well away from the substrate is moving back and forward with velocity amplitude v_0. We know, however, that viscous drag keeps the fluid directly in contact with the surface completely still, so we first ask how the displacement amplitude builds up as we move away from the surface. The answer to this question is given by the equations of fluid dynamics—actually a particular case of the Navier-Stokes equations—and it would take us too far away from our main interest to discuss this in detail. If x measures the distance away from the substrate surface, then the result is that

$$v(x,t) = v_0(1 - e^{-x/\delta})e^{j\omega t} \tag{4.2}$$

where δ is the thickness characterizing the viscous boundary layer, and is given by

$$\delta = \left(\frac{\eta}{\rho_F \omega}\right)^{1/2} = \left(\frac{\nu}{\omega}\right)^{1/2}. \tag{4.3}$$

The behavior of the velocity through the boundary layer is shown in Fig. 4.1(a). Note that the boundary layer is not sharply defined in this situation, as it may be for high-speed flow, but is simply a transition region.

To get some feel for magnitude, suppose that the fluid flow is generated by the wings of a wasp, beating at say 200 Hz in air. Then the boundary layer thickness is about 0.1 mm. For the disturbance generated by the fins of a fish in water, the characteristic frequency is perhaps 5 Hz, which gives a boundary layer thickness of about 0.2 mm. Any hair more than about 1 mm in length therefore experiences

Figure 4.1 (a) Variation of fluid velocity amplitude v with distance x away from a plane surface when the velocity amplitude in the bulk fluid is v_0. The boundary layer thickness is δ. (b) Variation of boundary layer thickness δ with distance from the leading edge of a plane obstacle over which fluid is flowing.

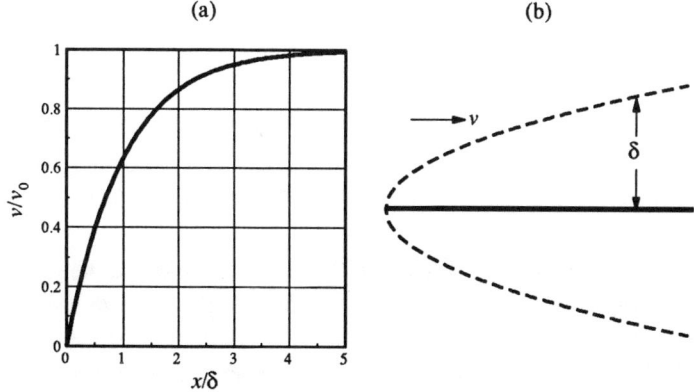

the force of the full flow velocity over most of its length, and we can really ignore boundary layer effects.

Another aspect of boundary layer behavior is seen in the case of a body moving steadily through a fluid—for example, a fish swimming through water. The fluid close to the moving surface is dragged along, but it takes some time for this disturbance to reach farther into the fluid. There is thus a characteristic length, or diffuse boundary layer thickness, which increases steadily from the front to the back of the moving body. This characteristic thickness is given by

$$\delta = \left(\frac{\eta x}{\rho_F v}\right)^{1/2} = \left(\frac{\nu x}{v}\right)^{1/2} \tag{4.4}$$

where x is the distance from the leading edge of the object and v is its velocity through fluid of kinematic viscosity ν. This is illustrated in Fig. 4.1(b). The behavior of velocity with normal distance z through the boundary layer is similar to that shown in Fig. 4.1(a). For small animals in motion, the boundary layer thickness is typically a few millimeters.

4.3 Viscous Force on a Hair

Some 140 years ago, Stokes calculated the force on an oscillating sphere in a fluid, and we shall need this result in our treatment of otoliths. For a sphere of radius a oscillating with velocity v at frequency ω in a fluid of density ρ_F and kinematic viscosity ν, the resultant force is

$$F = -3\pi a^2 \rho_F \left(\frac{2\nu}{\omega}\right)^{1/2}\left[1+\frac{1}{a}\left(\frac{2\nu}{\omega}\right)^{1/2}\right]v - j\left(\frac{4}{3}\pi a^3 \rho_F\right)\left[\frac{1}{2}+\frac{9}{4a}\left(\frac{2\nu}{\omega}\right)^{1/2}\right]\omega v \tag{4.5}$$

where the time variation $e^{j\omega t}$ is understood. This force has a real part tending to slow the motion, and an imaginary part representing an added mass, which is essentially that of the fluid carried along in the boundary layer around the sphere. In the limit of low frequencies and small radii, such that $\omega a^2/\nu \ll 1$, (4.5) reduces to the familiar form

$$F \approx -6\pi\eta a v. \tag{4.6}$$

This condition is usually at least marginally satisfied for the biological systems in which we are interested, so we can generally use the simple form (4.6). For the case of a sphere immersed in a fluid moving with velocity v_0, we simply replace v with $v - v_0$.

The viscous force acting on an infinitely long cylinder moving sinusoidally in a fluid was also studied by Stokes. The general problem is by no means simple, but fortunately the hairs with which we are concerned have very small radius a, and the frequency ω is also small, so that the parameter $\omega a^2/\nu$ appearing in the theory is very small, leading once more to considerable simplification. The force F acting on a length l of the cylinder, for a cylinder velocity $ve^{j\omega t}$, is

$$F \approx -\eta G l [2 + j(\pi^2/g)] v \qquad (4.7)$$

where

$$G = -g/[g^2 + (\pi/4)^2] \quad \text{and} \quad g = 0.58 + \ln\left[\frac{a}{2}(\omega/\nu)^{1/2}\right]. \qquad (4.8)$$

(Note that g has nothing to do with the gravitational acceleration!) This result (4.8) is approximately correct only for $\omega a^2/\nu \ll 1$. Once again the real part of F in (4.7) is a viscous drag force, while the imaginary part is equivalent to a mass added to the hair, representing the fluid trapped in the boundary layer surrounding it, a boundary layer which is, however, much thinner than that adjacent to a plane surface.

Substitution of numerical values in (4.7) and (4.8) shows that the viscous force is a slowly varying function of radius and frequency—almost logarithmic, in fact—and that the added mass is significant only in the case of extremely fine hairs a few microns in diameter. Except for such extremely fine hairs, the mass component can therefore be neglected. The slow variation of damping force with hair radius, and the fact that G is typically in the range 0.1 to 1 means that fine hairs are extremely highly damped. Indeed, comparing (4.7) with (4.6) shows that the total damping force on a fine hair is equal to that on a sphere with diameter about one tenth the length of the hair.

This treatment can again be applied to give the force acting on a long cylinder in an oscillating flow field v_0 by replacing v by $v - v_0$. It is something of an extrapolation to apply it to a relatively short hair, but this extrapolation is justified if the radius of the hair and the amplitude of the motion are both very small compared to the length of the hair, as is generally the case in practice. Even if the assumptions are not fully justified, the treatment will still give good approximate results.

4.4 Sensory Hairs

Sensory hairs vary widely in length, diameter and stiffness in different animals. Sometimes they are moderately flexible, but generally they are fairly stiff so that they transmit bending moments efficiently to the sensory cells at their base. The mounting itself is generally only moderately stiff, so as to transmit motion of the hair to the underlying sensory cells, but needs to be stiff enough that the hairs do not suffer too large an angular displacement during the motion of the animal through the surrounding fluid. This combination of features means that, from our discussion in Section 3.11, the resonance frequency of the fundamental mode is small compared with that of a rigidly clamped bar with the same properties, while the deflection associated with the fundamental mode is nearly linear with distance from the hair root, as given by (3.40). We could thus treat the motion as one of angular deflection, but it is simpler to use our general approach.

We can measure the dimensions of the hair microscopically, and we can calculate an upper limit to its resonance frequency in the fundamental mode from equations (3.36) and (3.37). The actual resonance frequency will be lower than this, as shown in Fig. 3.8, because of the angular compliance of the root. We might estimate this in a practical case by bending the hair from the tip and noting the ratio between the angular deflection of the tip and the difference in angle between tip and root ends (thus essentially giving the ratio $\mu SE\kappa^2/l$ used in the theory), but we cannot expect an accurate result because hairs are generally tapered. Armed with an estimate of this frequency, we can now proceed to calculate the expected response.

First consider a hair that can detect vibrations in the fluid while the animal is stationary. Provided the hair is more than about a millimeter in length, the flow field can be taken to be constant along its length, as discussed in Section 4.1. If the flow field normal to the hair has velocity v_0, then the viscous drag on the hair is given by (4.7) with $v - v_0$ substituted for v. The part of this expression involving v contributes the damping coefficient α, while the part in v_0 gives the forcing term. We can use the general results (3.23) and (3.24), along with the explicit expression (3.40) for the fundamental eigenfunction $\psi_1(x)$ of the hair to calculate its amplitude response $a_1\psi_1(x)$. The damping constant α is given by

$$\alpha = \frac{2\eta G}{\pi a^2 \rho_s} \tag{4.9}$$

where G is the factor from (4.8) and has a value between about 0.1 and 1 for typical hairs in either air or water. We need to compare this with $\omega_0/2$ to find the quality factor Q of the resonance, defined by (2.23) to be $\omega_0/2\alpha$. For a resonance frequency of order 100 Hz, we conclude that $Q \sim 10$ in air, while in water $Q \sim 0.1$. There is thus some influence of the resonance properties of the hair on its behavior in air, while in water hairs are usually much more than critically damped.

After a little algebra on (3.23) and (3.24), we find for a hair of length l in water the particularly simple approximate result

$$z(x,t) \approx \frac{3x}{4j\omega l} v_0 e^{j\omega t} \quad \text{or} \quad v(x,t) \approx \frac{3x}{4l} v_0 e^{j\omega t}. \tag{4.10}$$

This is not exact, since bending of the hair under the influence of the viscous forces, as represented by excitation of higher modes, has been neglected, but it is a good approximation. The angular velocity of the tilting motion of the hair in its root cell, which is presumably the quantity transduced to nerve impulses, depends on the ratio μ/μ_H between the compliance μ of the root mounting and that of the hair

SENSORY HAIRS AND OTOLITHS

itself μ_H, but, provided this ratio is large, this angular velocity is just $(3/4l)v_0$. A moderately short hair is therefore a more effective transducer than a very long hair, provided it is several times greater in length than the thickness of the boundary layer and also long enough to be well coupled to the fluid medium, as judged by having a Q value significantly less than 1.

For hairs in air or other gaseous medium, there is still significant influence from the hair resonance. After some algebra, we find the result

$$v(x,t) = \frac{j\omega\omega_0}{Q(\omega_0^2 - \omega^2) + j\omega\omega_0}\left(\frac{3x}{4l}\right)v_0 e^{j\omega t}. \qquad (4.11)$$

Again, assuming that the angular compliance of the clamp at the hair root is large compared with the bending compliance of the hair itself, the mechanical stimulus applied to the neural transducer is just the angular motion $\partial v/\partial x$. The response thus has a maximum value for $\omega = \omega_0$ that is equal to the response attained in the fully damped case $Q \to 0$. For frequencies greater or smaller than ω_0 the response falls but, since the Q value is not large, the bandwidth is quite broad. For hair-receptors in air, therefore, there is clearly an advantage if the mechanical resonance is tuned to the frequency of the stimulus that it is desired to detect. The general behavior is shown in Fig. 4.2.

For a hair of uniform cross-section, $\omega_0 \propto a/l^2$ from (3.36), while from (4.9) $\alpha \propto a^{-2}$ approximately. This means that $Q \propto a^3/l^2$, and this relation defines the

Figure 4.2 Response of a sensory hair to fluid motion as a function of the Q value of its resonance. Frequency is normalized to the resonance frequency and response to that of a greatly over-damped hair ($Q \to 0$).

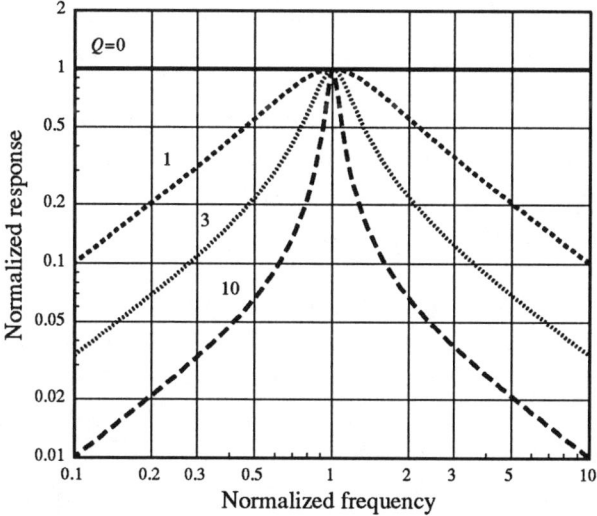

bandwidth ω_0/Q of the response. For a given resonance frequency, long thin hairs have a lower Q and broader response bandwidth than short thick hairs. However the hairs must be long enough to protrude well outside the boundary layer, and thick enough that they do not bend too much in traversing the boundary layer. We should also emphasize once again that this analysis assumes that the hair is stiff relative to the follicle in which it is mounted.

We could, if we wished, carry through a more detailed derivation for the case when the compliance at the hair root is not large compared with the bending compliance of the hair. The algebra is tedious, however, and we shall not give it here. We can easily see, in a general way, that each of (4.10) and (4.11) should be multiplied by a factor $\mu/(\mu+\mu_H)$, where μ is the angular compliance of the root clamp and $\mu_H = ES\kappa^2/l$ is the effective bending compliance of the hair, when calculating from them the mechanical stimulus to the transducer cell. This factor is nearly unity unless the hair is rather flexible or its mount rather stiff. When this happens to the extent that $\mu < \mu_H$, the hair length l nearly disappears from the result, and only a short length of hair near the root contributes to the stimulus.

A second type of sensory hair is one intended to detect the movement of the animal through the surrounding fluid and thus to give a measure of velocity. In this case the boundary layer thickness depends on the speed of motion of the animal and on the position of the hair on its surface according to (4.4). If we take $v \sim 0.1$ m s^{-1} as a typical velocity and suppose the hair to be about 10 cm from the leading edge of the animal, then the boundary layer thickness at the position of the hair is around 4 mm in air and 1 mm in water. This requires that the sensory hair be at least 1 cm in length if it is to be efficient as a velocity sensor. Of course if the animal is smaller so that the hair is closer to the leading edge, its length can be reduced. For an insect with the hair only 5 mm from the leading edge in air, the boundary layer thickness at 0.1 m s^{-1} is only 1 mm.

We could write down a solution to this problem of about the same complexity as those derived in the earlier part of this section, but it is good to look at a slightly more complex model that includes boundary layer effects explicitly. Because the motion in this case is steady, we can neglect time-varying terms in the equation of motion (3.29) so that, including the viscous force $F(x)$, it becomes

$$-ES\kappa^2 \frac{d^4z}{dx^4} = F(x) = -2\eta G v_0(1-e^{-x/\delta}). \tag{4.12}$$

The drag force itself is given by the real part of (4.7), with the velocity given by the boundary-layer equation (4.2), using (4.4) for the boundary layer thickness. Experience allows us to guess a general solution to (4.12) of the form

$$z(x) = A + Bx + Cx^2 + Dx^3 + \frac{\eta G v_0}{12ES\kappa^2}(x^4 - 24\delta^4 e^{-x/\delta}) \tag{4.13}$$

where A, B, C, D are new constants and the final term is chosen to balance the expression on the right-hand side of (4.12). The unknown constants in the general solution are fixed by the boundary conditions, which we take to be those discussed in Section 3.11 for an imperfectly clamped bar with clamp compliance μ. The algebra is straightforward but tedious, and the general solution is quite involved. In the limit when the hair is much longer than the boundary layer thickness, we find that the angular deflection θ at the root is given by

$$\tan\theta = \left[\frac{dz}{dx}\right]_{x=0} \approx \mu\eta G v_0 l^2. \qquad (4.14)$$

The length of the hair appears squared because a bending moment is involved. The angular deflection is clearly proportional to velocity and to the compliance μ of the mount, for small deflections, but sensibly saturates at 90° for large velocities.

Importantly from the point of view of sensory perception, we note that hair sensors of the types we have been discussing, and other hair sensors as well, are able to give information about the vibration direction, and thus about the direction of origin of the disturbance, provided they are connected to at least two sensor cells that respond to deflections in two orthogonal planes. Real hair sensors may, of course, have more than two transducer cells.

4.5 Sensory Thresholds

The response function for a sensory hair illustrated in Fig. 4.2 makes good sense from a physical point of view, since we may reasonably expect to be able to measure the velocity amplitude of the motion of the hair and relate it to the velocity amplitude v_0 in the moving fluid. In biophysical experiments, however, the measurement is usually of a different kind, and the response of the neural transducer cell is involved. Such cells generally have a discontinuous response—they either emit a pulse of fixed magnitude or they remain quiescent. Certainly the number of pulses emitted per second generally depends on the strength of the stimulus, but the dependence is far from linear. There is thus no obvious way in which to measure a curve such as that in Fig. 4.2 for comparison with theoretical predictions, and a different approach must be used.

Transducer cells are not completely quiescent when unstimulated, but fire randomly at a low rate. When a stimulus is applied, this rate increases, and we can define a stimulus threshold for detection to be that value of stimulus that increases the firing rate by a specified factor—say a factor 2—above the spontaneous rate. It turns out that if we adopt a different factor, say 10, then this makes relatively little difference to the form of the result, as we see later. If we determine the detection threshold v_T for the stimulus at each frequency, according to this criterion, then we can plot a curve that is called the sensory threshold curve.

In plotting such a curve we might measure the strength of the stimulus by giving, in the case of sensory hairs, the fluid velocity amplitude v_T that produces the specified increase in firing rate. Actually, however, it is more sensible to refer the threshold not to the amplitude but rather to the energy or power or intensity of the disturbance. We shall define these terms in detail in Chapter 6, and for the moment it is sufficient to note that each of them is proportional to the square of the disturbance amplitude, v_0^2. A threshold curve plotted in terms of v_T^2 would then be what we might call an absolute threshold curve. Often we are more interested in the shape of the curve than its absolute level, and it is then convenient to give a value of the relative threshold by plotting v_T^2/v_{ref}^2, where v_{ref} is an appropriate reference value for the disturbance. This might be taken, for example, to be the threshold at the most sensitive frequency. If we chose some physically defined value such as 1 m s^{-1} for the reference, then we would once more have an absolute threshold.

When we plot the sensory threshold as a function of frequency, it is convenient to use a logarithmic scale for the relative stimulus power v_T^2/v_{ref}^2, and we call the resulting function the threshold stimulus level $L_T(v)$. The defining equation for the stimulus level associated with a velocity amplitude v_0 is

$$L(v) = \log_{10}(v_0^2/v_{ref}^2) = 2\log_{10}(v_0/v_{ref}) \quad (4.15)$$

and the unit is called the bel, after the American inventor of the telephone, Alexander Graham Bell (1847-1922). The bel, which corresponds to a change in power by a factor 10, is rather too large for general use, and it has become customary to use a unit one-tenth as large, called the decibel. Applying this definition to (4.15) gives

$$L(v) = 20\log_{10}(v_0/v_{ref}) \quad \text{dB} . \quad (4.16)$$

We return to the subject of levels in decibels in Chapter 6 when we consider waves.

Figure 4.3 shows the response curves of Fig. 4.2 plotted as relative threshold curves, with the 0 dB level $v_0 = v_{ref}$ taken as the threshold at the frequency of maximum sensitivity. In this diagram the frequency axis is also drawn with a logarithmic scale and made relative to the resonance frequency ω_0. Curves of this sort, though with true frequency on a logarithmic scale rather than relative frequency, are a standard method of presenting the results of sensory experiments. Note that the form of the curves is related to those in Fig. 4.2 simply by reversing the diagram vertically.

This discussion has assumed that the sensor cell has no inherent frequency discrimination of its own, but responds equally over a broad frequency range. In reality the situation is usually more complex than this, and individual sensory cells often do have some form of internal tuning. This is generally compensated for by the fact that the neural transducer apparatus may have several sensory cells, each

SENSORY HAIRS AND OTOLITHS

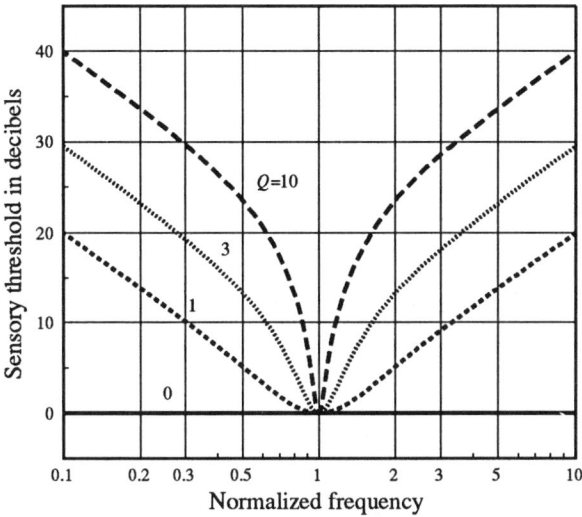

Figure 4.3 The velocity response curves of Fig. 4.2 replotted as sensory threshold curves, relative to the velocity level at peak sensitivity.

covering a different frequency band. Exploration of these complexities would take us outside the bounds set for this book.

4.6 Otoliths

An otolith detector consists of a dense grain of mineral material, supported upon one or more sensory hairs and immersed in aqueous fluid that moves with the animal. The otolith is generally internal to the animal, which usually is an aquatic species. The otolith is thus protected from external influences except for those associated with gravity or with the accelerated motion of the animal. From the point of view of response, these two influences are the same, except that the gravitational acceleration g is constant in time and direction. Both cases are important, but our primary interest here is with the time-varying acceleration produced by acoustic waves in water. The density of an aquatic animal is normally very close to that of water, so that it moves bodily to follow the acoustic displacement in the wave.

Let us calculate the behavior of a simple otolith, consisting of a small grain mounted on a single hair as shown in Fig. 4.4(a). The behavior of the multi-hair otolith shown in Fig. 4.4(b) is essentially similar, though the physical magnitudes involved are very different.

Suppose that the animal has an acceleration $f(t)$ in the z direction normal to the length of the hair. If the surrounding fluid, which moves with the animal, is liquid water, then its density is very nearly the same as that of the biological tissue

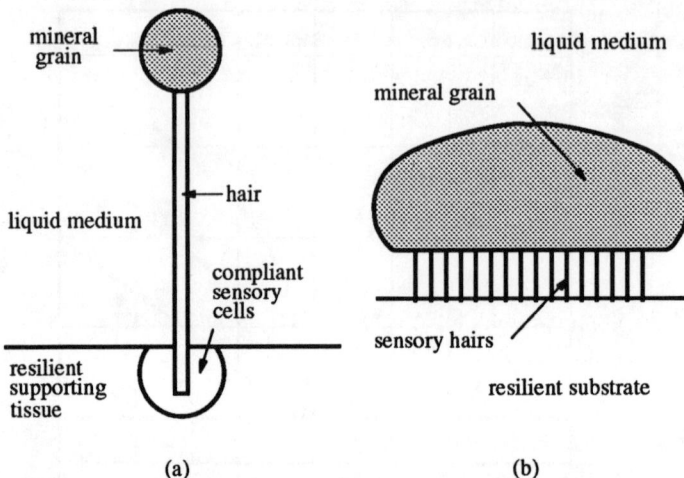

Figure 4.4 (a) Schematic drawing of a simple otolith detector. The hair root is surrounded by compliant neural transduction cells that distort and fire when the hair is deflected. (b) A larger otolith detector (not to the same scale) with multiple supporting hair cells.

of the hair, so that the acceleration force on the hair itself is the same as that on the fluid and there is no tendency for relative motion. The otolith, however, has a density $\rho_M \approx 3000$ kg m^{-3} characteristic of mineral material, which is very much greater than the density $\rho_F \approx 1000$ kg m^{-3} of the fluid. If we consider the motion from the point of view of coordinates moving with the animal, then the otolith experiences an inertial force associated with a negative acceleration $-f(t)$, and it tends to lag behind the motion of the animal and fluid. The acceleration of the system thus has just the same effect as a time-varying gravitational acceleration and, in calculating the net force on the otolith, we must allow for a buoyancy effect associated with the fluid displaced by the mineral grain. The effective acceleration force acting on the otolith is thus

$$F(t) = -m\left(\frac{\rho_M - \rho_F}{\rho_M}\right) f(t) \qquad (4.17)$$

and this force is localized at the tip of the hair, $x = l$. In addition there will be viscous forces acting on the hair that are proportional to its velocity relative to the surrounding fluid.

We discussed the behavior of such a mass-loaded bar in Section 3.12 and found in equation (3.47) an expression for the eigenfunction $\psi_1(x)$ for the fundamental mode. In the present case the calculation is quite easy, since the force is concentrated at the loaded end of the hair, $x = l$, which is also where the mass is concentrated. It does not matter how we assume the force to be concentrated over the internal coordinate x' for $l < x' < l + l'$, since the eigenfunction has the constant value $\psi_1(l)$ over this whole range. From (3.47) and (3.48) for large mass loading m the eigenfunction at the mass has the value

$$\psi_1(x') = (S\rho_s/m)^{1/2} x'/l \quad \text{for} \quad 0 < x' < l$$
$$= \psi_1(l) \quad \text{for} \quad l < x' < l + l'. \quad (4.18)$$

When we use this in the bar analog of the response equations (3.23) and (3.24), we find the complex amplitude a_1 of the fundamental mode, and thus the hair displacement $z_1(x)$ measured relative to the animal, to be

$$z_1(x) = a_1 \psi_1(x) \approx \left[\frac{-f(\rho_S - \rho_F)/\rho_S}{(\omega_1^2 - \omega^2) + 2j\omega\alpha(\omega)} \right] x \quad (4.19)$$

where ω is the frequency of the motion of the animal, and we still have to determine the damping coefficient α. For steady acceleration, or for a steady gravitational force, $\omega = 0$. The actual displacement of the mass at the tip of the hair, relative to the position of the root, is $z(l)$. The form of this result, using (4.19) in (4.18), is identical with what we would have obtained if we had simply assumed the hair and its mount to provide a spring return for the mass load and regarded it as a simple vibrator of the type discussed in Chapter 2. Indeed such an approach would have given us an adequately complex model, so that the detailed argument above is really unnecessary. It serves, however, to justify the simple treatment, shows how it can be extended for smaller values of the mass load or root compliance, and points up the nature of the approximations made.

From the form of the mechanical response (4.19), we can come to some conclusions about the sensory response. The force acting on the transducer cell in the root of the hair is proportional to its angular deflection, and thus to $z_1(l)/l$. The response of the cell could be proportional either to the hair deflection or to the rate of change of the deflection, which introduces another factor ω. For an acceleration of constant amplitude over the frequency range, the sensory response should have the form of a simple resonance curve, as illustrated earlier in Fig. 2.5, the velocity response curve applying to transducer cells sensitive to rate of change of force, and the displacement response curve to cells sensitive to steady forces. Sensory threshold curves for acceleration response are found by inverting these diagrams.

If we are interested in acoustic response rather than acceleration response, we should plot sensitivity for constant acoustic intensity over the frequency range, which is the same as the response for constant velocity amplitude. The threshold curves in this case have the forms shown in Fig. 4.5 for the two sorts of sensory

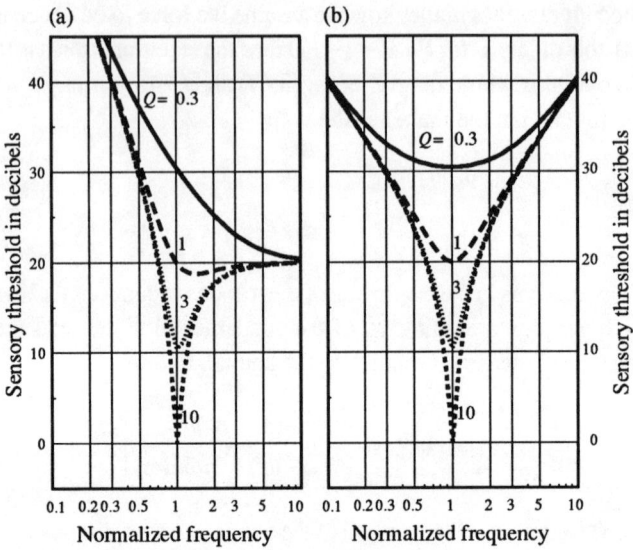

Figure 4.5 Sensory threshold curves for an otolith detector in an acoustic field, assuming sensory cells respond to (a) rate of change of hair deflection angle, or (b) hair deflection angle. The reference level is arbitrary.

cells. The curve for cells responding directly to hair deflection gives a conventional type of sensory threshold and is associated with a steady response to a constant gravitational acceleration, as illustrated in Fig. 2.5(a).

Damping behavior is clearly important in determining sensory threshold and frequency response. Equations from which the damping can be calculated are given in Section 4.3, and it is perhaps most helpful to simply calculate a representative example. Suppose that the otolith consists of a mineral grain 2 mm in diameter on the end of a hair 5 mm in length and 200 μm in diameter, immersed in water. We take the relevant time scale of the motion to be about 0.01 s for a small aquatic animal so that, in calculating the frequency-dependent drag on the hair, we can take $\omega \sim 1000$ s^{-1}. Using the numerical values for viscosity given in (4.1) we find, from (4.6), the drag force on the mineral grain to be about $2 \times 10^{-6} v$ while the real part of (4.7) gives the drag on the hair as about $2 \times 10^{-5} v$, where we have made allowance for the fact that, because it is pivoted at its root, the average velocity of the hair is only $v/2$. These results, which depend little on the frequency or the diameter of the hair, show that most of the drag acts on the hair rather than on the mineral grain. Writing the total drag force as γv, as in Chapter 2, gives $\gamma \approx 2 \times 10^{-5}$ N m^{-1}s. The mass of the grain and hair together, appropriately reduced by buoyancy effects, is about 10^{-8} kg, so that $\alpha = \gamma/2m \approx 10^3$. If the resonance frequency is about 100 Hz so that $\omega_0 \approx 600$ s^{-1}, this gives $Q \sim 0.3$. For this example, therefore, the damping is a little more than critical. The Q value is, however, very closely dependent upon the size of the mineral grain and the resonance frequency of the loaded hair.

Let us examine now the detection of a steady gravitational force, in which case $f = g$ and $\omega = 0$. The steady deflection of the otolith is then

$$z(l) \approx -\frac{g}{\omega_0^2}\left(\frac{\rho_S - \rho_F}{\rho_S}\right) = -\mu m g l^2 (\rho_S - \rho_F)/\rho_S \,. \qquad (4.20)$$

The natural frequency ω_0 in the absence of the surrounding medium is $\omega_0 = (\mu m l^2)^{-1/2}$ where μ is the angular compliance of the hair root, and this leads to the second expression in (4.20). No great sensitivity is required for this detector, since $g \approx 10$ m s^{-2}, and it would clearly be of advantage if it were critically damped. For lighter damping, sudden change of position would induce oscillations in the otolith that would take some time to die away, while heavier damping would make the response unduly slow. The calculation outlined above suggests that a simple otolith might indeed be nearly critically damped and thus serve the dual purpose of attitude detection and acoustic sensing. Clearly we require a set of at least two orthogonally related transducer cells in order to resolve the direction of deflection of the hair and thus to allow the animal to sense the vertical. Two such otolith detectors with different base planes would then suffice. Alternatively an animal might possess three or more appropriately oriented otolith detectors with unidirectional transducer cells.

The behavior of an otolith mounted on many hairs and perhaps attached by means of some form of viscous jelly, as shown in Fig. 4.4(b), is essentially similar to that of the simple otolith discussed above. We might term this a shear otolith, for the elastic restoring force acting on the stone derives from shear distortion of the supporting layer with its embedded hairs, and this can be best treated as a homogeneous elastic medium. Clearly the mineral grain itself might well have a mass 1000 times that discussed in our example above, but dynamic similarity can be retained if it is supported on 100 hairs, each only one-tenth as long as in the simple example. If a jelly-like attachment medium is present, then its elasticity and viscous losses must also be taken into account. While it would be possible for such a multitude of hairs to be organized into two families with sensitivity axes at right angles, it is equally possible that all hairs have the same sensitivity axis and that the animal possesses several independent otolith detectors.

4.7 Nonlinearity

The response theory outlined above is quite linear, in the sense that the amplitude of the angular motion of the hair in its socket is proportional to the amplitude of the exciting fluid velocity v, in the case of a sensory hair, and to the amplitude of the acceleration f in the case of an otolith. In fact, the response shows this linearity only for small signals, as we can easily see.

The first nonlinearity is a geometrical one. We have assumed that the hair protrudes normal to the surface on which it is mounted and that it has length l. In the case of a sensory hair, the force exerted by the moving fluid is proportional to l, while the same proportionality holds for the bending moment in the case of an otolith. When the hair has been bent away from the normal through an angle θ, however, the effective length becomes $l \cos \theta$, and this decreases steadily towards zero as $\theta \to 90°$. At large excitation amplitudes, therefore, the stimulus imparted to the sensory cell at the hair root tends to saturate. This geometrical nonlinearity thus has a fortunate protective effect on the transducer cells.

The second type of nonlinearity is a structural one, and depends upon the elastic behavior of the compliant mounting of the root of the hair. It is reasonably to expect this mounting to become stiffer for large amplitude deflections, but there is no available data on this point. Again, this will lead to gradual saturation of the stimulus at large amplitudes.

A nonlinearity of either of these two types will have a distorting effect on the stimulus waveform. A large-amplitude sine wave motion, for example, will be converted to a shape closer to a square wave. Whether or not this is important as an amplitude clue to the animal concerned is debatable.

References

Viscous losses: [12]
Otoliths: [31] Ch 3

Discussion Examples

1. The keratin of which a hair is made has a density of 1000 kg m^{-3} and a Young's modulus of 2×10^9 N m^{-2}. Calculate the fundamental frequency of a hair of length 3 mm and diameter 20 μm, assuming the hair to be rigidly clamped at its root. Find the approximate Q value in air.
2. Make a similar calculation if the length of the hair is 10 mm.
3. Make a similar calculation for a bristle of length 5 mm and diameter 200 microns in water.
4. The hair of example 1 is loaded with a sand grain of diameter 0.5 mm. Using the graph in Fig. 3.8, estimate the vibration frequency.
5. If the root of the hair in example 4 has an angular compliance of 2×10^{10} radian N^{-1} m^{-1}, estimate the first resonance frequency.

Solutions

1. From (3.36) and (3.37), $f_1 \approx 440$ Hz. From (4.8), $g \approx -2$, so $G \approx 0.4$. From (4.9), $\alpha \approx 50$, so $Q \approx 30$. The hair is only lightly damped.
2. Frequency varies as 1/length^2, so $f_1 \approx 40$ Hz. G is little changed ($G \approx 0.3$), but the frequency is much lower. $Q \approx 3$.
3. Frequency also varies directly as the diameter, so $f_1 \approx 1580$ Hz. The value of g given

by (4.8) is positive, and the assumptions underlying the result (4.7) are no longer valid, since the diameter is too large and the frequency too high. Our simple theory is inapplicable.
4. The mass m_B of the hair is about 10^{-9} kg and that m of the sand grain about 2×10^{-7} kg, so that $m/m_B \approx 200$. From Fig. 3.8, β_1 is about 0.1, so the frequency is reduced to $440 \times (0.1/0.6)^2 \approx 12$ Hz.
5. The compliance of the root is large compared with that of the hair, so from (3.45) $f_1 \approx 1$ Hz.

5 VIBRATION OF MEMBRANES, PLATES, AND SHELLS

SYNOPSIS. Vibrating membranes, plates, and shells are important components of biological acoustic systems. Like strings, membranes rely upon being stretched with appreciable tension over rigid frames, and their vibration modes have nodal lines or curves that divide the membrane into areas vibrating with opposite phase, as shown in Figs 5.1 and 5.4. Generally only the lowest mode, which has no nodal curves and a maximum amplitude near its center, is important in biological applications. The frequency of this first mode depends on the square root of the tension divided by the membrane thickness (or the elastic stress divided by the density) and is inversely proportional to the square root of the area of the membrane. It depends relatively little upon the exact shape of the membrane, provided it is reasonably equiaxial. This is illustrated for rectangular and circular membranes in Fig. 5.2.

An interesting exception to this generalization is the tapered membrane, illustrated in Fig. 5.5, which has some connection with the behavior of the basilar membrane in the mammalian cochlea. Analysis shows that the vibrations of the normal modes are confined towards the broad end of the membrane, but extend further towards its apex as the frequency is raised. The area of maximum amplitude similarly moves towards the apex for higher frequencies.

The response of a membrane to an external force, for example an acoustic pressure applied uniformly over its surface, can be expressed in a similar form to the case of a force acting on a string, or even a force acting on a simple vibrator. The response is greatest at the mode frequencies of the membrane, and is limited by the membrane damping.

In biological systems the membrane may be elastically braced, as shown in Fig. 5.6, to control its tension. It may also be loaded at its center by the necessary mechanical connection to the neural transducer mechanism. Such loading can affect the resonance frequencies of the membrane, lowering it significantly if the load is mass-like, and can also introduce significant damping.

If the thickness of a membrane is significant, its stiffness begins to control its vibrational behavior, and it is more appropriately regarded as a plate. The vibration behavior of a plate is broadly similar to that of a membrane, but it is self-supporting and does not require to be held under tension. Its mode frequencies are proportional to thickness and inversely proportional to plate area.

If the plate is curved to a shallowly dished shape it is called a shell. The vibration frequency of a shell depends on the ratio of its thickness to its dome height. If the thickness is very small, the fundamental resonance is proportional to dome height and inversely proportional to area. For thicker shells, the dome height has less influence. Because the natural frequencies of domed shells are higher than those of similar flat plates, they are useful as components of high-frequency acoustic systems.

Application of force to the center of a domed shell can cause it to buckle towards an inverted configuration. If the shell is ribbed so that its thickness is not uniform, then the buckling may proceed in a sequence of abrupt steps, thus generating a train of mechanical or acoustic pulses. Such a system is used in the tymbals of certain insects as a sound-producing mechanism.

MEMBRANES, PLATES, AND SHELLS

5.1 Extended Surfaces

The sound-detecting and sound-producing organs of nearly all animals are based upon the vibration of extended surfaces, which may be either simple stretched membranes, flat plates with appreciable stiffness, or curved shells. In this chapter we examine the vibrational properties of each of these components, building once again upon the foundations laid in Chapters 2 and 3. The details are certainly different, and a little more complicated than in the case of the string and the bar, but the formal mathematics is nearly identical. As before, we shall find that the normal modes and eigenfunctions play a vital part in our understanding of the system, and we shall meet some new mathematical functions, the Bessel functions, that have application in other parts of our development as well.

5.2 Vibration of Rectangular Membranes

The behavior of a membrane is closely related to that of a string, since it needs to be stretched on a rigid frame if it is to vibrate, and there is a major part played by the elastic stress or tension. Again we neglect motion in the plane of the membrane, and concentrate on the displacements in the normal direction. Suppose that the tension T is the same in all directions in the membrane—a common but by no means universal situation—then the equation of motion can be written down almost by analogy with that for the string, (3.3). We can write $T = \sigma d$, where σ is the elastic stress in the membrane and d is its thickness, and the equation of motion is then

$$\frac{\partial^2 z}{\partial x^2} + \frac{\partial^2 z}{\partial y^2} = \frac{\rho_s}{\sigma}\frac{\partial^2 z}{\partial t^2} \equiv \frac{1}{c_M^2}\frac{\partial^2 z}{\partial t^2} \tag{5.1}$$

where ρ_s is the density of the membrane material and we have introduced

$$c_M = \sqrt{\sigma/\rho_s} \tag{5.2}$$

by analogy with the case of the string. Clearly (5.1) is a two-dimensional wave equation of some sort, and we identify c_M with the speed of transverse waves on the membrane. Once again we shall defer discussion of wave propagation until a later chapter and concentrate on the vibrational behavior of a finite membrane.

If we assume that the behavior is oscillatory with frequency ω, then we can write the time dependence as $e^{j\omega t}$ and the equation for the transverse displacements $z(x, y)$ becomes

$$\frac{\partial^2 z}{\partial x^2} + \frac{\partial^2 z}{\partial y^2} = -\frac{\omega^2}{c_M^2}. \tag{5.3}$$

The solution to this equation has the form

$$z(x, y) = A \sin(k_x x + \theta_1) \sin(k_y y + \theta_2) \qquad (5.4)$$

where

$$k_x^2 + k_y^2 = \omega^2/c_M^2. \qquad (5.5)$$

Again as in the case of the string, if we suppose the membrane to be clamped over a rectangular frame of dimensions $l_1 \times l_2$, then the boundary condition $z(x, y) = 0$ all around the frame fixes $\theta_1 = 0$, $\theta_2 = 0$, and imposes the requirements

$$k_x = \frac{m\pi}{l_1} \quad \text{and} \quad k_y = \frac{n\pi}{l_2} \qquad (5.6)$$

so that the complete solution for the motion of a normal mode is

$$z_{mn}(x, y, t) = \psi_{mn} e^{j\omega t} \qquad m = 1, 2, 3, \ldots \qquad n = 1, 2, 3, \ldots \qquad (5.7)$$

where the eigenfunctions $\psi_{mn}(x, y)$ are

$$\psi_{mn}(x, y) = \frac{4}{l_1 l_2} \sin\left(\frac{m\pi x}{l_1}\right) \sin\left(\frac{n\pi y}{l_2}\right). \qquad (5.8)$$

From (5.5), the frequency ω_{mn} is given by

$$\omega_{mn} = c_M (k_x^2 + k_y^2)^{1/2} = c_M \pi \left[\left(\frac{m}{l_1}\right)^2 + \left(\frac{n}{l_2}\right)^2\right]^{1/2}. \qquad (5.9)$$

The normal modes for a rectangular membrane thus form a two-dimensional set, characterized by the two integers (m, n). The first few modes are shown in Fig. 5.1 for the case of a nearly square membrane, and the associated mode frequencies are shown in Fig. 5.2. For this nearly square case, the two modes with the same value of $m^2 + n^2$ have nearly the same frequency. If the membrane were exactly square the two frequencies would be equal, and they diverge as the membrane deviates further from squareness. For any rectangular membrane, the nodal lines of the mode (m, n) divide the membrane into rectangular sections. There are m such sections and $m - 1$ internal nodal lines in the x direction, and n sections and $n - 1$ nodal lines in the y direction. Adjoining sections of membrane vibrate with opposite phase.

In biological systems we shall be particularly interested in the frequency of the fundamental mode (1,1). From (5.9) this can be written

$$\omega_{11} = \frac{\pi c_M}{S^{1/2}} \left(\frac{l_2}{l_1} + \frac{l_1}{l_2}\right)^{1/2} \qquad (5.10)$$

where $S = l_1 l_2$ is the area of the membrane. The shape factor in brackets, which depends on the ratio l_2/l_1, varies only between 1.4 and 1.6 for $0.5 \leq l_2/l_1 \leq 2$, so that

MEMBRANES, PLATES, AND SHELLS

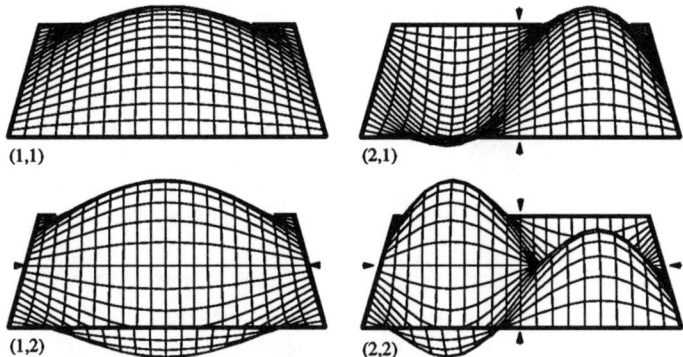

Figure 5.1 Normal mode shapes for a rectangular membrane with sides $l_1:l_2 = 1.1:1$. The modes are labeled with the numbers (m,n) from equation (5.8), and arrowheads indicate the nodal lines.

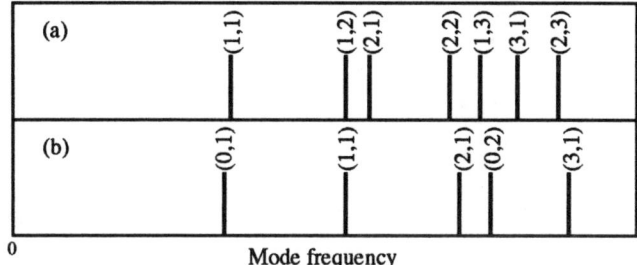

Figure 5.2 Mode frequency distribution (a) for the rectangle of Fig. 5.1 and (b) for a circle of same area.

precise shape is unimportant as far as the frequency of the fundamental of a nearly square membrane is concerned.

5.3 Vibration of Circular Membranes

The equation (5.1) also describes the vibration of membranes of shapes other than rectangular, but needs first to be transformed into a coordinate system in which we can express the boundary conditions reasonably simply. For a circular membrane this clearly involves the use of polar coordinates (r, ϕ). The transformed equation is

$$\frac{1}{r}\frac{\partial}{\partial r}\left(r\frac{\partial z}{\partial r}\right) + \frac{1}{r^2}\frac{\partial^2 z}{\partial \phi^2} = \frac{1}{c_M^2}\frac{\partial^2 z}{\partial t^2}. \tag{5.11}$$

The solution is found to be of the form

$$z(r,\phi,t) = \psi_{mn}(r,\phi)e^{j\omega t} \tag{5.12}$$

as usual, and the eigenfunctions $\psi_{mn}(r,\phi)$ are

$$\psi_{mn}(r,\phi) = M_{mn}J_m(k_{mn}r)\cos m\phi \qquad m = 0, 1, 2, \ldots \quad n = 1, 2, 3, \ldots$$

$$M_{0n} = [J_1(k_{0n}a)]^2 \qquad M_{mn} = \frac{1}{2}[J_{m-1}(k_{mn}a)]^2 \tag{5.13}$$

where the value of the constants M_{mn} comes from the normalization requirement that $\iint \psi_{mn}^2 r\, dr\, d\phi = 1$. $J_m(kr)$ is a special function called a Bessel function, which is discussed in some detail in Appendix A. The subscripts m, n are both integers, and the eigenvalues $k_{mn} = \omega/c_M$ are determined by fitting the boundary condition that $z = 0$ on the clamped boundary of the membrane at $r = a$. The k_{mn} are thus the solutions of the equation $J_m(k_{mn}a) = 0$. The Bessel functions are oscillatory, as shown in Fig. 5.3, but have rather more complex behavior than cosine or sine functions. The first few zeros occur for

$$m = 0: \quad k_{01}a \approx 2.405 \quad k_{02}a \approx 5.520 \quad k_{03}a \approx 8.654$$

$$m = 1: \quad k_{11}a \approx 3.832 \quad k_{12}a \approx 7.016 \quad k_{13}a \approx 10.173 \tag{5.14}$$

$$m = 2: \quad k_{21}a \approx 5.136 \quad k_{22}a \approx 8.417 \quad k_{23}a \approx 11.620.$$

The shapes of the first few modes are shown in Fig. 5.4 and the distribution of mode frequencies, given by

$$\omega_{mn} = k_{mn}c_M, \tag{5.15}$$

is illustrated in Fig. 5.2 for comparison with that for a rectangular membrane. The mode (m, n) divides the membrane with m nodal diameters and $n - 1$ internal nodal circles. Adjoining sections of the membrane vibrate in opposite phase.

Again we are interested primarily in the fundamental frequency, which is in this case ω_{01}. From (5.15) and (5.14) we can write

$$\omega_{01} \approx \frac{2.405 c_M}{a} \approx 1.36 \frac{\pi c_M}{S^{1/2}}. \tag{5.16}$$

The shape factor 1.36 for the circular membrane should be compared with the value between 1.4 and 1.6 for a moderately square rectangular membrane, as given in (5.10), which now allows us to expand our remark about the role of membrane shape in determining the frequency of the fundamental to the assertion that this frequency is almost independent of exact shape, provided the membrane is not too far from equiaxial. This allows us to neglect details of shape in biological membranes, which are usually roughly oval, and to use (5.16) with simply the area inserted. Fig. 5.2 illustrates this in relation to mode frequencies, and shows that the result holds only for the lowest few modes, though a weaker statement could

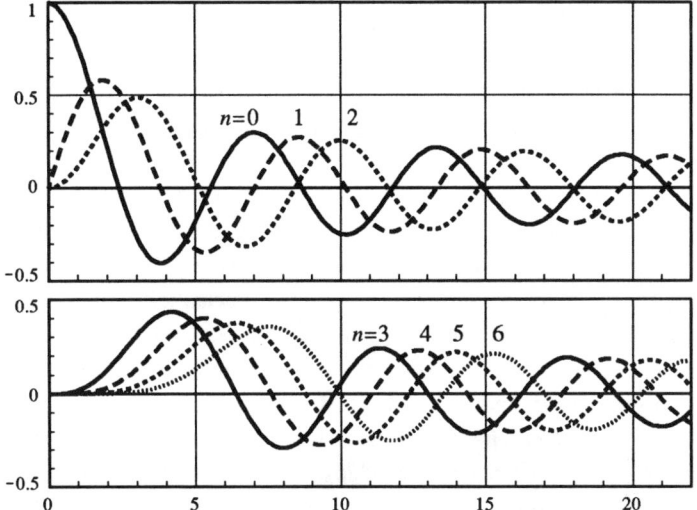

Figure 5.3 The Bessel functions $J_n(x)$ for various values of n. For $x \ll 3$, $J_n(x)$ behaves as $(x/2)^n$, while for $x \gg 3$, $J_n(x) \approx (2/\pi)^{1/2} x^{-1/2} \cos(x - n\pi/2 - \pi/4)$.

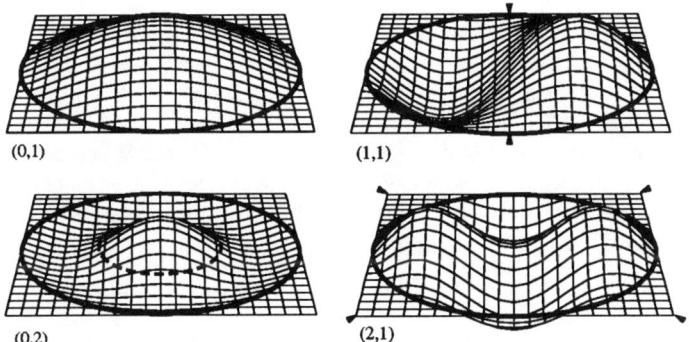

Figure 5.4 Mode shapes for a membrane, clamped around the heavy circle shown, labeled with (m,n) values from equation (5.13). Arrowheads indicate the nodal lines, and there is also a nodal circle for mode (0,2). Note the similarity of the fundamental and (1,1) mode shapes to the corresponding modes (1,1) and (2,1) for a rectangular membrane, as shown in Fig. 5.1.

be made about the average mode density distribution over the whole frequency spectrum.

We see also from Figs 5.1 and 5.4 that the shapes of the fundamental, and indeed of the next mode as well, are quite similar for a rectangular and a circular membrane. It is only for the higher modes that details of boundary begin to affect the shape.

5.4 Tapered Membranes

While most external biological membranes are oval in shape and nearly equiaxial, we do know that some important internal membranes, such as the basilar membrane in the mammalian cochlea, have a form more like an elongated triangle. Consideration of cochlear mechanics is reserved for Chapter 12, but it is of interest to examine the oscillatory behavior of an elongated tapered membrane, which may be a useful model for some simpler auditory transducers. This is not presented as a description of the motion of the basilar membrane in the cochlea, which is dictated largely by fluid dynamics rather than membrane mechanics, and which involves propagating waves, as we shall discuss in Chapter 12.

We can simplify the analysis by making use of our observation that fine details of shape affect only the higher modes. This allows us to replace an elongated symmetrical quadrilateral membrane by a similar membrane with curved ends, as shown in Fig. 5.5, secure in the knowledge that this will make very little difference to the result. This membrane, being bounded by concentric circles and radii, can be analyzed quite easily in the polar coordinate system used for the circular membrane. We suppose the angle included between the two straight sides is ϕ_0, the radius describing the broad end a, and the radius describing the narrow end b.

Figure 5.5 A tapered membrane, approximated as the circular sector between radii b and a. The displacement is shown for the first few modes of a complete sector ($b \to 0$) of included angle 30°, the frequency increasing from top to bottom in the diagram. The apex lies at O and the wide end at A. Arrowheads show the location of the principal maximum, which moves towards the apex with increasing frequency. Because there is very little motion of the membrane near its apex, the modes are very little changed if the sector is truncated at a radius b, provided $b \ll a$.

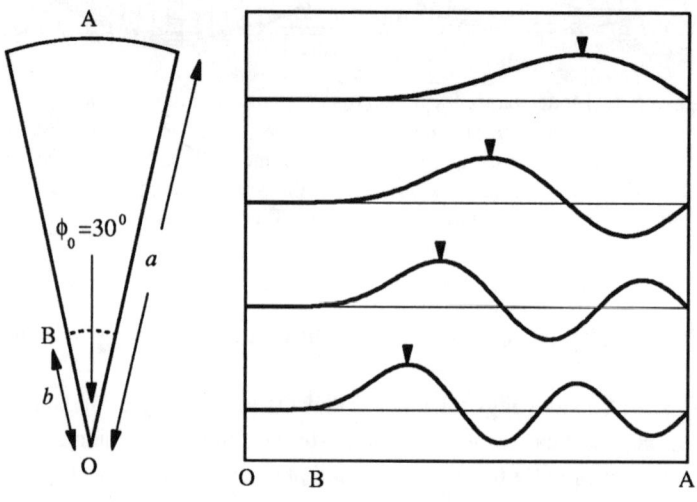

Although formal solution of this problem is quite straightforward, the numerical results are hard to evaluate, and we therefore adopt a further simplification by supposing that $b \ll a$, so that the membrane is nearly triangular. We can then make a calculation for the case $b = 0$ knowing that it will be at least approximately correct for the real membrane.

The solution for the sector of a circle is formally very similar to that for a full circle, as given by (5.12) and (5.13), though we write it a little differently for convenience as

$$\psi_{\mu n} = M_{\mu n} J_\mu(k_{\mu n} r) \sin \mu \phi. \tag{5.17}$$

If we suppose the sector to extend over $0 \le \phi \le \phi_0$, then the choice of a sine function ensures that $\psi = 0$ along the straight boundary $\phi = 0$. To make ψ vanish along the other straight boundary $\phi = \phi_0$ requires that

$$\mu = \frac{m\pi}{\phi_0} \qquad m = 1, 2, , 3, \ldots \tag{5.18}$$

so that μ is not an integer, and the order of the Bessel function is not integral either. This fact leads to numerical complication, for tables of Bessel functions are readily available only for integer or half-integer orders, but we can always approximate ϕ_0 by the closest angle π/s, where s is an integer, to get a good idea of the solution. Indeed in a semiquantitative manner we can do better than this by simply interpolating between the behavior of the Bessel functions in Fig. 5.3, noting that they all vanish at $r = 0$ except for J_0. The final boundary condition to be satisfied is that $\psi(ka) = 0$ which, translated to the form $J_\mu(k_{\mu n} a) = 0$, determines the eigenvalues $k_{\mu n}$.

We need not go into the details of this solution, but a few qualitative remarks are in order. The first is to note that, if the angle ϕ_0 is small, then the order μ of the Bessel function is large. Now, as is illustrated in Fig. 5.3, $J_\mu(kr)$ varies as $(kr)^\mu$ for small kr, so that this means that there is relatively little motion of the membrane at its narrow end unless k is very large, which corresponds to a high exciting frequency. Equally we note from Fig. 5.3 that the successive maxima in $J_\mu(kr)$ decline in amplitude as $(kr)^{-1/2}$, the first maximum being the largest. In fact the position of the first maximum for $J_\mu(k_{\mu n} r)$ occurs for $k_{\mu n} r \approx \mu + 0.8 \mu^{1/3}$. We have seen that μ is fixed by the angle of taper of the membrane, and is typically a number of order 10, so that we can neglect the $\mu^{1/3}$. The peak amplitude of mode n therefore occurs near $r^* \approx \mu/k_{\mu n}$. The wave number $k_{\mu n}$ increases linearly with frequency by (5.15), so that the position r^* of the maximum amplitude moves steadily towards the apex of the membrane as the frequency increases.

If we examine descriptions of the basilar membrane in the mammalian cochlea, then we find a very similar behavior to that described here. There is, however, a potential source of confusion. Mechanical vibration is transmitted to the cochlear fluid through a flexible window in the large end, or base, of the cochlea. The basilar membrane is narrow at this end and widens towards the remote end or apex of the cochlea. We must therefore associate the apex of the membrane in our discussion above with the base of the cochlea, and vice versa.

A rather similar analysis can be carried out for the case in which the membrane is long in one direction and the tension transverse to that direction is not constant but increases from one end of the membrane to the other. We shall not examine this problem in detail, but simply remark that the effect on the modes is rather similar to that of tapering the membrane width. Somewhat similar effects occur if the density varies along the membrane, and, of course, all these things can happen together in a real biological system.

Again we emphasize that the mechanics of the basilar membrane in the cochlea is a much more complex subject than the mechanics of a simple tapered membrane. In the cochlea the membrane is immersed in watery fluid, to which its motion is closely coupled, and we must consider the fluid dynamics as an integral part of the vibration problem. Sharp modal resonances are also generally undesirable, so that the cochlear system is highly damped. Nevertheless, there is some resemblance between the behavior of the actual system and that of the very simple structure investigated here.

5.5 Sinusoidally Driven Membranes

The behavior of a driven membrane can be analyzed in a similar way to that of a driven string, though there are problems associated with the fact that a finite point force applied to a perfect membrane gives an infinite displacement at the point of contact, a problem that did not occur for the string. Fortunately this case is of little interest in the present context, in which membranes are generally driven by sound waves and therefore by continuously distributed forces.

To get sensible answers when the exciting frequency coincides with a mode frequency of the membrane, we must include damping, both from internal losses and, more importantly, from interaction with the air on both sides of the membrane. We defer most of this discussion to a later chapter, and simply note the result for the case in which the damping has a simple frequency-dependent form specified by the parameter $\alpha(\omega)$ as in previous discussions.

For the particular case in which the force per unit area due to varying air pressure p in a sound wave is applied uniformly and with constant phase across the surface of the membrane, we can generalize the result (3.25) for a uniformly excited string to write

$$z(x,y,t) = \left(\frac{p}{\rho_s d}\right) \sum_{m,n} \left[\frac{\psi_{mn}(x,y) \int_S \psi_{mn}(x',y') \, dS'}{(\omega_{mn}^2 - \omega^2) + 2j\omega\alpha(\omega)} \right] e^{j\omega t} \quad (5.19)$$

MEMBRANES, PLATES, AND SHELLS

where the integral is over the surface S of the membrane using the dummy variables x', y', and the eigenfunctions ψ_{mn} are given by (5.4) or (5.13). This result applies also for a circular membrane. All we have to do is to perform the integral over a circle rather than a rectangle, for which polar coordinates are more convenient. It is important to note that the integral in (5.19) is large for the fundamental, since its eigenfunction has the same sign everywhere, and much smaller for the higher modes, for which the eigenfunctions have partially cancelling positive and negative regions. In the case of a circular or rectangular membrane, certain modes integrate identically to zero and therefore are not excited at all. These are all modes for which the angular index $m \neq 0$ in the case of a circle and all modes for which either m or n is even for a rectangle. These symmetry properties are overlaid on the general resonance denominator, which accentuates the response of those modes with frequencies close to the forcing frequency ω.

While (5.19) is correct and accurately applicable to membranes of reasonable area and thickness, we must exercise care in applying it to biological situations in which the membrane may be very thin. The expression implies that, as the membrane thickness d approaches zero, so the response amplitude becomes infinite, which is clearly wrong physically. The reason is that we have neglected the loading effect of the air on the other side of the diaphragm, which necessarily moves with it. We discuss this problem properly in a later chapter.

If the force itself is not evenly distributed across the membrane, then this brings an angular term into the integral and allows otherwise forbidden modes to be excited. This is, however, not generally important for small membranes. Indeed we shall find that in nearly all biological problems we can concentrate on the fundamental and ignore higher modes.

5.6 Elastically Braced Membranes

The tympanic membranes of many vertebrate ears are not simple flat membranes but are elastically braced by a central link that is pulled tight by one or more special muscles. It is interesting to look briefly at this situation. Suppose the brace is connected to a small central section of the membrane of radius b, the radius of the membrane being a. This is illustrated in idealized form in Fig. 5.6. We would like to know the shape of the membrane, and any influence the brace has upon its behavior.

The static shape of the membrane is given by the solution of its equation of motion, (5.11), with the time-varying term omitted and the angular term omitted for reasons of symmetry. This leaves only the radial derivative term, and the solution is

$$z(r) = A \ln(r/a) \tag{5.20}$$

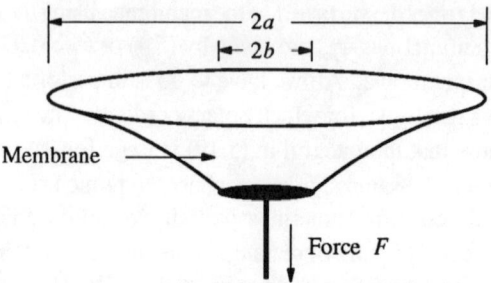

Figure 5.6 The shape of an elastically braced membrane, such as that found in the mammalian tympanum. The neural transducer is usually connected to the same central region.

where A is a constant. Clearly $z = 0$ at $r = a$ as required by the edge-clamping conditions. Now the force balance on the central section of radius b requires that

$$F = 2\pi bT \left[\frac{dz}{dr}\right]_{r=b} \qquad (5.21)$$

where T is the membrane tension. This fixes the value of A and gives

$$z(r) = \frac{F \ln(b/a)}{2\pi T} \ln(r/a). \qquad (5.22)$$

This analysis assumes that the original tension T in the membrane is large and the force F relatively small, so that the deflection $z(b)$ of the central part of the membrane is small compared with its radius a. Under these conditions the deflection has little effect on the tension, but the vibration behavior of the membrane depends upon the mechanical impedance of the brace, as we discuss in the next section.

We could, of course, have the opposite situation, in which the original tension in the membrane is small. This may often be the case in real biological systems. Exact solution of the elastic problem is rather complicated, so we shall be content with an outline. Suppose that the force F communicated by the brace deflects the central portion of the membrane by an amount z. Then an analysis of the curved geometry of the membrane shows that it is stretched by an amount proportional to z^2, provided z is very much less than the radius of the membrane. This means that the radial component of the tension, which may not be uniform across the membrane, is proportional to z^2. The slope dz/dr of the membrane at the edge of the connection circle of the brace ($r = b$) is proportional to z, so that by (5.21) the component of tension balancing F is $2\pi bT[dz/dr]_{r=b}$, which is proportional to z^3.

This means that the deflection z is proportional to $F^{1/3}$ and the membrane tension to $F^{2/3}$. The force applied by the brace thus directly controls the tension of the diaphragm, and so its mode frequencies. A mechanism such as this, with appropriate feedback to the tensioning muscle, could allow an animal to keep the tympanic membrane of its auditory system in a state of appropriate adjustment.

5.7 Loaded Membranes

Since a tympanic membrane itself does not usually contain sensory transducer cells, it is necessary to make some connection between it and the neural transduction organ. The obvious, and usual, way to do this is through a connection to the center part of the membrane. It is quite likely that this connection will apply a significant mechanical load to the membrane and thus will affect its vibration behavior.

Suppose that the central connection is made over a circular area of radius b. The membrane itself then occupies the annular region from $r = b$ to $r = a$, and we need to calculate its behavior. The membrane is clamped at its outer edge $r = a$, but its inner edge $r = b$ moves with the attached load. This load can be characterized by its mechanical impedance Z_L, which may have spring-like, massive and resistive components. The motion of the membrane is described by the same equation as before, (5.11), but the solution needs to be rather more general than (5.13), involving not only the ordinary Bessel functions J_n but also a second kind of Bessel function N_n, called a Neumann function (sometimes confusingly denoted by Y_n, a symbol normally used in physics for a spherical harmonic). These Neumann functions did not appear in the solution for a complete circular membrane because they diverge to infinity for $r = 0$. Since $r = 0$ is not within the present annular membrane, we must include them, and in fact they are necessary for the solution. They are discussed further in Appendix A.

The eigenfunctions for the annular membrane thus have the form

$$\psi_{mn}(r, \phi) = [A_{mn} J_m(k_{mn} r) + B_{mn} N_m(k_{mn} r)] \cos m\phi \tag{5.23}$$

where A_{mn} and B_{mn} are constants. We need three equations to determine these two constants and the mode frequency parameter k_{mn}. The first equation is simply the normalization requirement on the eigenfunction, $\iint \psi^2 r \, dr \, d\phi = 1$, and the second comes from the clamping condition $\psi(a, \phi) = 0$ at the outer edge of the membrane. The third describes the coupling of the membrane to the load at $r = b$. We write this down just for the case of centrally symmetric modes for which $m = 0$, but the principle is the same for the others. The force acting on the load is $2\pi b T [d\psi/dr]_{r=b}$ and the velocity of the load is $[d\psi/dt]_{r=b}$, so that the boundary condition is

$$\frac{2\pi b T}{j\omega \psi_{mn}(b)} \left[\frac{d\psi_{mn}}{dr} \right]_{r=b} = Z_L(\omega). \tag{5.24}$$

It is not particularly difficult to solve these equations, but the algebra is a little involved so we shall not write it down here. The general conclusion is that, if the impedance of the load is comparable with that of the membrane itself, measured over the attachment circle $r = b$, then the load has little effect and the attachment moves with the center of the membrane. At the other extreme when the impedance of the load is high, motion of the load is very small and the membrane behaves almost as though its center were clamped. The real situation in biological systems is closer to the first of these possibilities than the second, but the load generally adds both mass and resistance, so that the frequency and Q value of the fundamental membrane resonance are both lowered. Higher modes are affected rather differently, and generally less severely, as with the loaded bar.

One simple example serves to show the behavior of the fundamental in an extreme case when the added load is mass-like and large compared with the mass of the membrane itself. The frequency of the motion is then sufficiently below that of the first resonance of the free diaphragm that we can use static analysis for the membrane tension as in Section 5.6. Suppose that the mass is spread over a contact circle of radius b at the center of the membrane. Then the analysis of Section 5.6 shows that, when the deflection of the center section of the membrane is z, the restoring force exerted by the diaphragm tension is $2\pi z T/\ln(b/a)$. If we write this as the equation of motion of the added mass m and neglect the mass of the diaphragm, then

$$m\frac{d^2z}{dt^2} = -\frac{2\pi z T}{\ln(b/a)}. \tag{5.25}$$

The frequency of this simple oscillator, which is now the frequency of the fundamental mode for the diaphragm, is

$$\omega = \left(\frac{2\pi T}{m \ln(a/b)}\right)^{1/2}. \tag{5.26}$$

This frequency depends not only upon the magnitude of the added mass, but also to a small extent upon the degree to which it is concentrated. It is not a very realistic formula if $b \ll a$, since the assumptions upon which it is based then cease to hold, the membrane begins to stretch significantly, and its tension changes. It does show, however, that an added mass load can have a large effect on the frequency of the fundamental mode. As with the loaded bar, the frequencies of higher modes are much less affected, but these are not biologically significant.

In real systems involving a tensioned and loaded membrane, the same central attachment may be used for both, effectively combining the results of this and the previous section. If the membrane and load are both geometrically symmetric relative to the membrane, then only symmetric membrane modes can transfer

MEMBRANES, PLATES, AND SHELLS

energy to the transducer. The fundamental is the most important and most readily excited of these symmetric modes.

5.8 Slack Membranes

Sometimes in biology we encounter a membrane that is stretched with nearly zero tension over a rigid frame. Since the vibration of such a membrane presents several interesting features, it is worthwhile to give it brief attention.

Suppose the frame is rectangular, with dimensions $l_x \times l_y$. For a deflection $z(x,y)$ of appreciable magnitude relative to the frame dimensions, the membrane is stretched. Since an element dx becomes one of length $(dx^2 + dz^2)^{1/2}$, the new length of the membrane in the x direction becomes

$$l_x' = \int_0^{l_x} \left[1 + \left(\frac{dz}{dx}\right)^2\right]^{1/2} dx \approx l_x + \frac{1}{2}\int_0^{l_x} \left(\frac{dz}{dx}\right)^2 dx \tag{5.27}$$

and similarly in the y direction. If E is the Young's modulus and d the thickness of the membrane, then the tension in the x direction becomes

$$T_x \approx \frac{Ed}{2l_x} \int_0^{l_x} \left(\frac{dz}{dx}\right)^2 dx . \tag{5.28}$$

The tension in the y direction is found similarly. The membrane adjusts by motion in its plane so that the tension is essentially uniform across its whole surface. Thus, to a first approximation, when the membrane is vibrating with amplitude a it behaves like a simple membrane with a tension given by the average value of T_x from (5.28). The integral has the average value $a^2 l_x/4$ when both space and time averages are considered, together with an oscillating part of frequency 2ω which we ignore, so that $\overline{T}_x \approx Eda^2/8$ and the average tension stress is $\overline{\sigma} \approx Ea^2/8$.

From (5.2) the transverse wave velocity c_M on the membrane is thus proportional to the amplitude a and, by (5.9), all the mode frequencies are therefore proportional to a. This is a simple example of nonlinear behavior, the restoring force being proportional to $T\partial^2 z/\partial x^2$ and thus to z^3. More detailed discussion of the motion shows that each mode is accompanied by harmonics of its own frequency, as outlined in Section 2.9, and there is also coupling between the modes. Such a cubic-nonlinear vibrator can also show chaotic vibrational behavior if it is lightly damped, and this can be heard in some bird calls, as discussed further in Chapter 13.

5.9 Vibration of Plates

The exact equations for the vibratory motion of a plate are very complex, so we content ourselves with writing down the first approximation for the transverse motion, which is closely analogous to the equation of motion (3.29) for a bar, extended to two dimensions. For a rectangular plate we naturally use rectangular coordinates (x, y), and the equation is

$$d\rho_s \frac{\partial^2 z}{\partial t^2} = \frac{Ed^3}{12(1-\gamma^2)}\left(\frac{\partial^2}{\partial x^2}+\frac{\partial^2}{\partial y^2}\right)^2 z \equiv \frac{Ed^3}{12(1-\gamma^2)}\left(\frac{\partial^4 z}{\partial x^4}+2\frac{\partial^4 z}{\partial x^2 \partial^2 y}+\frac{\partial^4 z}{\partial y^4}\right). \quad (5.29)$$

Here d is the plate thickness and γ is Poisson's ratio for the plate material, usually about 0.3. The elastic stiffness of the plate is $Ed^3/12(1-\gamma^2)$. Note that the differential operator expression in brackets in the first way of writing the right-hand side is just the expression from the membrane equation (5.1), but that expression is then squared. The edges of the plate, in this simplest approximation, can be either free, hinged, or clamped, as for the bar. The equation in polar coordinates, as appropriate for a circular plate, is similar in form, except that the differential operator expression in brackets in the first way of writing the right-hand side is replaced by the appropriate expression in polar coordinates from the membrane equation (5.11).

It would take us into too much mathematical complication to solve the plate equation for either of these two cases, but we can remark on the general form of the solution. For the rectangular case the eigenfunctions resemble those of the bar, discussed in Section 3.9, in that they contain cosine, sine, hyperbolic cosine and hyperbolic sine terms in proportions depending on the boundary conditions. They also resemble the membrane eigenfunctions (5.8) in that each term in x is multiplied by a similar term in y. In the case of a circular membrane the solutions involve both Bessel functions and modified Bessel functions, which behave as analogs of the hyperbolic cosine and hyperbolic sine. The case with which we are usually concerned in biological applications is one in which the edges of the plate are clamped into a stiffer structure. The shapes of the normal modes are then very similar to those of the membrane, shown in Figs 5.1 and 5.4 for rectangular and circular shapes respectively.

Just as the first few normal mode frequencies of a bar are more widely separated than those of a string, so the first few mode frequencies of a plate are more widely separated than those of a membrane. For a circular plate of area $S = \pi a^2$ and thickness d, clamped around its edge, the mode frequencies are given by

$$\omega_{mn} = \frac{\pi^3 d}{S}\left[\frac{E}{3(1-\gamma^2)\rho_s}\right]^{1/2} \beta_{mn}^2 \qquad \beta_{mn} \approx n + \frac{m}{2}. \quad (5.30)$$

The close resemblance in form to the equations (3.36), (3.37) for the mode frequencies of a bar is clear. We have written (5.30) in terms of plate area since, once again, the frequencies of the first few modes are not greatly influenced by the exact shape of the plate, provided it is nearly isometric.

5.10 Vibration of Shells

A shell is similar to a plate, in that it is a sheet-like structure in which the thickness of the material from which it is made is small compared with its lateral extent. It differs from a plate in that it is not planar, but curved. The simplest sort of shell is cylindrical, and can be made from a rolled-up plate, thus having curvature in one direction only. While this sort of shell is important in biology, since it is a good approximation for parts of the exoskeleton of insects and crustaceans, the shells met in acoustic systems are generally curved in two directions. To a reasonable approximation, such a shell can be regarded as a shallow cap cut from a spherical surface. Both cylindrical and spherical shells are very complicated to analyze, and only partial results are available in the literature. We shall therefore simply sketch their vibration behavior and quote a few useful results.

The vibrations of a shell can be divided into two classes. In the first class, called inextensional, the motion is such that any line drawn on the surface of the shell retains a constant length during the motion, at least to first order. An example of an inextensional vibration is the distortion of a circular cylinder to elliptical cross-section, or the deformation of a spherical cap with a free edge to an ellipsoidal shape. In such a vibration the restoring force is provided by the bending stiffness of the shell material, which varies as d^3, where d is the thickness of the material. The mass of the shell per unit area is proportional to d, so the vibration frequency, which varies as the square root of the stiffness divided by the mass per unit area, is proportional to the material thickness d. This result, and the argument leading to it, is just the same as used in the calculation for vibration of a plate in Section 5.9.

The second class of vibrations is called extensional and refers to motion in which the lengths of at least some lines drawn on the shell surface vary to first order (i.e. linearly) with the vibration. An example of such motion is the simple change in radius of a cylindrical or spherical shell. In this case the restoring force is primarily due to the elongation or compression of the shell material, and the elastic force that this involves, termed a membrane force, is simply proportional to the shell thickness. When we divide by the mass per unit area, we find that the frequency of extensional vibrations is independent of shell thickness. In most cases this is an oversimplification, because there is generally some bending associated with the extensional vibration, but the variation of frequency with shell thickness is small, provided that the thickness is small compared with the curved dimension of the shell.

In biological acoustic systems we can meet both kinds of vibrations, but those of principal interest are the axially symmetric vibrations of shallow spherical-cap shells. These vibrations have shapes similar to the $(0, n)$ modes of a flat circular membrane, illustrated in Fig. 5.4. Because parts of the shell move closer to or further away from the center of curvature, such modes are extensional, and we expect them to have frequencies nearly independent of shell thickness. In fact, if the radius of the circular edge of the spherical cap is a and the height of the dome of the cap is H, it can be shown that the frequency of the first axially symmetric mode is

$$\omega_0 \approx 2\left(\frac{E}{\rho}\right)^{1/2}\frac{H}{a^2}. \qquad (5.31)$$

This result is very closely accurate if the dome H is more than about 25 times the shell thickness. For thicker shells, Reissner has shown that the frequency of this mode is given to a better approximation by

$$\omega_0 \approx \frac{3d}{a^2}\left[\frac{E}{\rho(1-\gamma^2)}\right]^{1/2}\left\{1+(1+\gamma)[0.9-0.2(1+\gamma)]\frac{H^2}{d^2}\right\}^{1/2} \qquad (5.32)$$

where d is the shell thickness and γ is Poisson's ratio for the shell material. The contributions of both stretching and bending terms are clearly evident in this result.

If we substitute numerical values into these equations for a typical thin shell, as is done in the discussion examples at the end of the chapter, we see that the shell curvature greatly increases the effective stiffness of the shell, relative to that of a flat plate of similar dimensions, and consequently greatly raises the frequency of the first resonance. The frequencies of all axially symmetric modes will be similarly raised, since they are all extensional, but the frequencies of modes (m, n) with angular dependence (i.e. $m \neq 0$) are not changed greatly from their flat-plate values. Shallow circular-cap shells are thus good candidates for use as diaphragm elements in biological systems in which a high resonance frequency is required, in contrast to the use of very thin membranes under high tension.

5.11 Buckling of Shells

A nonlinear phenomenon of importance in shells in some biological systems is their propensity to buckle when a large force is applied. This is the basis of sound production in some insects, the shells of the tymbals being repeatedly buckled in and out by powerful thoracic muscles.

It is easy to see that a shell, or indeed a curved beam, can be made to buckle to a new shape by the application of a sufficiently large force. The situation is illustrated in Fig. 5.7. A force F applied to the center of the arch tends to flatten the curve there and to increase its curvature partway towards the edge. As the force is increased, the curvature of the central section decreases through zero and the center becomes concave instead of convex. Depending upon the thickness d of the shell in relation to its dome height H, and thus upon the ratio between stretching and bending forces, the situation may now be unstable, so that the central region of reversed curvature spreads without any increase in force. Analysis of this buckling condition is difficult but, for dome heights between about 2 and 10 times the shell thickness, the theoretical buckling force is about

MEMBRANES, PLATES, AND SHELLS

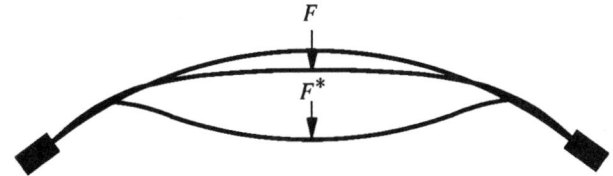

Figure 5.7 Buckling of a uniform shallow spherical shell under the action of a concentrated central force.

$$F^* \approx EH^2 d^2/a^2 \qquad (5.33)$$

where E is the Young's modulus of the shell material and a the radius of its circular edge. For the tymbal of a large insect, such as a cicada, insertion of numerical values suggests that $F^* \sim 0.1$ N, which is large, but not unreasonably so. In fact, however, F^* has its maximum value for a perfect shell, and any geometrical irregularities can reduce it by up to a factor 10 at least.

The distance that the buckled section of reverse curvature extends towards the edge of the shell depends on shell thickness and on the way in which the edge of the shell is restrained. An increase in force extends the area of the buckled region. In very thin shells the buckled configuration may be elastically stable, particularly if the shell edge is unrestrained. An opposite force is then required to return the shell to its original shape. If the shell thickness is larger, the bending terms become more important and a shallow shell will spring back to its original configuration once the force is removed. In highly domed shells the stresses may exceed the elastic limit of the material, resulting in plastic deformation or rupture.

While a uniform shell buckles in a relatively simple manner, it is not surprising that a shell with thickness variations behaves with more complexity. The tymbals of some insects are ribbed, and so have regular thickness variations which, as well as reducing the initial buckling force F^*, affect the extension of the buckled zone. If the force is increased steadily, or if the center of the shell is displaced smoothly, the buckled zone jumps quickly through each thin section of the shell and pauses at the next rib until the force builds up sufficiently to cause further buckling. In this way the buckling process can be made to yield a train of regular mechanical or acoustic pulses that constitute the characteristic song of the insect. We return to this in Chapter 12 when considering active acoustic systems.

References

Vibration of membranes: [1] Ch 5; [8] Ch 4; [10] Ch 3
Vibration of plates: [1] Ch 5; [8] Ch 4; [7] Ch 8; [10] Ch 3
Vibration of shells: [13]
Bessel functions: [46] Ch 9; [47] Ch 8

Discussion Examples

1. A mammalian eardrum has a diameter of 8 mm and a thickness of 20 μm. What must be its tension if it is to have a first resonance at 1000 Hz? What is the elastic stress in the membrane?
2. The eardrum of an insect is an oval patch of thin cuticle in its exoskeleton. If the dimensions of this patch are 0.8 mm by 0.6 mm, its thickness 1 μm, and its internal tension essentially zero, calculate its approximate resonance frequency in the fundamental mode, given that the cuticle has Young's modulus 4×10^9 N m^{-3}.
3. If the eardrum in question 2 is not flat but rather a shallow shell with arch height 100 μm, calculate its fundamental resonance frequency.
4. A tapered membrane comes to a point at one end and is bounded by a circular arc at the other. The angle between the two straight sides of the membrane is 30° and their length is 30 mm. The thickness of the membrane is 30 μm and its tension 0.03 N m^{-1}. By consulting a table of Bessel functions, find the first three resonance frequencies.
5. For the membrane of question 4, find, by consulting a table of Bessel functions, the position of maximum vibration amplitude for the first, second, and third modes.

Solutions

1. From (5.14), $ka = 2.405$ for the first mode, so $k \approx 600$ m^{-1}. But $k = \omega/c_M$ so if $\omega = 2000\pi$, then $c_M \approx 10.5$ m s^{-1}. Since $\rho_M \approx 1000$ kg m^{-3}, then from (5.2) the stress is $\sigma \approx 1.1 \times 10^5$ N m^{-2}, so $T \approx 2.2$ N m^{-1}.
2. Using (5.28) for $m = 0$, $n = 1$, gives $f_{01} \approx 16$ kHz.
3. $H/a = 100$, so we can use (5.29), giving $f_{01} \approx 80$ kHz.
4. Wave speed $c_M = 1$ m s^{-1}. Since $\phi_0 = 30° = \pi/6$, the eigenfunctions are $J_6(k_n r)$ with $J_6(k_n r) = 0$ for $r = 30$ mm. The zeros of J_6 occur for $kr = 9.9, 13.6, 17.0$, so $k = 3300, 4500, 5700$ m^{-1}. Frequencies are therefore 520, 710, 900 Hz.
5. The first maximum of $J_6(kr)$ is at $kr = 7.5$, so maxima are at $30 \times 7.5/9.9 = 22.7, 16.5$, and 13.2 mm from the apex respectively.

6 ACOUSTIC WAVES

SYNOPSIS. Acoustic waves will propagate in any medium that possesses some form of elasticity. In solids, which have both a bulk modulus (the reciprocal of the compressibility) and a shear modulus (or rigidity) wave propagation is complex, involving both longitudinal and transverse waves. Liquids and gases have no rigidity, and acoustic waves are simply longitudinal. Plane sound waves propagate in fluids with a speed equal to the square root of the bulk modulus divided by the density. This gives a wave speed of about 340 m/s in air and about 1480 m/s in water, though both values depend on temperature and, in the case of sea water, on salinity. The main parameters for a sinusoidal wave are illustrated in Fig. 6.2.

The acoustic pressure is the principal variable characterizing a sound wave, and it is convenient to define a logarithmic scale, called the sound pressure level (SPL), measured in decibels relative to a standard pressure of 20 micropascals r.m.s. that is approximately the threshold of human hearing, as shown in Fig. 6.3. A change of 10 dB in SPL corresponds to a change by a factor $10^{1/2} \approx 3.2$ in sound pressure. Intensity measures the power carried by an acoustic wave and is measured in watts per square meter. Again we define an intensity level (IL) in decibels relative to a standard intensity of 10^{-12} W/m². Sound pressure level and intensity level are numerically the same for a plane wave, but whereas sound pressures simply add (bearing in mind relative phase) when waves are superposed, intensities are directional and can add or subtract.

It is helpful to define a wave impedance z, equal to the pressure divided by the acoustic particle velocity for a plane wave, that is characteristic of the medium. At a boundary between two media of different wave impedances there is partial reflection and partial transmission, the extent of each depending upon the ratio of the two wave impedances. For air $z = 415$ rayl while for water $z = 1.48 \times 10^6$ rayl and the power transmission from one medium to the other is only 0.001, a transmission loss of 30 dB.

As a wave expands spherically from a small source, its intensity decreases by 6 dB for every doubling of distance. This dominates the behavior at small distances, while at large distances losses caused by viscosity and thermal conduction in the medium become important. This is illustrated in Fig. 6.5. Attenuation losses increase as the square of the frequency and amount to about 1 dB/km for air but only 0.05 dB/km for water, both at 1000 Hz. Sound propagation is thus possible to great distances in the ocean.

Another biologically important form of wave propagation occurs at the surface of water. The wave velocity is rather slow, typically less than 1 m/s, and depends upon frequency. These waves carry information of value to surface-feeding aquatic insects. Similar but more complex waves propagate at the surface of large solid bodies such as the earth, but their wave speed is high, several thousands of meters per second. Wave propagation along smaller solid bodies, such as spider webs, twigs, and leaves, can be approximated by the theory developed in Chapters 3 and 5 for strings, rods, and plates respectively.

In biological applications we are often interested in the propagation of sound near an obstacle, such as the head of an animal. When the sound wavelength is large compared with the diameter of the obstacle, the sound pressure is nearly uniform over the whole surface. As the frequency increases, however, the side remote from the sound source is increasingly shadowed and the sound pressure on it reduced. Another important case is the interaction of sound with an air bubble, or the swim bladder of a fish, under water. Because the air in the bubble is compressible, the bubble can be set into radial oscillation by a sound wave near its resonance frequency, and this has important implications for hearing in aquatic animals.

6.1 Waves

Waves are a very common phenomenon in the physical universe, and indeed are the most important means by which energy, and with it information, is transferred from one place to another. Electromagnetic waves propagating in free space, particularly those in the visible frequency range we call light, are perhaps the most important, but they are not a part of the theme of this book. Our concern here is with mechanical waves of one sort or another, which require the presence of a material medium for their propagation. As our discussion will show, such waves can propagate in any medium that possesses a mass density and some sort of elastic modulus. More complicated propagation phenomena occur in solid materials, which have both an elastic bulk modulus, opposing volume changes, and a shear modulus that resists changes in shape. In liquids and gases there is no elastic resistance to changes in shape, though there may be a viscous resistance, and a single elastic bulk modulus describes their wave-propagation behavior.

In biological acoustics of the sort we are discussing in this book, we are concerned mostly with sound propagation in air or water, though we have already met with some of the wave-propagation properties of solid materials formed into strings, bars, and plates. Fortunately the general analysis of wave propagation is just the same for all fluid media, and in particular for air and water, so that we can adopt a unified presentation and later simply insert the relevant values of density and bulk modulus for the particular medium. The initial discussion will concern waves in unbounded media, for this is both the simplest and most important case, but we shall also give brief attention to the types of waves that can propagate at the boundaries between media, the most familiar of which are the ripples spreading across the surface of a pond, or waves on the ocean.

6.2 Plane Waves

When waves are generated by a vibrating object, they spread out in all directions, as we discuss in the next section. If we observe these waves over a small region far enough from their source, however, they simply propagate in one direction and have no variation at right angles to that direction. We call such waves plane waves, and examine them first since they represent the simplest possible case.

Referring to Fig. 6.1, suppose that ξ measures the displacement of the fluid during passage of the sound wave, so that the element of fluid ABCD of thickness dx moves to A'B'C'D'. Taking S to be the area normal to the propagation direction x, the volume V of this element changes to

$$V + dV = S\, dx \left(1 + \frac{\partial \xi}{\partial x}\right). \tag{6.1}$$

ACOUSTIC WAVES

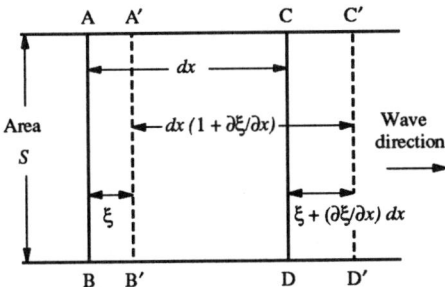

Figure 6.1 During propagation of an acoustic wave, the volume element ABCD, of cross-sectional area S, moves to A'B'C'D'. The acoustic displacement associated with the wave is denoted by ξ.

Now suppose that p_0 is the total pressure in the fluid. Then the bulk modulus K is defined by

$$dp_0 = -K \frac{dV}{V}. \tag{6.2}$$

We call the small varying part dp_0 of p_0 the sound pressure or acoustic pressure and write it simply as p. Comparison of (6.2) with (6.1), noting that $V = S dx$, then gives

$$p = -K \frac{\partial \xi}{\partial x}. \tag{6.3}$$

Now the motion of the volume ABCD must be described by Newton's equations so that, setting the pressure gradient force in the x direction equal to the mass of the element times its acceleration,

$$-S\left(\frac{\partial p}{\partial x} dx\right) = \rho S \, dx \frac{\partial^2 \xi}{\partial t^2} \quad \text{or} \quad -\frac{\partial p}{\partial x} = \rho \frac{\partial^2 \xi}{\partial t^2} \tag{6.4}$$

where ρ is the density of the fluid. Then, from equations (6.3) and (6.4),

$$\frac{\partial^2 \xi}{\partial t^2} = \frac{K}{\rho} \frac{\partial^2 \xi}{\partial x^2} \tag{6.5}$$

or, differentiating (6.5) with respect to x and using (6.3),

$$\frac{\partial^2 p}{\partial t^2} = \frac{K}{\rho} \frac{\partial^2 p}{\partial x^2}. \tag{6.6}$$

If we introduce the quantity c, defined by

$$c^2 = \frac{K}{\rho} \tag{6.7}$$

then we can rewrite (6.6) as

$$\frac{\partial^2 p}{\partial t^2} = c^2 \frac{\partial^2 p}{\partial x^2}. \tag{6.8}$$

Equations (6.8) and the equivalently rewritten (6.5) are two different versions of the one-dimensional wave equation, (6.8) referring to the acoustic pressure p and (6.5) to the acoustic displacement ξ. If we differentiate (6.5) with respect to t, we get a third similar equation for the acoustic velocity $v = \partial \xi/\partial t$. These equations apply to any fluid if appropriate values are substituted for the bulk modulus K and the density ρ.

It is easy to verify, by differentiation, that possible solutions to the wave equation (6.8) have the general form

$$p(x,t) = f(x - ct) + g(x + ct) \tag{6.9}$$

where f and g are completely general continuous functions of their arguments. As discussed in Section 3.2 and illustrated in Fig. 3.2, $f(x - ct)$ represents a pressure wave traveling in the $+x$ direction with wave speed c, while $g(x + ct)$ represents a different pressure wave traveling in the $-x$ direction, also with speed c. The two waves do not interact with each other, but their pressures simply add where they overlap.

To calculate the sound speed c for a particular case we need to evaluate the bulk modulus K. For sound waves in an infinite medium it turns out that the frequency is so low, and the wavelength consequently so long, that there is essentially no conduction of heat from the regions of compression to the regions of rarefaction, so that the behavior is adiabatic. We must therefore use the adiabatic bulk modulus rather than the isothermal one, which is somewhat smaller.

For water at about 20° C, $K \approx 2.2 \times 10^9$ N m^{-2}, so that $c_W \approx 1480$ m s^{-1}. The sound speed in sea water is about 3 percent higher. The sound velocity in water of salinity s in parts per thousand, over the range 0 to 20° C, is given approximately by

$$c_W \approx 1450 + 4.6\Delta T + 1.4(s - 35) \quad \text{m s}^{-1} \tag{6.10}$$

where ΔT is the temperature in degrees Celsius. More accurate empirical relations can be found in the literature.

In the case of gases, the adiabatic bulk modulus can be derived from the equation

ACOUSTIC WAVES

$$pV^\gamma = \text{constant} \tag{6.11}$$

where γ is the ratio of the specific heats at constant pressure and constant temperature and has the value 1.4 for a diatomic gas such as air. If we take the logarithm of both sides of this equation and differentiate, then we find from the definition (6.2) that $K = \gamma p_0$, so that the sound speed is given from (6.7) as

$$c = \left(\frac{\gamma p_0}{\rho}\right)^{1/2}. \tag{6.12}$$

Since the static pressure p_0 follows the equation $p_0 V = n k_B T$, where k_B is Boltzmann's constant, T is the absolute temperature, and n is the number of molecules in the volume V, we see that c varies as $T^{1/2}$ for gases. For air at temperature ΔT degrees Celsius and 50% relative humidity,

$$c_A \approx 332(1 + 0.00166\Delta T) \quad \text{m s}^{-1} \tag{6.13}$$

in a temperature range around room temperature. A convenient general value to use is 343 m s^{-1} at 20° C. In the following development we shall generally use ρ and c without subscripts to denote the density and sound velocity for air, unless a subscript A is desirable for clarity.

Because, as we have already seen, objects left to themselves vibrate in combinations of their normal modes, it is natural to consider a wave consisting of a regularly repetitive sinusoidal vibration $a \sin \omega t$, which we write in the complex form $a e^{j\omega t}$. From (6.9) this leads to waves with the general form

$$p(x,t) = a e^{j(\omega t - kx)} \tag{6.14}$$

where

$$k = \frac{\omega}{c} \tag{6.15}$$

is called the propagation number or angular wave number of the wave. Since the phase of the wave repeats when x changes by $2\pi/k$, we call this distance the wavelength λ of the wave, and note that

$$k = 2\pi/\lambda. \tag{6.16}$$

This is illustrated in Fig. 6.2. Usually k is simply called the wave number, which is a little ambiguous since this term is more usually used in optics for the number of wavelengths per meter, $\sigma = 1/\lambda$. However k naturally goes with ω just as σ goes with f, and these symbols are used universally, so there is usually no confusion. In this book we use only k for the wave number and never σ. Note that the wave going in the $+x$ direction is $a e^{j\omega t} e^{-jkx}$, which is one of the reasons physicists prefer to use $i = -j$.

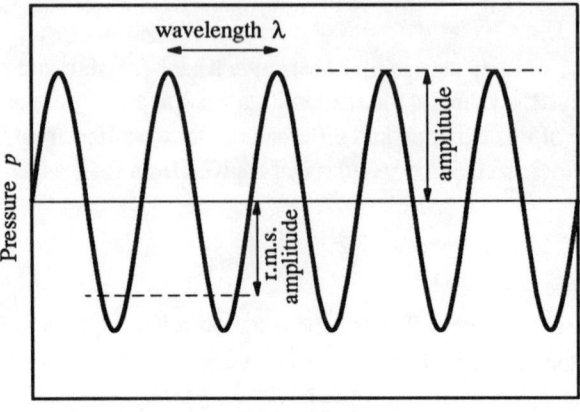

Figure 6.2 In a simple sinusoidal acoustic pressure wave, the important physical quantities are the propagation velocity c, the angular frequency ω, the wavelength λ and the amplitude p. The root-mean-square (r.m.s.) amplitude is equal to $0.707p$.

For sinusoidal pressure waves $p = p e^{j(\omega t - kx)}$, we see from (6.4) that $v = p/\rho c$, independently of frequency. This leads us to define a quantity z, called the wave impedance (or sometimes the specific acoustic impedance) of the medium, by

$$z = p/v = \rho c. \tag{6.17}$$

As we see in a moment, knowledge of the wave impedance tells us a great deal about propagation in the medium. The units for wave impedance are clearly Pa s m^{-1}, but the common name is the rayl, after Lord Rayleigh (1842-1919) who did so much to establish the mathematical basis of acoustics. Approximate values for our two common media, air and water, are

$$z_A \approx 415 \text{ rayl} \quad \text{and} \quad z_W \approx 1.48 \times 10^6 \text{ rayl}. \tag{6.18}$$

6.3 Sound Pressure and Intensity

In the study of vibrations the easiest quantity to measure is the acceleration of the object at a point, by use of a piezoelectric accelerometer. This can be integrated to give the velocity or displacement. The easiest quantity to measure in a fluid medium, however, is the acoustic pressure. This measurement is carried out with a microphone or a hydrophone, which can be appropriately calibrated. Most of our discussion therefore regards acoustic pressure as the primary variable, and we shall find it generally convenient to use acoustic impedances rather than acoustic admittances.

Acoustic pressure can be specified by giving either the pressure amplitude or the root-mean-square (r.m.s.) amplitude. This second quantity, which is $1/\sqrt{2} = 0.707$ times the first for a sine wave, is the more usual in practical applications, for it is directly related to the power in the wave in the case of more complex waveforms. A similar convention is used in electrical applications where 110 volts, for example, refers to the r.m.s. value of the normal mains voltage.

The range of sound pressures to which animals can respond, from those that are at the threshold of detectability to those that are so large as to cause discomfort, is very great—a factor of about one million for the human auditory system. Our own experience also shows us that we tend to use a roughly logarithmic response scale—the phrase "twice as loud", for example, meaning about the same thing for soft and for loud sounds. For both these reasons it is appropriate to define a logarithmic scale for sound pressure, and we refer to this as the sound pressure level $L(p)$. Because of the logarithmic nature of the scale, we need also a reference pressure p_{ref}, and we can then write $L(p) = A \log(p/p_{ref})$, where A is a constant yet to be determined.

As we discussed earlier in Section 4.5, it is conventional to define the logarithmic scale so that it makes sense in terms of power, rather than amplitude. Power is proportional to the square of the amplitude. We therefore take logarithms to the base 10, for convenience, and define the level change associated with a factor 10 in power to be a unit called the bel, after Alexander Graham Bell of telephone fame. This unit is a bit large, so one a tenth as big, called the decibel (dB), is always used. We therefore have the definition

$$L(p) = 10 \log_{10}(p^2/p_{ref}^2) = 20 \log_{10}(p/p_{ref}) \quad \text{dB}. \tag{6.19}$$

We still have to fix the standard reference pressure p_{ref}. For sound in air it is convenient to take this to approximate the threshold of human hearing in its most sensitive range, 1–3 kHz. The actual adopted value, the rationale for which will be explained when we discuss acoustic intensity, is 20 µPa (or 0.0002 microbars) r.m.s., so that to use the definition (6.19), p should also be an r.m.s. value. Fig. 6.3 shows a set of equal-loudness contours for human hearing, known as Fletcher-Munson curves and initially measured for the telephone industry. The region of maximum sensitivity, from about 500 to 5000 Hz, corresponds to the region of maximum information content in human speech, in terms of the distinction between different vowel and consonant sounds. The equal-loudness contours are labeled with a loudness level in phons, which for each curve is the sound pressure level of a pure tone of frequency 1000 Hz that gives the same subjective impression of loudness. We shall not give further attention to these curves just now, except to remark that a major aim of this book is to explain how such curves arise.

For much psychophysical and noise-monitoring work it is usual to use a sound-level meter equipped with a filter to produce a sensitivity curve close to that of the human auditory threshold. Such a filter is called an A-weighting filter, and

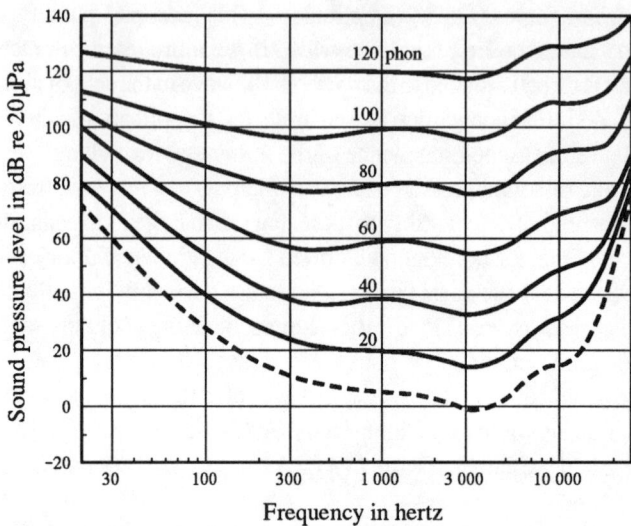

Figure 6.3 Equal-loudness contours for average human hearing in young adults (Fletcher-Munson curves). The loudness level in phons is equal to the sound pressure level of a pure tone of frequency 1000 Hz that gives the same subjective loudness as the test tone. The broken curve shows the average auditory threshold level. With advancing age, sensitivity to high frequencies decreases. The threshold of discomfort is about 130 dB.

sound pressure levels measured with it in use are called A-weighted sound pressure levels and specified in dB(A). Clearly a measurement in dB(A) discounts the contribution of sounds with frequency above about 10 kHz and, much more importantly, of sounds with frequencies below about 200 Hz. To give some idea of typical magnitudes, a quiet room in the evening has an A-weighted sound pressure level of perhaps 30 dB(A), ordinary conversation is about 70 dB(A), and a rock music concert may reach more than 110 dB(A). Industrial legislation generally limits allowable factory noise levels to the equivalent of 90 dB(A) continuous, unless hearing protectors are worn. Two other weighting networks, B and C, are occasionally used. These correspond roughly to equal-loudness contours at higher levels (about 60 and 100 phons respectively) and have progressively flatter response.

In the case of sound transmission through water, the pressure levels are generally much higher than in air because of the higher wave impedance of water. It is therefore usual to choose the reference pressure p_{ref} in (6.19) to have the value 0.1 Pa (1 microbar). In cases of possible ambiguity it is customary to quote the reference pressure by writing, for example, "20 dB re 0.1 Pa".

The other important quantity that we can define for a plane wave is the acoustic intensity I, which is the power per unit area carried by the wave, measured in watts per square meter. Since the wave impedance z of the medium is essentially real, the energy transported by a wave of pressure amplitude p is just

$$I = \frac{1}{2}pv = \frac{1}{2}p^2/z = \frac{1}{2}p^2/\rho c \quad (6.20)$$

per unit area, the factor 1/2 arising since p is an amplitude rather than an r.m.s. value. Once again it makes practical sense to use a logarithmic scale and to define a quantity $L(I)$, called the intensity level, by

$$L(I) = 10\log_{10}(I/I_{ref}) \quad \text{dB.} \quad (6.21)$$

The reference intensity I_{ref} is taken as 10^{-12} W m^{-2}, which is almost identical to the intensity of a wave with r.m.s. pressure $p_{ref} = 20$ µPa. For a plane sound wave in air, therefore, the intensity level is numerically essentially equal to the sound pressure level.

We must note, however, one major difference between sound pressure and intensity that is often overlooked. Sound pressure and sound pressure level are both quantities defined at a point, and are the physical entities to which an ear or a microphone responds. If there are several sound waves present, their sound pressures simply add. Intensity, however, is a vector quantity with orientation parallel to the propagation direction of the wave. If we were to measure the sound intensity close to an acoustically hard surface, a measurement requiring two closely spaced microphones and some computation, then we would find it to be nearly zero, for the energy carried by the reflected wave would almost cancel that of the incident wave. For obvious reasons, most of our discussion will deal with the biologically more relevant quantities sound pressure and sound pressure level.

To give some feeling for magnitudes, we can note that a typical biological sound source such as a singing bird has an acoustic power of a few milliwatts. If we measure the intensity at a distance of 1 m, then this power is spread over $4\pi \approx 12.6$ m^2, which, for a 1 mW source, gives an intensity of 8×10^{-5} W m^{-2}. This translates to an intensity level or sound pressure level of 78 dB. We can easily scale this figure for different powers or distances. Another convenient reference point to remember is that 1 Pa r.m.s. corresponds to an SPL of 94 dB.

6.4 Reflection and Transmission at a Boundary

To illustrate the importance of wave impedance, let us consider a sound wave falling normally upon a plane interface separating two media with wave impedances z_1 and z_2, the incident wave being in medium 1. An example important in biology is the transmission of sound from air to water, or from water to air. Suppose that the amplitude of the incident pressure wave is p_I, that of the wave reflected from the interface p_R, and that of the transmitted wave p_T, as shown in Fig. 6.4. If

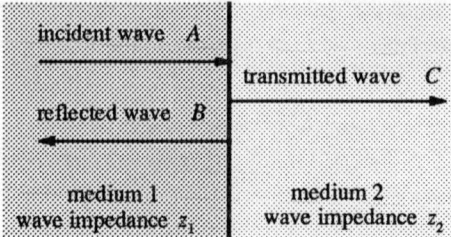

Figure 6.4 Reflection at a plane interface between two media with wave impedances z_1 and z_2 respectively. A, B, and C are the pressure amplitudes of the incident, reflected, and transmitted waves respectively.

the frequency of the wave is ω and its wave number $k_i = \omega/c_i$ where the subscript i refers to the medium, then the three waves can be written

$$p_I(x,t) = A e^{j(\omega t - k_1 x)} \qquad p_R(x,t) = B e^{j(\omega t + k_1 x)} \qquad p_T(x,t) = C e^{j(\omega t - k_2 x)}. \quad (6.22)$$

The acoustic velocities associated with these waves are, by the definition (6.17) of wave impedance,

$$v_I = \frac{A}{z_1} e^{j(\omega t - k_1 x)} \qquad v_R = -\frac{B}{z_1} e^{j(\omega t + k_1 x)} \qquad v_T = \frac{C}{z_2} e^{j(\omega t - k_2 x)}. \quad (6.23)$$

Note the change in sign for the velocity of the reflected wave. If we take the interface to be at $x = 0$, then we recognize that the two media must move in the same way on either side of the interface, so that the velocities match, and the pressures must also match, otherwise the molecules at the interface would have a large net force on them. We can write these two conditions in terms of the waves as

$$A + B = C \qquad \frac{(A-B)}{z_1} = \frac{C}{z_2} \qquad (6.24)$$

from which a little algebra gives

$$\frac{B}{A} = \frac{z_2 - z_1}{z_1 + z_2} \qquad \frac{C}{A} = \frac{2 z_2}{z_1 + z_2}. \qquad (6.25)$$

Reflection and transmission coefficients R and T are more usefully defined in terms of energy flow or intensity than in terms of pressure amplitudes. Again a little algebra shows that

$$R \equiv \frac{B^2}{A^2} = \frac{(z_2 - z_1)^2}{(z_1 + z_2)^2} \qquad T \equiv \frac{C^2 z_1}{A^2 z_2} = \frac{4 z_1 z_2}{(z_1 + z_2)^2}. \qquad (6.26)$$

Necessarily $R + T = 1$, since energy is conserved. These equations tell us that the extent of reflection at the interface depends upon the mismatch in wave impedance between the two media. From the form of the expression for the transmission coefficient given in (6.26), it is clear that energy transmission is the same in either direction, another example of the principle of reciprocity. For the important practical case of transmission between air and water, $z_A = 415$ rayl and $z_W = 1.54 \times 10^6$ rayl, so that the mismatch is very large and reflection is almost complete. The transmission coefficient is only 0.001, which corresponds to a transmission loss of 30 dB. These formulae are changed a little when the wave falls obliquely on the interface, but the fact remains that there is a large transmission loss for sound transmission between air and water, and ambient noise is likely to obscure all but the very loudest signals coming from the other medium.

Another important phenomenon occurring at an interface is refraction, which is completely analogous to the optical case, with the ratio of sound velocities taking the place of refractive index. The propagation direction of an obliquely incident sound wave is bent towards the normal to the interface when it passes from a region of high sound velocity to a region of lower sound velocity. Such interfaces occur in the ocean between water masses of different temperature or salinity, and in the air across temperature inversions. Similar but more complicated effects occur when the fluid is in motion. More generally, when the properties of the medium change gradually with position, the propagation direction will also change.

6.5 Wave Attenuation

We have not yet given any attention to loss processes in the propagation of a sound wave, but it is clear that such losses must occur. For one thing, we came to the conclusion that sound propagation is essentially adiabatic but, since there are temperature gradients caused by the compressions and rarefactions in the wave, there must be some heat flow and this will cause energy loss. In a more subtle way, the changes in temperature necessitate changes in the excitation of the internal vibrations of the gas molecules in all but monatomic gases, and this too causes energy loss. Finally, since acoustic velocity is aligned in one direction only, there is a necessary change in the shape of small volumes of the fluid as the wave passes, and this is resisted by viscous forces, causing still further energy loss.

Since all the gradients rise linearly with frequency, we expect the energy loss associated with all of these mechanisms to increase as the square of the wave frequency. If we write the wave number k of the wave as a complex quantity $k = (\omega/c) - j\alpha$, then the wave decreases in amplitude as $e^{-\alpha x}$. For air, the value of α, which is called the attenuation coefficient, depends considerably upon humidity, reaching a maximum of about 3 times the dry-air value for relative humidities in the range 10 to 20 percent. The behavior is somewhat complex, but an approximate value for air of 50% relative humidity at about 20° C is

$$\alpha_A \approx 1 \times 10^{-10} f^2 \quad \text{m}^{-1} \tag{6.27}$$

where f is the frequency in hertz. This is equivalent to an attenuation of about 1 dB/km at 1000 Hz or about 100 dB/km at 10 kHz. This is in addition, of course, to the decrease in intensity caused by spreading of waves in the real world, which we discuss in the next section.

A similar relation holds for sound propagation in water, though the mechanisms are rather different and depend upon the solute concentration in the case of sea water. To a moderate approximation for sea water at 5° C,

$$\alpha_W \approx (1 \text{ to } 10) \times 10^{-12} f^2 \quad \text{m}^{-1} \tag{6.28}$$

the higher values of the coefficient applying at low frequencies. The attenuation is actually about 0.001 dB/km at 100 Hz, 0.05 dB/km at 1000 Hz and 1 dB/km at 10 kHz. As far as attenuation is concerned, therefore, sound propagation in the ocean is relatively loss-free compared with sound propagation in the atmosphere. The effect is exaggerated by the fact that the oceans are relatively shallow, so that energy is spread over a cylindrical rather than a spherical surface as the wave expands.

6.6 Spherical Waves

In many cases we are not far from the source generating the sound wave, so that it is necessary to examine the way in which the wave spreads. As the simplest case, consider a wave generated at a point source and expanding in a spherical manner. The three-dimensional form of the wave equation in Cartesian coordinates (x, y, z) is written in obvious generalization of (6.8) as

$$\nabla^2 p \equiv \left(\frac{\partial^2}{\partial x^2} + \frac{\partial^2}{\partial y^2} + \frac{\partial^2}{\partial z^2} \right) p = \frac{1}{c^2} \frac{\partial^2 p}{\partial t^2}. \tag{6.29}$$

To treat our problem it is clearly best to use spherical polar coordinates (r, θ, ϕ). The general expression is unnecessarily complex and we can simplify by realizing that for a simple spherical wave there will be no dependence upon either of the angles θ or ϕ. For this spherically symmetrical case we can rewrite the differential operator ∇^2 and thus the wave equation (6.29) as

$$\frac{1}{r^2} \frac{\partial}{\partial r} \left(r^2 \frac{\partial p}{\partial r} \right) = \frac{1}{c^2} \frac{\partial^2 p}{\partial t^2}. \tag{6.30}$$

If now we write $p = p'/r$, then the equation becomes

$$\frac{\partial^2 p'}{\partial r^2} = \frac{1}{c^2} \frac{\partial^2 p'}{\partial t^2} \tag{6.31}$$

which is just an ordinary one-dimensional wave equation with solutions $p' = f(r \pm ct)$. The pressure wave therefore has the simple form

$$p(r,t) = \frac{A}{r} f(r - ct) + \frac{B}{r} g(r + ct) \tag{6.32}$$

corresponding to an outgoing spherical wave f and an incoming wave g. The amplitude of the outgoing wave decreases as $1/r$ while the incoming wave increases in amplitude similarly. At any fixed distance, the time variation of the wave is reproduced without distortion—an important feature in communication. If we confine our attention to sinusoidal waves of frequency ω, then we can write the outgoing wave, which is our primary interest, as

$$p(r,t) = \frac{A}{r} e^{j(\omega t - kr)}. \tag{6.33}$$

The decrease in amplitude as $1/r$ is clearly necessary from considerations of conservation of energy since, once the wave is well away from the origin, its intensity is proportional to p^2 and it is spread over a spherical surface of area $4\pi r^2$. Expressed in decibels, we find that the intensity decreases by 6 dB for every doubling of the distance r. For small distances, up to about α^{-1} where α is the attenuation coefficient for the medium, this decrease caused by spreading is the dominant influence on the wave intensity, while for larger distances the attenuation of the medium becomes more important. This is illustrated in Fig. 6.5.

The velocity associated with a spherical wave has a rather more complicated behavior than that in the plane-wave case. From Newton's equation of motion (6.4), written for the radial coordinate r, we see that

$$\rho \frac{\partial v}{\partial t} = -\frac{\partial p}{\partial r} \quad \text{or} \quad v = \frac{j}{\omega \rho} \frac{\partial p}{\partial r} \tag{6.34}$$

which, with (6.33), leads to

$$v(r,t) = \frac{A}{\rho c r} \left(1 - \frac{j}{kr}\right) e^{j(\omega t - kr)}. \tag{6.35}$$

At distances very much greater than a wavelength, $kr \gg 1$ and the term j/kr can be neglected. The behavior is then essentially the same as that of a plane wave. Within about one wavelength of the origin, however, we cannot make this approximation, and indeed the term $1/kr$ dominates the velocity behavior near the origin. This is called the near-field region, and in it the velocity is always much greater than would be expected from the magnitude of the acoustic pressure. This circumstance aids animals, such as insects, that use sensory hairs to detect acoustic signals produced by the buzzing wings of predators. The great increase in velocity amplitude within about one sixth of a wavelength—a distance of 50 cm or so for a wing-beat fre-

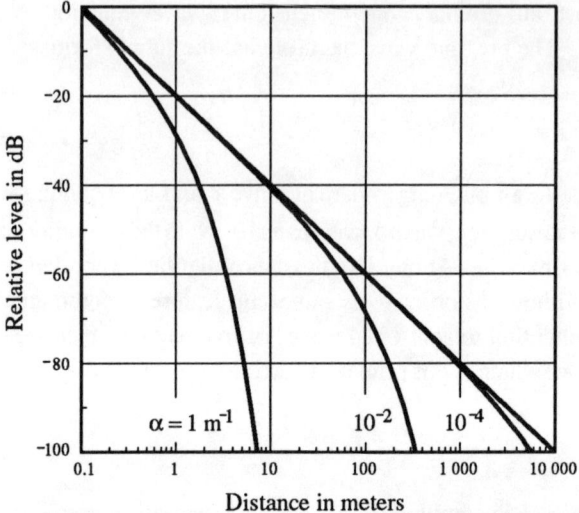

Figure 6.5 Decrease in intensity with distance for a spherical wave propagating in a medium with attenuation coefficient α. The intensity decreases by 6 dB for every doubling of distance in the absence of losses in the medium, while attenuation causes a more rapid decrease at large distances. The range of values illustrated is characteristic of sound propagation in air. The curve for $\alpha = 1$ corresponds to a frequency of about 100 kHz, and each decrease in frequency by a factor 10 is associated with a decrease in α by a factor 100.

quency of 100 Hz—gives a signal that stands out well from the background. Indeed, when we come to consider the dipolar nature of the wing-beat source in the next chapter, we see that the near-field effect is even more extreme than that for a simple spherical wave.

We can define a sort of wave impedance $z_s = p/v$ for a spherical wave as a function of the parameter kr. From (6.33) and (6.35) its value is

$$z_s(kr) = \frac{\rho c}{1 - (j/kr)} = \rho c \left(\frac{k^2 r^2}{1 + k^2 r^2} \right) + j\rho c \left(\frac{kr}{1 + k^2 r^2} \right). \tag{6.36}$$

This wave impedance is a complex quantity, with the imaginary part dominating near the origin. The magnitude of the impedance is

$$z_s(kr) = |z_s(kr)| = \rho c \frac{kr}{(1 + k^2 r^2)^{1/2}} \tag{6.37}$$

which tends to zero as we approach the origin. We shall find this to be important in the next chapter.

6.7 Surface Waves on Water

Another major type of wave-borne acoustic signal that is of significance in biology is that used by insects living just below the surface of water, and hunting prey trapped in the surface by capillary forces. All surface waves are characterized by the fact that their influence extends away from the surface only for a distance comparable with their wavelength, and this has considerable influence on their propagation behavior. There are two types of simple wave motion possible on the surface of a relatively deep body of water. The first, called a capillary wave, derives its restoring force from the surface tension of the liquid, which tends to keep the surface as flat as possible; the second, called a gravity wave, depends upon gravitational restoring forces to level the surface. In general, of course, both restoring forces must be considered. We shall not discuss these waves in detail, but a brief survey may be useful.

Suppose that σ is the surface tension of water, g the gravitational acceleration, ω the frequency, and c_S the wave velocity of the sinusoidal surface disturbance with which we are concerned. Then we can define a wave number $k = \omega/c_S = 2\pi/\lambda$, where λ is the wavelength, in terms of which the wave velocity can be shown to be

$$c_S = \frac{\sigma k}{\rho_W} + gk = \frac{2\pi\sigma}{\rho_W \lambda} + \frac{g\lambda}{2\pi} \qquad (6.38)$$

where ρ_W is the density of the liquid. If the frequency is high, so that k is large and λ is small, the second term can be neglected and only the surface tension affects the propagation velocity, which rises with increasing frequency. Such waves are called capillary waves, and their velocity is given approximately by

$$c_C \approx (\sigma\omega/\rho)^{1/3}. \qquad (6.39)$$

On the other hand, if the frequency is low so that k is small and λ large, we can neglect the first term and only gravitational restoring forces enter. The waves are then called gravity waves, and their propagation velocity increases with increasing wavelength (decreasing frequency) approximately as

$$c_G \approx g/\omega. \qquad (6.40)$$

In the border between these two regimes where the two terms in (6.37) are nearly equal, the surface wave velocity has its minimum value.

To put some figures into these equations, we note that $\sigma \approx 0.07$ J m^{-2} for water, and $g \approx 10$ m s^{-2}, so that the boundary between capillary and gravity waves occurs for a wavelength of about 17 mm. The wave velocity is then about 0.23 m s^{-1} and the frequency about 14 Hz. For capillary waves of frequency 100 Hz, $c_C \approx 0.35$ m s^{-1}, while for gravity waves of frequency 1 Hz, $c_G \approx 1.7$ m s^{-1}.

Propagation in which wave speed depends on wavelength is called dispersive, and has much more complex properties than the nondispersive propagation characteristic of sound waves in bulk air or water. Clearly for a liquid surface any transient disturbance will have its waveform distorted as it propagates, for the high- and low-frequency components will arrive before those in the range 10 to 20 Hz. In addition to this, energy propagates at a rate characterized by the group velocity, $\partial \omega / \partial k$ rather than the wave velocity, ω/k. Since the group velocity is normally less than the wave velocity, the waves inside a transient pulse envelope move through it from the trailing to the leading edge and then disappear. Some of these phenomena can be observed when a stone is thrown into a pond. The impulse of the stone generates a pulse with duration of a few tenths of a second, the time for creation and collapse of the depression in the water surface. From the Fourier transform theory discussed in Chapter 15, this impulse can be decomposed into a superposition of waves with median frequency of a few hertz, which are thus primarily gravity waves. These waves propagate independently with velocities given primarily by (6.40), the waves of lowest frequency and longest wavelength spreading most rapidly, so that after a little time the disturbance has waves of long wavelength at its edges and short-wavelength waves near its center. Similar phenomena occur for the waves generated by an insect trapped by surface tension, the characteristic signature of the wave telling something about the nature of the source. Surface-dwelling predators must obviously make use of this information in identifying the nature of the prey they are hunting.

One other observation can be made generally about surface waves. If we set aside the attenuation caused primarily by viscous effects, then the wave amplitude must decrease with distance r from a small source as $r^{-1/2}$ in order that the energy flux may remain constant. This is equivalent to a reduction in intensity, defined here as power per unit length in the surface, by 3 dB for every doubling of distance from the source. A similar relation holds for ordinary acoustic waves traveling very large distances in the ocean, for they are essentially confined between the surface and the sea floor and spread cylindrically rather than spherically.

6.8 Surface Waves in Solids

As indicated in the introduction, elastic wave propagation in solids is a very complex subject. The waves may be of longitudinal type, like sound waves in fluids, with the displacement parallel to the direction of propagation, or of transverse type, with the displacement normal to the propagation direction. Generally both types of wave will be present, traveling with different velocities, and they may convert partially from one type to the other when scattered or reflected. Some animals, such as ants or worms, live inside solids and may derive information from the different character of vibrations reaching them through the surrounding earth. If the bulk modulus of the solid is K and its shear modulus, or modulus of rigidity,

ACOUSTIC WAVES

is μ, then longitudinal (compressive) waves travel with speed c_L and transverse (shear) waves with speed c_T where

$$c_L = \left[\frac{(K+4\mu/3)}{\rho}\right]^{1/2} \qquad c_T = \left(\frac{\mu}{\rho}\right)^{1/2}. \tag{6.41}$$

These speeds are quite high for most materials—several thousands of meters per second—and always $c_T < c_L$.

Many other animals live on the surfaces of solids, either the earth or vegetation, and sense the vibratory signals transmitted through the substrate. For solid layers that are thick relative to the wavelength concerned, for example the earth's surface or the trunk of a tree, some types of waves can be confined to the immediate vicinity of the surface. These waves are relatively easily excited by surface impacts and convey important information to many animal species. There are two types of surface waves in simple elastic solids. Assuming the surface to be horizontal, the first, called a Rayleigh wave, has the material displacement confined to a vertical plane containing the direction of propagation, so that it is essentially a coupled longitudinal/vertical-transverse wave. The speed of Rayleigh waves is approximately $0.9c_T$. The second type of surface wave is called a Love wave, and in it the material motion is horizontal and transverse to the propagation direction. Propagation of Love waves is dispersive, but their velocity is close to c_T and is a little greater than that of Rayleigh waves.

When the object in which the vibrations propagate is not large, it may be possible to treat it as a flat plate (as in the case of a large leaf), as a thin rod or bar (as in the case of a twig or grass stem), or as a taut string (as in the case of a spider web). We can then use the expressions derived in Chapters 3 or 5 to estimate the propagation velocity. Propagation along strings is nondispersive, but propagation along rods or plates is dispersive with $c \propto \omega^{1/2}$.

6.9 Scattering by Solid Objects

In many biological applications we shall need to calculate sound pressures at the ears of an animal, and to take account of the fact that the ears may be located on the surface of a much larger object, the head or body of the animal. The general problem of interaction of waves with obstacles is very complicated, even if the obstacle is taken to have a rather simple shape, so we shall be content with a semiquantitative discussion of a few typical examples.

The simplest case to think about, though by no means the easiest to calculate, is the scattering of sound by a thin solid disc of radius a placed at right angles to the sound incidence direction. We can solve this problem by superimposing on the original plane wave, taken to fill all of space, a diverging wave centered on the disc and representing the scattering. The fact that the disc is solid introduces the requirement that the acoustic particle velocity must vanish on its surface, so that the scattered wave must be just the wave that would be generated by the disc alone if it were to vibrate with the same velocity amplitude as the incident wave, but with opposite phase. We discuss the radiation from a vibrating disc in the next chapter. At low frequencies, for which the wavelength is much greater than the disc diameter and $ka \ll 1$, the radiated intensity is small and has an angular distribution like a figure 8. This causes a small intensification of the pressure field just in front of the disc and a small reduction just behind it, but the effect is not large and the whole surface of the disc experiences very nearly the free field pressure p_0. At higher frequencies, for which the wavelength is less than the disc diameter and $ka > 1$, the radiation is increasingly concentrated into two beams normal to the two sides of the disc and with opposite phases. The backward-propagating or reflected beam nearly doubles the sound pressure in front of the disc and creates a standing-wave pattern, while the forward-propagating beam nearly cancels the original wave, leaving a sound shadow. In semi-quantitative terms, the sound shadow extends for a distance behind the disc of order a^2/λ, where λ is the sound wavelength. This rule can be applied in the same approximate spirit to flat obstacles of other shapes.

Of more relevance to our discussion is the scattering of sound by a spherical obstacle of radius a, for this is a first approximation for something like a human head. The calculation method is essentially similar, with the final sound field being a superposition of the original wave and a spherical wave centered on the sphere. The complication arises because of the thickness of the sphere in the original wave-incidence direction, for this means that there must be phase shifts between the vibration of different parts of the spherical surface in order to cancel out the velocity of the incident wave. The final result is rather complicated, but it takes on a simple form if the sphere is small compared with the sound wavelength, so that $ka \ll 1$. In this limit, the sound pressure at the surface of the sphere at a point whose location makes an angle θ with the direction of incidence can be shown to be approximately

$$p(a, \theta) \approx \left(1 - \frac{3}{2} jka \cos\theta\right) p_0. \tag{6.42}$$

Interestingly, this has a magnitude that is symmetric about the equatorial plane of the sphere, so that the sound pressure is the same at ipsilateral and contralateral positions. As with the disc, as ka is increased this symmetry vanishes and the sound pressure in the direction of the source increases, while that behind the sphere decreases. Once $ka > 1$ the sound pressure on the sphere in a direction facing the sound source, $\theta = 0$, becomes very nearly $2p_0$, while that in the opposite direction is nearly zero. These results allow us to make some estimates of head-shielding effects in real situations.

Another case that is of biological relevance is scattering by a cylinder, since this is a first approximation to the shape of the pinnae of animals like rabbits. The case of normal sound incidence on an infinite cylinder can be analyzed exactly, as for the sphere, and the sound pressure on the cylinder surface for $ka \ll 1$ has the angular variation

$$p(a,\theta) \approx (1 - 2jka\cos\theta)p_0. \tag{6.43}$$

Again the pressure amplitude distribution is symmetrical for $ka \ll 1$ and becomes increasingly asymmetric as ka increases until, for $ka > 1$, the sound pressure on the side facing the sound source is nearly $2p_0$ and that on the opposite side nearly vanishes.

6.10 Scattering by Bubbles

The related problem of the scattering of sound by a bubble in a liquid is of importance in the hearing process for some aquatic animals, not because bubbles themselves are important but rather because some animals have air-filled swim bladders in their bodies that act as quite efficient sound scatterers because of the large acoustic impedance mismatch between water and air. The body of the animal, in contrast, is a relatively poor scatterer since its density differs little from that of water and it does not have a high shear modulus. The shape of a swim bladder may be complex, but it will be an adequate first approximation to examine the scattering behavior of a spherical bubble, since sound wavelengths are relatively long in water.

It would take us too far afield to consider this problem in detail, and we shall be content with outlining the main results of importance to our present application. There are two different effects contributing to the scattering and to the acoustic field close to the sphere. In the first place, by matching the scattered wave and the incident field in such a way that the requirements of the impedance mismatch are met, it can be shown that the bubble scatters predominantly in the backwards direction, and that it actually moves in the field with an amplitude that is just about three times the displacement amplitude of the surrounding liquid. Clearly this displacement effect will be experienced also by any acoustic sensors close to the bubble and will have the effect of increasing their sensitivity by as much as a factor 3, or 10 dB. Because the bubble moves in the direction of the acoustic displacement, this near field contains directional information so that a fish with a swim bladder is at a considerable advantage in the matter of acoustic information gathering.

The other phenomenon associated with scattering from a bubble is one of resonance, when the frequency of the incident wave is close to the characteristic frequency ω_0 for radial oscillations of the bubble. This frequency depends upon pressure p_B in the bubble, and thus upon the depth below the surface. To a reasonable approximation

$$\omega_0 \approx \frac{1}{a}\sqrt{\frac{3\gamma p_B}{\rho_W}} \tag{6.44}$$

where a is the radius of the bubble, $\gamma = 1.4$ is the ratio of specific heats for air, and ρ_W is the density of water. For a bubble of radius 1 cm, which might be typical of the volume of the swim bladder of a small fish, $\omega_0 \approx 2000$ s^{-1}, which corresponds to about 300 Hz. The elastic properties of the surrounding membranes may raise this frequency somewhat and damp the resonance, which otherwise may have a Q value higher than 10.

A bubble resonating in this manner radiates a great deal of sound in a spherically symmetric pattern, and is thus a very efficient scatterer. From a sensory point of view, since the resonance may occur within the auditory range of the animal, it may also provide a very low auditory threshold near the resonance frequency. The auditory behavior is, naturally, dependent upon the shape of the swim bladder, its location relative to the otoliths or other sensors, and the nature of any direct anatomical connections.

References

Wave equation: [1] Ch 5,6; [6] Ch 2; [8] Ch 5,7; [10] Ch 6; [52]
Spherical waves: [1] Ch 6; [8] Ch 7; [10] Ch 6
Reflection and transmission: [8] Ch 6
Wave attenuation: [8] Ch 9; [29]; [30]
Surface waves on water: [52] Ch 12, 13
Surface waves on solids: [7] pp. 463–467
Underwater acoustics: [8] Ch 15; [33] Ch 13
Scattering: [1] Ch 7

Discussion Examples

1. Calculate the wavelengths corresponding to the upper and lower limits of human hearing (20 kHz and 20 Hz respectively).
2. Calculate the wavelengths in water for sounds of the same frequencies as in example 1.
3. Calculate the acoustic displacement amplitude for a plane sound wave of (a) moderate loudness (74 dB re 20 µPa at 1000 Hz), and (b) threshold loudness (0 dB at 1000 Hz) in air. Compare the result (a) with the displacement amplitude for a wave with the same pressure level (0 dB re 0.1 Pa) in water.
4. A fog-horn produces a pressure level of 100 dB re 20 µPa measured at the surface of still ocean. Calculate the pressure level re 0.1 Pa just below the water surface.
5. A bird singing in a high tree has an acoustic output of 2.5 mW. Calculate the intensity level and the sound pressure level generated by its song at a distance of 10 m.
6. If the bird sings with the same acoustic power while sitting on the ground, what is the intensity level at 10 m?

Solutions

1. 20 kHz corresponds to a wavelength of 17 mm, 20 Hz to 17 m.
2. 20 kHz in water is 74 mm wavelength, 20 Hz is 74 m.
3. (a) pressure = 0.1 Pa r.m.s., amplitude = 38 nm r.m.s. or about 0.06 µm peak, (b)

pressure = 20 µPa r.m.s., amplitude ≈ 0.76×10^{-11} m r.m.s. or about 0.1 Å peak. For 0.1 Pa in water, the amplitude ≈ 10^{-11} m r.m.s. or only about 0.14 Å peak.

4. 100 dB SPL corresponds to a pressure amplitude of 2 Pa r.m.s. This includes the pressure of the reflected wave, so the incident wave is only 1 Pa. From pressure balance at the interface, the transmitted wave has pressure amplitude 2 Pa r.m.s. or a level of 26 dB re 0.1 Pa.

5. The acoustic power is spread over a sphere so the intensity is 2 µW m^{-2}, which gives an intensity level of 63 dB re 10^{-12} W m^{-2}. The SPL is the same in decibels, since we have free radiation.

6. The power is spread over only a hemisphere so the level is increased by 3 dB to 66 dB. (Note output power assumed constant, not source strength or volume flow.)

7 ACOUSTIC SOURCES AND RADIATION

SYNOPSIS. When a surface vibrates in a fluid, acoustic waves are generated. The simplest case to consider is that of a sphere of radius a pulsating in volume at frequency ω. The wave number $k = \omega/c$ in the combination ka is the appropriate general parameter to describe the behavior. If $ka \ll 1$ the sphere is a very inefficient radiator of sound, but it improves in efficiency as the square of the frequency until $ka = 2$. Above this frequency the efficiency, measured as radiated power per unit area for given surface velocity, is constant. This behavior can be expressed in terms of an acoustic radiation impedance, and in particular its real part, the acoustic radiation resistance, as illustrated in Fig. 7.1. For a given volume flow generated by the source, the radiated acoustic power is proportional to radiation resistance. A small spherical source is often a reasonable model for a more complex small radiator.

The efficiency of radiation from a given vibrating source is directly proportional to the wave impedance of the medium in which it is immersed. Since the wave impedance of water is some 3000 times greater than that of air, vibrating sources radiate particularly efficiently under water. This fact, together with the relatively small sound attenuation in water, is important for acoustic communication in aquatic animals.

When a simple spherical source is located a small distance above a flat surface, as for example an insect singing near the ground, then the total radiated power is increased by a factor of as much as 2 compared with that of the same source in free air. The larger value is approached when the source is within about one-sixth of a wavelength of the surface. The directional pattern is complicated, as illustrated in Fig. 7.2.

A vibrating rigid disc of radius a is a reasonable first model for the sound generated by a vibrating wing. Because the sound waves generated from the two surfaces of the disc are in opposite phase, they cancel exactly in the plane of the disc, and sound radiation is concentrated along its axis in both directions. The radiation efficiency is very low if $ka < 1$, as illustrated by the real part of the acoustic impedance in Fig. 7.3, but increases as the fourth power of the frequency and reaches the same constant value as for the sphere if $ka > 2$, when both sides of the disc are taken into account.

A panel vibrating in a more complex pattern, with regions of opposite phase, is more difficult to analyze. The fundamental mode radiates relatively efficiently, provided the panel is reasonably large and we make allowance for the dipolar character of the radiation. In the higher modes, however, the contributions of adjacent regions of opposite phase tend to cancel, so that their radiation efficiency is small. The irregular shape of biological objects ensures that cancellation is not exact, but higher modes do not approach the radiation efficiency of the fundamental.

Radiation from the open end of a circular pipe of radius a is a good first model for radiation from open mouths in many biological species. Again the radiation efficiency is small if $ka \ll 1$, as shown in Fig. 7.6, but reaches the standard constant value when $ka > 2$. The directional behavior of the radiation from an open pipe is illustrated in Fig. 7.5. The radiation intensity is uniform in direction at low frequencies, but becomes increasingly concentrated along the pipe axis as ka becomes greater than unity.

A discussion of a general principle called the reciprocity theorem shows that the behavior of acoustic systems as generators of sound is closely related to their behavior as sound detectors. In particular, the directional behavior is the same in both cases.

ACOUSTIC SOURCES AND RADIATION

7.1 Sound Generation

In the early chapters of this book we considered the behavior of vibrating bodies, and in Chapter 6 we discussed the propagation of acoustic waves in air and water. We now come to link these two subjects together by examining the way in which vibrating bodies couple to the fluid in which they are immersed to generate acoustic disturbances. This is potentially a rather complicated subject, and we begin as usual from the simplest possible idealizations, expanding these gradually to encompass more complicated and realistic situations.

For convenience once again we take air as the fluid medium of concern, but all the results and formulae can be applied immediately to water or to other fluids simply by inserting appropriate values for the density and bulk modulus. Indeed, all the behavior is conveniently described in terms of the non-dimensional number ka, where $k = \omega/c$ is the propagation number or wave number $2\pi/\lambda$ and a is a characteristic dimension of the vibrating source. This quantity ka is a useful universal parameter for describing wave behavior, and indeed we have already met it in the formulae of Chapter 5.

We remark here also on the useful fact, embodied in the reciprocity theorem, that the behavior of any acoustic system as a receiver of sound is essentially the same as its behavior as a transmitter. This remark requires a little interpretation which we come to later on, but it underlines the fact that our discussion in this chapter turns out to be applicable to auditory systems as well as to vocal systems.

7.2 Simple Spherical Source

The simplest possible acoustic source is a sphere of radius a that pulsates at a frequency ω. We suppose the radial velocity of the surface to be $u = da/dt$. If this source generates any sort of acoustic wave, it must have spherical symmetry and constitute the sort of spherical wave discussed in Section 6.6. To determine the amplitude of the wave generated, all we have to do is match its particle velocity to the velocity of the sphere at $r = a$, since there is no restriction on the pressure acting on the surface. Now by (6.35) the velocity at radial distance r in a spherical wave has the form

$$v(r,t) = \frac{A}{\rho c r}\left(1 - \frac{j}{kr}\right) e^{j(\omega t - kr)} \tag{7.1}$$

so that the condition $v(a) = u$ fixes the constant A. In terms of the acoustic pressure p, the wave becomes, by (6.33),

$$p(r,t) = \frac{\rho c a u}{(1 - j/ka)r} e^{j(\omega t - kr)}. \tag{7.2}$$

If we define the source strength q to be the volume flow created by the source, so that $q = 4\pi a^2 u$, we see that

$$p(r) = \frac{j\rho c k}{4\pi r} \frac{q}{(1+jka)} e^{j(\omega t - kr)} \tag{7.3}$$

If the sphere is small, so that $ka \ll 1$, we can neglect the term ka in the denominator, and the result then becomes independent of source size. This leads to the concept of an acoustic point source of strength q. Such a point source is a good approximation to the behavior of many real sources at low frequencies, and gives at least a good semiquantitative idea of the behavior of more complex radiators, as we see presently.

We can calculate the power P radiated by a spherical source by taking the time average of the real (physical) part of the product of the pressure and the acoustic volume flow. The time average introduces a factor $1/2$ since p and u are amplitudes rather than r.m.s. values—the time average of $\cos^2 \omega t$ is $1/2$. The result is

$$P = \frac{1}{2}\operatorname{Re}[p(a)q] = \frac{2\pi \rho c k^2 a^4 u^2}{1 + k^2 a^2} \tag{7.4}$$

$$\to 2\pi \rho c a^2 u^2 = \frac{\rho c q^2}{8\pi a^2} \quad \text{if} \quad ka > 1 \tag{7.5}$$

$$\to 2\pi \rho c k^2 a^4 u^2 = \left(\frac{\rho c q^2}{8\pi a^2}\right)(ka)^2 \quad \text{if} \quad ka < 1. \tag{7.6}$$

This shows that, for a given surface velocity u, the radiated acoustic power per unit area, $P/4\pi a^2$, increases as $(ka)^2$ while $ka < 1$ but then saturates for $ka > 1$. Alternatively, if we take the radius a and source strength q to be constant, then the total radiated power increases as the square of the frequency while $\lambda > 2\pi a$ and becomes independent of frequency for $\lambda < 2\pi a$, λ being the sound wavelength in air. Note that the result in (7.6) is independent of a if q is fixed, so that very small sources can be defined by their volume flow q independently of their size.

7.3 Mechanical and Acoustic Impedance

In many applications it is important to know not only the nature of the wave generated by a vibrating source, but also the interaction of the medium back on the source. The simplest quantity to calculate for a pulsating spherical source is the mechanical radiation impedance $Z_R^{(M)}$. As usual, this is simply defined as $F/v(a)$, where $F = 4\pi a^2 p(a)$. From (7.2)

$$Z_R^{(M)} \equiv R_R^{(M)} + jX_R^{(M)} = \rho c S(r_R + jx_R) \tag{7.7}$$

where $S = 4\pi a^2$ is the area of the sphere and

$$r_R = \frac{(ka)^2}{1+(ka)^2} \qquad x_R = \frac{ka}{1+(ka)^2}. \tag{7.8}$$

This way of separating real and imaginary parts displays the behavior of the radiation mechanical resistance and the radiation mechanical reactance, which are shown graphically in Fig. 7.1 as functions of the parameter ka. The resistive term accounts for energy lost from the source as acoustic radiation, while the reactive term represents the mass of a blanket of air moving along with the surface of the source. The thickness of this blanket of air is either $\lambda/2\pi$ or a, whichever is smaller. Note that, for $ka < 1$, $r_R \approx (ka)^2$ and $x_R(ka) \approx ka$, while for $ka > 1$, $r_R \approx 1$ and $x_R \approx (ka)^{-1}$.

Another concept, which turns out to be of even more importance, is the acoustic impedance $Z_R^{(A)}$ seen by the vibrating sphere. This is the first time we have met this quantity, which is defined as the ratio of the acoustic pressure p to the acoustic volume flow $U = Sv$. From (7.7) this is simply

$$Z_R^{(A)} = \frac{p}{U} \equiv R_R^{(A)} + jX_R^{(A)} \equiv \frac{\rho c}{S}(r_R + jx_R) \tag{7.9}$$

Figure 7.1 Real and imaginary parts R_R and X_R of the acoustic radiation impedance on the surface of a vibrating sphere of radius a. The behavior of the mechanical radiation impedance is identical except that the units are $\rho c S$.

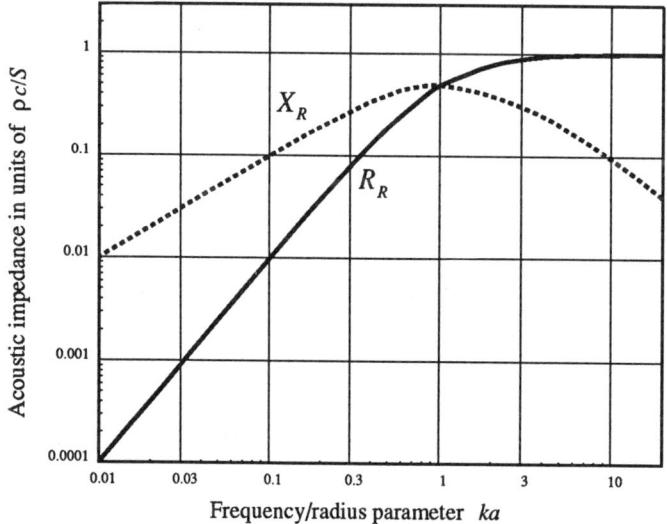

where r_R and x_R are still given by (7.8). The relation

$$Z^{(M)} = S^2 Z^{(A)} \tag{7.10}$$

holds quite generally. The behavior of $Z_R^{(A)}$ as a function of ka is similarly shown in Fig. 7.1. We should keep the general form of these curves in mind, since the behavior of more complicated radiating systems is very similar. Note that, since we shall generally use acoustic impedances rather than mechanical impedances in our discussion, we usually drop the superscript (A) for convenience.

The SI units for mechanical impedance are clearly newtons per (meter per second) and those of acoustic impedance pascals per (cubic meter per second), but it is usual to refer to these simply as mechanical ohms and acoustic ohms, by analogy with their electrical counterparts.

7.4 Source near a Reflector

Since many animals generate sound in a position quite close to the ground, it is useful to see what effect this has on radiated power and intensity. To simplify matters as much as possible we assume that $ka \ll 1$ so that the source is effectively a point source of strength q, and that it is located a distance d above an infinite plane reflector or baffle. Since there can be no flow into the surface, the whole arrangement can be simulated by imagining that there is an image source of the same strength and phase located a distance d below the baffle. The flow is then automatically zero in the plane of the surface and it makes no difference if the baffle is removed. We can then simply sum the radiation pressures and velocities from the two sources. Calculation of the radiation field at the point with coordinates (r, θ) gives a complicated expression unless $r \gg d$, in which case, from (7.3),

$$p(r,\theta) = \left(\frac{j\rho ckq}{4\pi r}\right) e^{-jkr}(e^{jkd\cos\theta} + e^{-jkd\cos\theta}) e^{j\omega t}$$

$$= \left(\frac{j\rho ckq}{2\pi r}\right) e^{-jkr} \cos(kd\cos\theta). \tag{7.11}$$

The two terms summed in brackets include the phase difference in propagation from the two sources to the point (r, θ) and sum to the angular function shown.

The directional dependence of the radiated intensity, which is proportional to the square of $p(r, \theta)$, is shown in Fig. 7.2. The directional pattern is complicated, and there is a null in the sound intensity, from (7.11), when $kd\cos\theta = (2n-1)\pi/2$, for $n = 1, 2, 3, \ldots$. These nulls appear only if the height d is greater than one quarter of a wavelength, and are not perfect unless the surface is ideally flat and completely reflecting. This effect accounts, however, for the difficulty often experienced in locating singing insects.

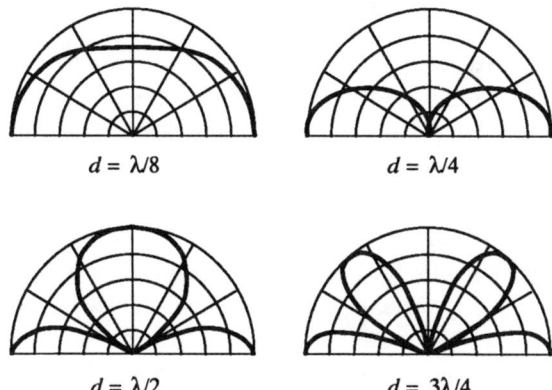

Figure 7.2 Acoustic intensity pattern radiated by a point source at a distance d above a reflecting plane, when the acoustic wavelength is λ. The radial scale is linear, and is normalized to the power radiated in a direction tangential to the plane.

The total power radiated by the pair of sources can be found by squaring the pressure, dividing by the wave impedance ρc, and integrating over a spherical surface. The final result for the power radiated by the single real source is just half this, namely

$$P = \frac{\rho c k^2 q^2}{8\pi c}\left(1 + \frac{\sin 2kd}{2kd}\right) \qquad (7.12)$$

with q as the r.m.s. source strength. From the behavior of the term in brackets, the animal gains a power advantage of a factor 2 by singing at a height for which $kd < 1$, or less than about one-sixth of a wavelength above the ground. Since the solid angle into which the song radiates is also halved to 2π instead of 4π, the animal gains an intensity advantage of a factor 4, or 6 dB.

This analysis can be extended to the situation in which we have a line of identical sources, separated by equal distances. The sound intensity at a large distance is always a maximum in the plane normal to the line of the sources and passing through its midpoint, for the contributions of all the sources then add. For a line with many sources, this maximum is quite sharp, particularly if the line is more than one wavelength in extent. There are subsidiary maxima at angles away from this plane and, once the line length exceeds the sound wavelength another principal maximum, equal in strength to that on the median plane, enters. Such line sources do not seem to occur in biological systems, but there is a possibility that the line-organ of fishes is a receptor system of this distributed type, and the directional behavior is the same.

7.5 Radiation from a Vibrating Disc

When we consider extended sources, we need to be careful about their shape and vibration behavior. The simplest example is a small rigid circular disc of radius a, vibrating normal to its plane with frequency ω. This is a moderately good model for a vibrating wing, though we recognize that the wing actually pivots rather than vibrates, and its up-and-down strokes are not symmetrical, so that aerodynamic lift is generated. These features can be accommodated semi-quantitatively by taking an average normal velocity to account for pivoting, and by simply adding a steady flow normal to the wing in a downwards direction. This flow has no acoustic effect.

A proper treatment of radiation from a vibrating disc involves complex mathematics and we shall not even outline it here. The resulting acoustic impedance, considering both sides of the disc, is again given by an expression of the form $Z_R \equiv R_R + jX_R = (\rho c/S)(r_R + jx_R)$ and the behavior of R_R and X_R is shown in Fig. 7.3. To a good approximation, we can write

$$r_R \approx 0.125(ka)^4 \quad \text{for} \quad ka < 2$$

$$\approx 1 \quad \text{for} \quad ka > 2$$

$$x_R \approx 0.8ka \quad \text{for} \quad ka < 2$$

$$\approx 4(ka)^{-1} \quad \text{for} \quad ka > 2. \tag{7.13}$$

Figure 7.3 Real and imaginary parts R_R and X_R of the acoustic radiation impedance on both sides of a disc of radius a vibrating in free air. The behavior of the mechanical radiation impedance is identical, except that the units are $\rho c S$.

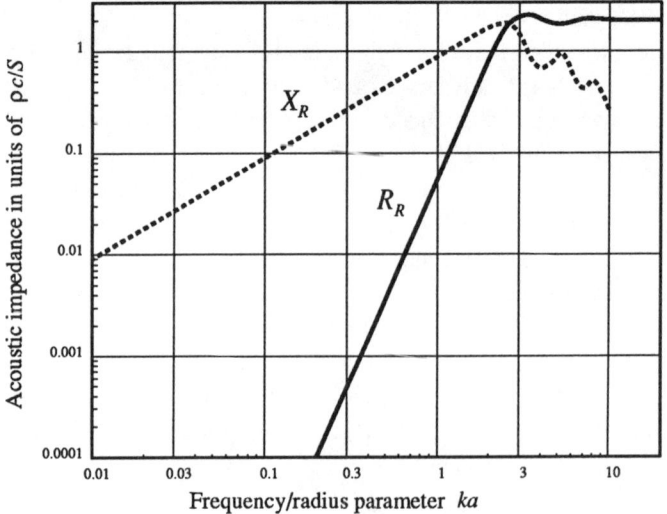

ACOUSTIC SOURCES AND RADIATION

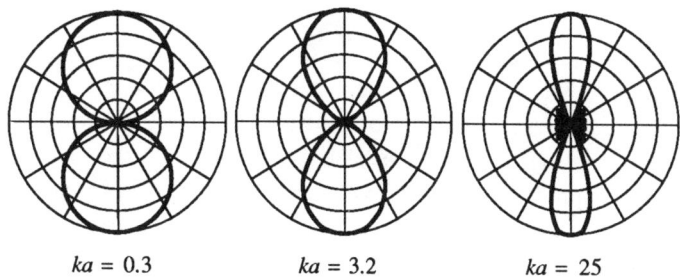

$ka = 0.3$ $ka = 3.2$ $ka = 25$

Figure 7.4 Acoustic intensity pattern radiated by a disc of radius a vibrating in free air. The radial scale is linear, and is normalized to the intensity radiated normal to the plane of the disc.

The contrast with Fig. 7.1 for a pulsating sphere should be noted. For low frequencies ($ka < 1$) the radiated power rises with increasing frequency at 12 dB per octave, until for $ka > 2$ the radiated power becomes independent of frequency. In this limit R takes the value 2 because of radiation from the two sides of the disc. The reactive mechanical load $jX\rho cS$ is an added mass, corresponding to the air moving passively along with the disc. This load is very similar to that on a pulsating sphere.

The directional behavior of the radiation from a vibrating disc is illustrated in Fig. 7.4. It is clear that there is a null in the plane of the disc, for pressure waves from the two sides arrive exactly out of phase. At low frequencies the intensity distribution is like $\cos^2\theta$, where θ is measured from the axis normal to the disc. At higher frequencies the radiation becomes increasingly concentrated along the axis, and for very high frequencies, for which $ka \gg 1$, there may be nulls and minor side-lobes at other angles as well.

The velocity behavior is also interesting, and relevant to our present subject because the sensory hairs of some insects have the function of detecting the velocity field generated by the wings of predators. We consider this in a little more detail in Section 7.8. For the present we simply note that there is a near-field component in the velocity field of a dipole, the amplitude of which varies as $k^{-2}r^{-3}$ and can therefore dominate the behavior by providing a large disturbance when r is small and $kr \ll 1$. The velocity signal near a dipole source may thus be easier to detect than the pressure signal. For a wing-beat frequency of 100 Hz, $kr = 1$ at $r \approx 50$ cm, so that this consideration can be significant for insects such as caterpillars, which may use sensory hairs to detect the approach of a predator such as a wasp.

7.6 Radiation from a Vibrating Panel

A more complex source often met in biological applications is a membrane or panel vibrating at its fundamental, or in one of its higher modes so that different parts of its surface move in opposite phase. It is difficult to treat this complicated system in any detail, but it is relatively easy to see the general trend of the behavior and to derive semiquantitative expressions for the acoustic output. If we suppose the panel to be divided into rectangular sections by its node lines, then our first task is to estimate a source strength for one such element. Suppose that the mode velocity, measured at the center of the element where it is a maximum, is u. Then to a reasonable approximation, for an element with hinged or simply supported edges or for an element surrounded by other similar elements, the mode shape over the panel is

$$u(x, y) \approx u \sin\left(\frac{\pi x}{a}\right) \sin\left(\frac{\pi y}{b}\right) \qquad (7.14)$$

where the dimensions of the element are $a \times b$. We can integrate this expression over the surface of the element to find the average velocity $\bar{u}$ associated with the mode, with the result

$$\bar{u} = (4/\pi^2)u \approx 0.4u. \qquad (7.15)$$

If the panel as a whole is larger in size than the wavelength, or if it is set in a baffle, then we can treat the radiation from the element to a reasonable approximation as equivalent to that from a free point source of strength $q = 0.8ab u$ located at its center, an extra factor 2 arising because the source is radiating into only the half-space on one side of the panel.

With this as background, let us look at the radiation behavior of an array of point sources in which nearest neighbors have opposite phase. We have already discussed the case of a single source, so the next most complex is a pair of sources corresponding to a mode of the form (1,2) or (2,1) in the notation of Section 5.2 and Fig. 5.1. Choose the first case, so that the two sections of the panel are separated by a distance b. Analysis of the radiation behavior is then nearly identical with that for two similar sources carried out in Section 7.4, except that we change the plus sign in the bracket of (7.11) to a minus and replace d by $b/2$. The source is now, in fact, a finite dipole with total radiated power

$$P = \frac{\rho c k^2 q^2}{4\pi}\left(1 - \frac{\sin kb}{kb}\right) \qquad (7.16)$$

and an intensity distribution like $\sin^2(\frac{1}{2} kb \sin \theta)$ where θ is measured from the normal to the plane.

In a biological system, the transverse wave velocity in the membrane or plate is always much less than the velocity of sound in air, so that the distance b between antiphase sections is always much less than half of the wavelength in air, and $kb \ll 1$. The factor $(1 - \sin kb/kb)$ in (7.16) is therefore always very much less than the value 1 characteristic of a single source and thus of the panel vibrating in its fundamental mode.

This analysis is, however, too simplistic, for it is virtually impossible, in biological systems, to find the degree of symmetry necessary to ensure that the two antiphase sources have exactly the same strength. In a real system, therefore, we should consider the mode (2,1) as consisting of a two sources of slightly unequal strengths q_1 and $-q_2$. These then constitute an exactly balancing antiphase pair of strength $\pm(q_1+q_2)/2$ and a co-phase pair of strength $(q_1-q_2)/2$. At low frequencies the radiation is primarily generated by the co-phase pair with modifying bracket $(1+\sin kb/kb)$ while the antiphase pair almost cancel as discussed above. The radiation pattern is moderately simple for the co-phase component, as shown in Fig. 7.2 for $kd \equiv kb/2 \ll 1$, but more complex for the antiphase sources, as discussed above.

A similar analysis can be carried out for more complex situations and modes of the type (m,n). The quantity kb does not vary very much from one mode to the next, since as k increases b decreases, and the radiated power for the exactly balanced case varies as kb raised to a power of about $\log_2 N$ where N is the number of antiphase segment pairs. If both the indices m and n are odd, however, there will be one more area of one sign than the other, and so a net monopole contribution of at least a fraction $1/mn$. More generally, the segments near the edge of the panel radiate rather differently from those near the middle, and there will always be a significant out-of-balance monopole that will contribute most of the radiation. It is clear, however, that the fundamental is generally a much more efficient radiator of sound than are any of the higher modes. This has significance in the operation of membrane or panel-based radiating systems.

The possibilities are too varied for it to be worthwhile to discuss directional patterns in any detail, except to remark that there are basically two classes of panel radiators. In the first, which might for example be a tymbal or other flexible structure on the surface of the body of an animal, only one side of the vibrating section is free to radiate sound. The low frequencies then radiate in a nearly omnidirectional manner, with increasing concentration in a direction normal to the vibrating panel at higher frequencies. In the second class, for example vibrating panels in the wings or wing covers of insects, both sides of the panel are free to radiate. The lower surface must always move in antiphase to the upper at each point so that, no matter what the distribution of panel modes, the whole vibration will have a dipole character normal to the plane of the panel. As with a simple dipole radiator, this will lead to an angular variation as $\cos\theta$, where θ is measured from the normal to the panel, and nearly complete cancellation in the plane of the panel itself.

7.7 Radiation from an Open Pipe

One of the most important radiating systems in biology is the end of an open pipe, as representative of open mouths in animals producing sound by aerodynamic means. If we omit the complications, the obvious simple model to investigate is the radiation of sound from the open end of a circular cylindrical pipe of radius a.

We shall see later that, provided the frequency is not too high, the air in the pipe simply moves parallel to the axis and there is essentially no variation in amplitude across the pipe. Certainly viscous forces ensure that the air in contact with the walls does not move but, as we saw in Section 4.2, the viscous boundary layer across which the motion builds up to its axial value is typically only about 0.1 mm in thickness. We can therefore think of the radiation problem from an open pipe as equivalent to that of radiation from a vibrating plane piston set in the end of the pipe, though there are actually some complications from higher modes in the pipe, which we mention briefly in Section 10.8.

The problem as stated is very difficult to solve exactly, and most texts solve the much simpler case of a vibrating piston set in an infinite plane baffle. The answers are not very different, but it is more useful for our discussion to deal only with the more realistic case, even if we cannot display the mathematics of the solution. Basically what we have to do is to replace the vibrating piston by a distribution of simple sources across its whole surface and then solve the problem for these sources subject to the condition that the normal flow vanishes on the outer surfaces of the pipe. We can leave the inner region of the pipe out of consideration. A more exact treatment places the piston some way down in the pipe and allows for nonuniformity of flow at the mouth, not caused by viscous effects but rather by the fact that outflow conditions are different for elements at the edge of the mouth and for elements near its center.

The outcome is expressed, as before, in terms of the real and imaginary parts of the radiation impedance, either in mechanical or acoustic terms. The exact expressions are complicated, but the calculated form is shown in Fig. 7.5. The very close similarity to the behavior of a spherical source, as shown in Fig. 7.1, should be noted. It is again useful to have approximate expressions for the real and imaginary parts of the acoustic impedance, defined as $Z_R = (\rho c/S)(r_R + jx_R)$, and these are readily seen from the figure to be

$$r_R \approx 0.25(ka)^2 \quad \text{for} \quad ka < 2$$
$$\approx 1 \quad \text{for} \quad ka > 2$$
$$x_R \approx 0.6ka \quad \text{for} \quad ka < 1.5$$
$$\approx 1.35(ka)^{-1} \quad \text{for} \quad ka > 1.5. \tag{7.17}$$

The only differences if the piston happens to be set in a plane baffle are that the coefficients for small ka become 0.5 and 0.85 respectively. The corresponding coefficients for a spherical source are both 1. The power radiated can be evaluated immediately from these expressions, given the acoustic flow velocity $U = Sv$ at the mouth of the pipe, from the relation

$$P = \langle R_R (\text{Re } U)^2 \rangle_{\text{Av}} = \tfrac{1}{2} R_R U^2. \tag{7.18}$$

ACOUSTIC SOURCES AND RADIATION

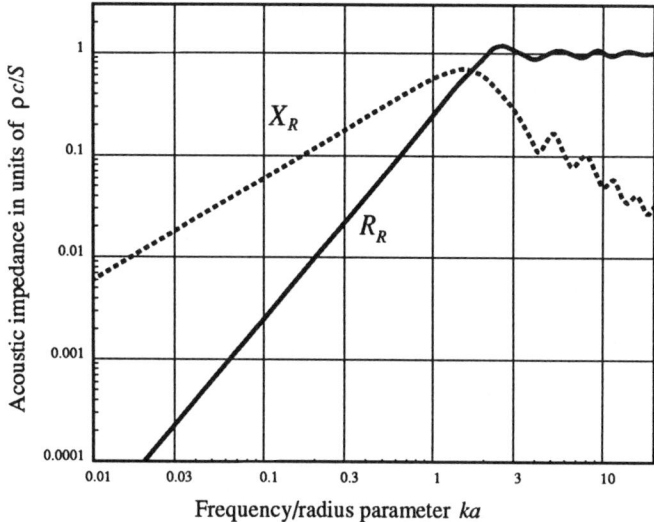

Figure 7.5 Real and imaginary parts R_R and X_R of the acoustic radiation impedance on the surface of a plane piston of radius a closing the end of a cylindrical pipe. The behavior of the mechanical radiation impedance is identical, except that the units are $\rho c S$.

It is important to examine the directivity of our simple open-pipe source since, as we remarked in Section 7.1, the reciprocity theorem implies that this is also the directionality of an aperture at the end of a pipe when used as a receiver of sound. The result for a pipe with a large plane baffle surrounding its mouth has a simple analytical form, and we can write the radiated pressure and intensity as

$$p(r,t) = \tfrac{1}{2} j\rho c k a^2 v \left[\frac{2 J_1(ka \sin\theta)}{ka \sin\theta}\right] \frac{e^{j(\omega t - kr)}}{r} \tag{7.19}$$

$$I(r) = \frac{\rho c a^2 (ka)^2 v^2}{8r^2} \left[\frac{2 J_1(ka \sin\theta)}{ka \sin\theta}\right]^2 \tag{7.20}$$

where $J_1(x)$ is a Bessel function of order 1, as discussed in Section 5.3 and Appendix A. The form of the radiated intensity is shown in the lower part of Fig. 7.6. The function in square brackets in (7.19) and (7.20) is unity at $\theta = 0$ and tends to unity for all θ if $ka \ll 1$. It becomes increasingly concentrated near $\theta = 0$ for $ka > 1$ and falls off steadily, with intervening nulls, for larger angles. The first null occurs for

$$\sin\theta = 3.8/ka. \tag{7.21}$$

We quote these results, not for their own importance, but rather because the radiated intensity from a simply open pipe, shown in the upper part of Fig. 7.6, for

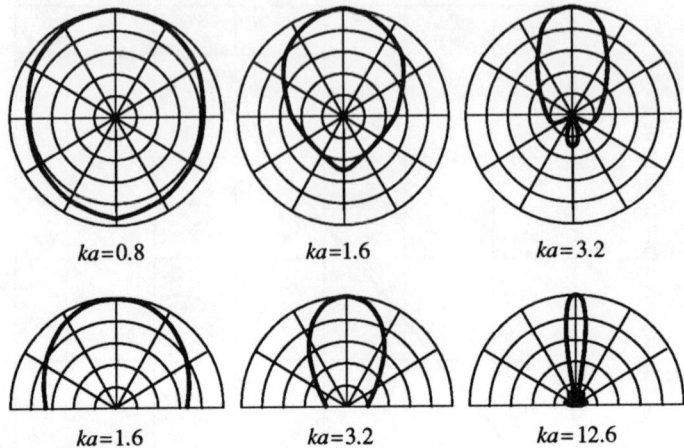

Figure 7.6 Acoustic intensity pattern radiated by a piston of radius a set in the end of a cylindrical pipe (upper row) or set into an infinite plane baffle (lower row). The radial scale is linear, and is normalized to the intensity radiated normal to the plane of the disc.

which there is no simple functional expression, has very much the same form at high frequencies, at least in the forward half-plane $\theta < 90°$. At low frequencies, radiation from the unbaffled pipe spreads nearly uniformly in all directions, rather than simply in the forward half-plane, and is consequently 3 dB lower in intensity than for the baffled pipe. A pipe with a small baffle, which is how we might approximate an animal's mouth, behaves in an intermediate fashion.

7.8 The Near Field

Our discussion above has been mostly concerned with the acoustic waves radiated by a small source, which in a bulk medium spread out spherically from the source, though their intensity is not necessarily the same in all directions. Once we are far away from the source, in the sense that $kr \gg 1$ or $r \gg \lambda$, where r is the distance from the source, and provided also that r is very much greater than the largest dimension a of the source, then these radiated waves decrease in amplitude as r^{-1} or in intensity as r^{-2}. This law applies to both the acoustic pressure and the acoustic particle velocity and is called the far-field or radiation behavior of the source.

In biological applications we are usually dealing with the far field, so that this approach is entirely adequate. It is important to recognize, however, that the behavior of the source can be dramatically different in the near field where either $kr < 1$ or $r < a$. Such situations can occur, for example, when the sound source has rather a low frequency, like the beating wings of a predator, and it is approaching its prey. The discussion is also relevant to any communication between two animals in close contact.

A detailed discussion of near-field behavior is complicated by the fact that every different source behaves differently, but we can draw some useful general conclusions by considering some idealized cases. The first of these is the simple point source, or monopole, discussed in Section 6.6. From equations (6.33) and (6.34), the pressure and velocity behave like

$$p_{monopole} = \frac{A}{r} \qquad v_{monopole} = \frac{B}{r} - j\frac{B}{kr^2} \tag{7.22}$$

where actually $B = 1/\rho c$, but we are not interested in this level of detail. There is thus no near-field variation in the acoustic pressure, but there is a near-field addition to the acoustic velocity of magnitude

$$v_{monopole}^{(NF)} = \frac{B}{kr^2} \tag{7.23}$$

which dominates the behavior when $kr < 1$. From (1.22) we see that this near-field velocity differs in phase by 90°, a factor j, from the acoustic pressure. The load it imposes on the source is therefore purely reactive and it carries away no energy. Nevertheless it can provide a strong acoustic stimulus in the near field for a detector, such as a sensory hair, that responds to acoustic particle velocity rather than to acoustic pressure.

The situation with an acoustic dipole is even more extreme. Such a dipole consists of a pair of simple sources of strength q and opposite phase, set very close together. The strength of the sources is made inversely proportional to the distance dz between them so that the dipole strength $q\,dz$ remains constant. This is an idealization of a vibrating disc or sphere. If we think of the pressure field generated by such a dipole, we can see that it is just the difference between the fields of two simple sources separated by a small distance dz along the z axis. Since the source strength is proportional to $1/dz$, we can see that the prescription for calculating the pressure field amounts simply to differentiating the expression for a monopole field with respect to the coordinate z of the dipole axis. Now $z = r\cos\theta$, so this introduces an angular variation as well as introducing factors of either k, from the exponential in (6.33), or $1/r$ into the expression for the pressure. The pressure field generated by a dipole therefore has the form

$$p_{dipole} = \left(\frac{A'k}{r} - \frac{jA'}{r^2}\right)\cos\theta \tag{7.24}$$

where θ is the angle between the measuring direction and the dipole axis z. To calculate the acceleration, and from it the velocity field, we need to find the gradient of this pressure field. This again involves a differentiation, and so introduces either a factor k, which cancels in the transformation between acceleration and velocity, or else a factor $1/r$, together with angular terms. The result is complicated to write down, since the velocity is no longer radial, but the important thing is that there is a dominant near-field term of magnitude

$$v_{dipole}^{(NF)} = \frac{B}{k^2 r^3} F(\theta) \tag{7.25}$$

where $F(\theta)$ is an angular function. This near-field term dominates all others when $kr \ll 1$. The velocity field that it describes varies in direction with the angular coordinate θ.

We need to modify our conclusions somewhat when dealing with the realistic case of a source that is not infinitesimal is size. The main effect is that there is a lower limit a, corresponding to the dimensions of the source, below which r cannot be decreased. The near-field pressures and velocities do not therefore rise without limit, but saturate at values appropriate for $r \sim a$. In addition, when the observation point is very close to the source, then the source geometry begins to influence the result. General calculations are difficult, but simple cases, such as a circular piston in a plane baffle, can be worked out in detail. For this case there is a series of pressure maxima and minima on the axis of the piston, and a near-field pattern extending for several times the piston diameter. Even more complex effects can be expected with less simple sources.

7.9 The Reciprocity Theorem

As a final comment on this initial discussion of acoustic radiation, it is useful to have a formal statement of the reciprocity theorem, which has already been referred to several times. Specifically, we remarked in Section 3.8 that the transfer admittance $Y_{xx'}$ between any two points on an extended body is unchanged if we interchange the excitation and measurement points x and x', and we observed in the present section that the transmission behavior of acoustic generators is very similar to their behavior as receptors.

The reciprocity theorem actually applies to any system in which magnetic fields do not play a part, but it will suffice to state it for the case of acoustic radiation in a form suitable for our present interests. The theorem states that, for any bounded or unbounded medium and for any configuration of passive objects within that medium, if a point source of strength q located at position $\mathbf{r}$ produces an acoustic pressure p at another point $\mathbf{r}'$, then the same source q located at $\mathbf{r}'$ will produce the same pressure p at $\mathbf{r}$.

In most of our applications we are concerned with vibrating surfaces, either as transmitters or receivers, and it is important to see how the reciprocity theorem can be applied to them. Suppose we are concerned with a vibration mode defined by the normalized eigenfunction $\psi(\mathbf{r})$. Then essentially what the reciprocity theorem shows is that if this mode with velocity amplitude numerically equal to ψ gives rise to a pressure $p(\mathbf{r}')$ at an external point $\mathbf{r}'$, then a unit source placed at $\mathbf{r}'$ will produce a force driving the amplitude of mode ψ that is numerically equal to p.

We may not need to apply this theorem in detail, but it is clear that it has very important consequences. For example, if a complex vibrating system radiates power preferentially in a given direction, then the same mode of the system used as a detector will be maximally sensitive to radiation coming from that direction. Equally, if the radiating system has a null in a particular direction, then the same system used as a receiver will be deaf to sound coming from that direction. In practice, of course, we need to examine the response of each mode of the system in this way.

7.10 Acoustic Sources and Power

We shall consider acoustic sources in biology in a later chapter, but it is useful here to have some idea of the range and magnitude of possible sources. We have already noted a few relevant points in Chapter 6, namely that the dynamic range of human hearing, which is not very different from that of other animals, extends from a threshold sound pressure of about 20 µPa in its most sensitive frequency range of 1 to 3 kHz to an upper limit of about 20 Pa. This reflects an evolutionary adaptation to the range of sound pressures occurring in the natural environment. If we consider sources in the open air, this corresponds to an intensity range from 10^{-12} W m^{-2} to 1 W m^{-2}. At one end of the range, the wing-beat of a mosquito produces a sound at a few hundred hertz that is audible above night-time background (say 20 dB) at a distance of about 30 cm. This corresponds to an acoustic source power of order 10^{-10} W. At a more usual level, a typical song bird might produce a level of 70 dB at a distance of 10 m, which indicates a source power of about 1 milliwatt. This acoustic power is also typical of the human voice in ordinary speech and of a small portable radio. Some extremely vocal insects, such as certain species of cicada, produce a comparable acoustic power. The air-powered vocal apparatus of vertebrates has a very flexible capacity, however, and a crowing domestic cock may produce a peak acoustic output approaching 1 watt.

As a further extension of this scale, if we set aside large-scale phenomena such as thunder or waterfalls, we must go to human artifacts. A large orchestra produces a peak output of a few tens of watts, and the same is true of a really large electronic music system. The apparent disparity between this figure and the rated power of the system arises from the fact that electromechanical transducers such as loudspeakers typically have conversion efficiencies of only a few percent. Unfortunately the conversion efficiency of devices like jet engines is lower than this by a factor of only 10 to 100, so that a jet transport engine may produce nearly a kilowatt of unwanted acoustic power.

7.11 Underwater Sources

All the analysis in this chapter is generally applicable in any medium, but our examples have all been for the case of sources in air. It is worthwhile to consider briefly the situation for sources in water. We note that all the radiation formulae are of the form (7.9), with an initial factor $\rho c/S$, which is just the wave impedance of the medium divided by the source area. If we increase the wave impedance $z \equiv \rho c$, keeping the vibration amplitude constant, then the radiated power increases with z. We recall from (6.18) that $z_A \approx 415$ rayl for air while $z_W \approx 1.5 \times 10^6$ rayl for water, so that a source of given vibrational amplitude radiates some 4000 times as much power under water as it does in air. There is, of course, an energy penalty involved in producing the same amplitude with the mass loading of the liquid, but sound radiation is a much more efficient process under water than in air. The same is true, as we saw in Chapter 6, for sound propagation in water, which has quite low attenuation with distance. These facts lead to all sorts of technological consequences for humans, but for naturally aquatic animals they mean that acoustic communication can be particularly efficient.

References

Radiation and sources: [1] Ch 7; [6] Ch 4,5; [8] Ch 7; [10] Ch 7
Directionality: [1] Ch 7; [6] Ch 4
Radiation impedance: [6] Ch 5; [9] Ch 2
Propagation in natural environments: [30]

Discussion Examples

1. An insect produces sound by muscular vibration of a tymbal membrane set into its body. If the tymbal is approximately circular with radius 3 mm, the vibration amplitude is 0.1 mm, and the song frequency 3 kHz, calculate the radiated acoustic power.
2. An insect produces sound by mechanical excitation of its wing covers with a file on its hind legs. If the covers have total area 30 mm² and radiate as a simple dipole source at a frequency of 4.5 kHz, producing a total radiated power or 0.1 mW, estimate their vibration amplitude.
3. What would be the necessary amplitude if the frequency were only 1.5 kHz?
4. The wings of a wasp have total area 50 mm² and vibrate at 200 beats per second with an amplitude of 2 mm at their tips. Estimate the magnitude of the radiated acoustic power.

Solutions

1. Area $S \approx 3 \times 10^{-5}$ m², $\omega \approx 2 \times 10^4$ s^{-1}, $ka \approx 0.2$. From Figs 7.1 or 7.5, $R_R \approx 0.02 \rho c/S \approx 2 \times 10^5$ acoustic ohms. Volume flow $U \approx 6 \times 10^{-5}$ m³s^{-1} so radiated power $\frac{1}{2} R_R U^2 \approx 0.4$ mW. Compare with the two following examples.
2. Treat the vibrator as a disc with radius $a \approx 3$ mm. Then $ka \approx 0.2$. (Use mechanical calculation for variety.) From Fig. 7.3, the mechanical radiation resistance is $10^{-4} \rho c S \approx 10^{-6}$ mechanical ohms, so the radiated power for average vibration amplitude x is $10^{-6} x^2 \omega^2 \approx 10^3 x^2$ W. This gives $x \approx 3 \times 10^{-4}$ m or 0.3 mm r.m.s.
3. From Fig. 7.3 for $ka \ll 1$, the radiation resistance varies as $(ka)^4$. A decrease in frequency by a factor 3 therefore decreases the radiation resistance by a factor 1/80, requiring an increase in amplitude by a factor 9, giving about 3 mm r.m.s. This is unrealistic, and the insect can radiate only a smaller power.

4. Area $S = 5 \times 10^{-5}$ m², equivalent radius $a \approx 4$ mm, angular frequency $\omega \approx 1200$ s⁻¹, so $ka \approx 0.016$. From Fig. 7.3 extrapolated, $R_R \approx 8 \times 10^{-9} \rho c/S \approx 0.05$ acoustic ohms. Average velocity $v \approx 0.001$ m s⁻¹ so $U = uS \approx 6 \times 10^{-5}$ m³s⁻¹. Radiated power = $\frac{1}{2} R U^2 \approx 10^{-10}$ W.

8 LOW-FREQUENCY NETWORK ANALOGS

SYNOPSIS. When a mechanical or acoustic system is small in dimensions compared with the wavelength of the vibrations or sounds concerned, it is possible to use particularly simple treatments to analyze its behavior. Many biological systems meet this condition in the low-frequency part of their operating range, though many require the more sophisticated techniques that we develop later.

A particularly useful technique is to exploit the analogy between mechanical and electrical systems. If we recognize voltage as analogous to force and current as analogous to velocity, then we find that an inductor is analogous to a mass, a capacitor to a spring, and a resistor to a viscous resistance. This analogy extends to numerical values as well, provided a consistent set of units is used. Circuits for these elements are shown in Fig. 8.1. Except for the voltage generator, the current generator, which corresponds to an impressed vibration velocity, and the free or fixed terminations, all are four-terminal elements which can be fitted together like dominos to form an electric circuit analog for the whole mechanical system.

Figure 8.2 shows two examples—a mass supported on a spring and driven by a sinusoidal force, which is our simple vibrator, and a free mass driven through a spring, which constitutes a sort of vibration isolator. This second circuit is also a simple model for the vibration detector system incorporated in the subgenual organs of some insects, in which the biosensor responds to the extension of the spring. It has greatest sensitivity near the resonance frequency of the system.

Figure 8.3 shows an electrical analog for a simple lever, such as is found in the coupling between the eardrum and the cochlea in mammals. The analog for a perfectly light and stiff lever is an ideal transformer with a turns-ratio equal to the reciprocal of the length-ratio of the lever arms. To make the transformer ideal, the number of turns on each winding is then made infinite. A real lever needs to be supplemented by an inductor and a capacitor in each branch, representing the mass and compliance of the separate lever arms. Analysis of this circuit, when the lever is driven by an oscillating force at one end and connected to a mechanical load at the other, is illustrated in Fig. 8.7.

We can proceed similarly for the analysis of acoustic systems that are small compared with the sound wavelength. The electric voltage is an appropriate analog for the acoustic pressure and the electric current for the acoustic volume flow. As shown in Fig. 8.4, an inductor is the analog for a short open pipe, a capacitor for an enclosed volume, and a resistor for a pipe or opening obstructed by wool or other fibrous or perforated material. Particular attention needs to be given to an open termination, for it is loaded by the radiation impedance. It is shown that the radiation reactance at the end of an open pipe effectively increases the length of the pipe by 0.6 times its radius. Similarly, an aperture in a plane baffle behaves like a short open tube of acoustic length 0.7 times the square root of the aperture area.

A Helmholtz resonator consists of a closed volume with a single aperture, generally in the form of a tubular neck as shown in Fig. 8.5. The electric analog circuit for this system, as shown in the figure, is exactly the same as that for a simple mechanical vibrator. The resonance frequency is determined by the enclosed volume and the length and area of the neck. A resonator of this type with a small auxiliary tube that can be placed in the ear forms a simple tuned filter, since the pressure response is greatly increased near the resonance frequency. If a Helmholtz resonator is driven internally by a small piston or tymbal set into its wall, as happens in the sound-production system of some insects, then the acoustic output from the open neck is greatly enhanced at the resonance frequency.

A tympanum, consisting of a resonant membrane stretched across a tube, is a common component of auditory systems. A four-terminal element representing a tympanum is shown in Fig. 8.6. The values for the analog elements can be calculated from the dimensions, resonance frequency, and Q value of the bare tympanum. Generally only the fundamental vibration mode is considered, but higher modes can be included as shown. The impedance associated with higher modes is much greater than that of the fundamental, so they are usually not important. If the tympanum is connected mechanically to an auditory capsule or cochlea, as is usually the case, then the mechanical impedance of this element can be included in the acoustic impedance of the tympanum as shown.

The behavior of complex systems, represented by complex electrical networks, can be analyzed as shown in Fig. 8.7. Circulating currents are drawn through all the elements of the network and, around each circuit, the driving voltage, if any, is set equal to the sum of the voltage drops caused by currents flowing through each component in the circuit. This leads to a set of equations equal in number to the number of currents. These equations can be solved algebraically for relatively simple systems, or numerically, using a general computer program, for more complex systems.

8.1 Electric Analogs

We have now examined most of the mechanical components that go to make up a typical acoustic system in biology, and we have looked briefly at the ways in which vibrating mechanical components can generate or be influenced by sound waves. We have not yet considered acoustic components such as long tubes or horns but, even before doing this, we can begin to understand the behavior of a variety of biological systems at low frequencies. In this context, low frequencies implies frequencies low enough that the dimensions of the system are small compared with the sound wavelength involved, say less than about one-tenth of a wavelength. Real biological systems often meet this requirement in the low frequency part of their range, though more complicated concepts are required to deal with behavior at higher frequencies.

Our purpose is not just to understand the behavior of systems in a qualitative manner, but rather to develop methods that will allow acoustic behavior to be calculated quantitatively with reasonable accuracy. We have treated the individual mechanical components and their interaction with sound radiation in this manner in preceding chapters, and we now need a convenient means for calculating behavior when several components are assembled to make a system. It would be possible simply to write down the differential equations for each element, allowing for its interaction with other elements, and then to solve the whole array of equations algebraically, but this approach is laborious and generally fails to show up the contribution of individual components in a physically meaningful way.

The approach we adopt is based on an analogy between the equations describing mechanical and acoustic elements and those describing elements in electrical circuits such as inductors, capacitors, and resistors. We can then replace the real acoustic system, for the purposes of calculation, by a network of interconnected electric analog components. Calculation of the response of the network is quite straightforward, as we shall see, and the results can then be interpreted back into acoustic terms. The real power of this approach comes from the fact that the calculation has the same form for all networks, and therefore for all acoustic systems. We can write down solutions for the simpler systems directly, and those for which the algebra becomes too involved can easily be solved numerically using a general computer program.

In this chapter we shall be concerned with the construction of electric analogs for mechanical and acoustic elements, and with an introduction to the solution of the electric network problem, as a preparation for the discussion of real biological systems in Chapter 9.

8.2 Analogs for Mechanical Components

There are two basic variables in mechanical systems, the applied force and the resulting velocity. Similarly, there are two basic variables in electrical systems, the applied potential and the resulting current. It is natural to make an analogy between force and potential and between velocity and current, and this is the convention we adopt. Similarly there are three basic mechanical components—a mass, a spring, and a viscous loss resistance - and three basic electrical components—an inductor, a capacitor, and a resistor. Let us see how we can extend the analogy.

The behavior of a mass m under the influence of a force $F(t)$, generally varying in time and conveniently represented by the complex quantity F, is defined by the equation of motion

$$F = m \frac{dv}{dt} \tag{8.1}$$

where v is its velocity. Similarly, the relation between the potential V across an inductance L and the current i flowing through it is given by

$$V = L \frac{di}{dt}. \tag{8.2}$$

If the quantities are all taken to vary sinusoidally in time with frequency ω, then these two equations can be written as

$$F = j\omega m v; \qquad V = j\omega L i \tag{8.3}$$

with the factor $e^{j\omega t}$ understood throughout. A mass is therefore formally analogous to an inductance, the electrical impedance of which is $j\omega L$.

Similarly, the behavior of a spring of stiffness β is given by the relation

$$F = \beta x = \beta \int v \, dt \qquad \text{or} \qquad F = \frac{\beta}{j\omega} v \tag{8.4}$$

LOW-FREQUENCY NETWORK ANALOGS

while that of a capacitor C is given by

$$V = \frac{Q}{C} = \frac{1}{C}\int i\,dt \quad \text{or} \quad V = \frac{1}{j\omega C}i \tag{8.5}$$

where Q is the charge on the capacitor. Again these equations are seen to be formally identical provided that we identify the capacitance C with the reciprocal of the spring stiffness β, which is just its elastic compliance C. The analog impedance is just $1/j\omega C$.

Finally, a viscous resistance R in a mechanical system induces a force proportional to the velocity, in just the same way as an ohmic resistance in an electric circuit has a potential drop proportional to the current through it.

$$F = Rv \qquad V = Ri. \tag{8.6}$$

Note that, in each of these equations (8.1) through (8.6), the analogy extends to the numerical values of all the quantities involved, provided that we use SI units, or some other consistent system, throughout. It is this numerical analogy that makes the whole procedure so useful. It may sometimes be necessary to distinguish between electrical analogs for mechanical quantities and for acoustic quantities, which we consider next. To make this clear, we can affix a superscript (M) to the mechanical analogs.

A small amount of skill is required in setting up electrical analog circuits for mechanical systems, but we shall try to make this as automatic as possible by introducing some graphical conventions. The difficulty is that some mechanical elements, such as a spring, clearly have two ends, while an element such as a mass does not. What we shall do is to construct block elements, each of which effectively has two ends, so that they can be joined together like dominos. This is illustrated for the three basic elements in Fig. 8.1. Each analog element has four terminals, grouped into two pairs, and within the block representing the element we have given the conventional representation of an electrical inductor, capacitor, or resistor. It is unfortunate that the symbol for an inductor looks very much like a spring, but once this anomaly is recognized it is easy to avoid confusion. In the case of the mass and resistance analogs, it does not matter whether the electrical element is placed in the upper or the lower link. Two special two-terminal elements are included, one representing a rigid support (open circuit) and one a completely free termination (short circuit).

To these elements we need to add analogs for the excitation of the system. These can be of two kinds: either we apply a sinusoidal force $Fe^{j\omega t}$ at some point, or else we impose a velocity $ve^{j\omega t}$ upon some element. The electric analogs are a voltage generator $Ve^{j\omega t}$ and a current generator $ie^{j\omega t}$, as shown in Fig. 8.1. The voltage generator has zero internal impedance, so that the voltage on its upper terminal, relative to the lower, is always $Ve^{j\omega t}$, whatever the impedance of the circuit to which it is connected; the current generator has zero internal admittance,

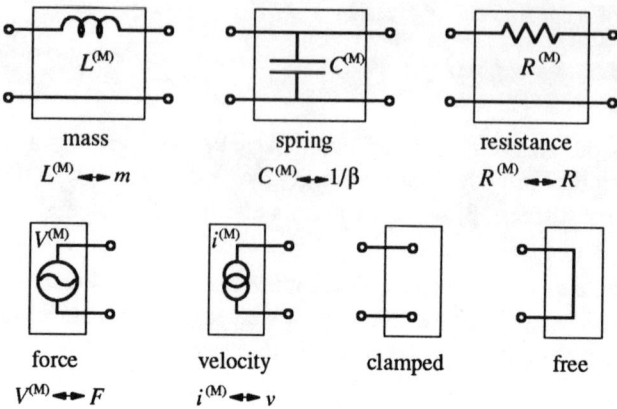

Figure 8.1 Electric analogs for mechanical elements. In both the force and velocity analogs, the upper terminal is taken to define the sign. All other analogs are non-polar.

so that the current flowing out of its upper terminal is always $ie^{j\omega t}$, whatever voltage might be applied to it.

The convention guiding construction of a network is that these blocks must be simply connected end-to-end (except in rather exceptional circumstances), and free terminals must be connected to an appropriate one of the two-terminal blocks. The free termination simply completes the circuit, while the clamped termination is an open circuit and prevents current flow. It is easy to see that a force driving a mass, with no restriction on the mass motion (free termination), simply gives us an inductance connected across the generator. A similar situation applies for a resistive load. In the case of a force driving a spring, it is important that the remote end of the spring is clamped, using the clamped analog termination. The analog circuit then appears as a capacitor connected across the generator terminals. These are all exactly what we expect.

In practice there are viscous losses associated with the motion of any mass, unless it happens to be in a vacuum, so that we might usefully take the impedance analog for such a "real" mass to consist of an inductor and a small resistor in series. Similarly, there are internal losses associated with the compression of a real spring, and particularly with springs of biological material. The analog for such a real spring therefore consists of a capacitor with a small resistor added in series with it in the vertical circuit path of the analog circuit of Fig. 8.1.

When we consider a combination of a mass and a spring, there are two possibilities, as shown in Fig. 8.2. In the first case the force is applied to the mass, which rests upon a spring, the remote end of which is clamped. This is our familiar simple vibrator. The electric analog circuit in the figure shows an inductor and a capacitor in series across a generator. If the mass moves in a medium with viscous

LOW-FREQUENCY NETWORK ANALOGS

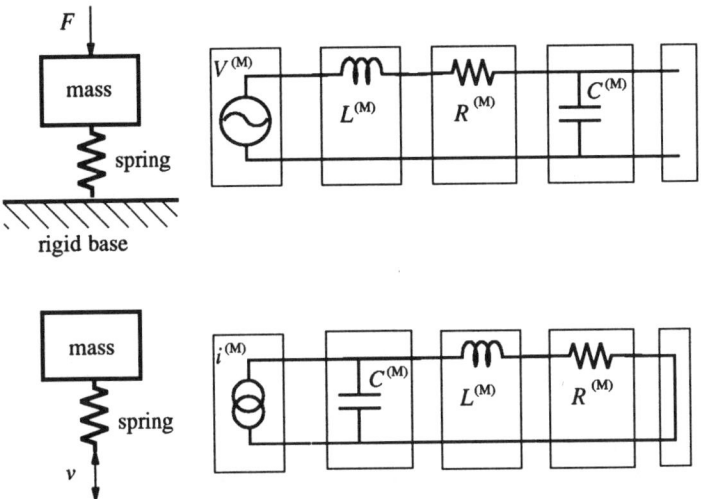

Figure 8.2 Physical realizations and analog circuits for simple oscillator, consisting of a mass supported on a spring and with force applied to the mass, and for a vibration detector or vibration isolator, consisting of a mass supported on a spring with a velocity signal applied to the spring.

friction, then we should add a resistive element to the chain, immediately following the inductor, as discussed above and shown in the figure. The total impedance in this simple circuit is the sum of the three individual impedances, so that the electric circuit equation for current flow gives, after a little rearrangement,

$$i^{(M)} = \frac{j\omega V^{(M)}}{L^{(M)}\omega^2 + 1/C^{(M)} + j\omega R^{(M)}}. \tag{8.7}$$

When we transform back from electrical to mechanical quantities by simple substitution, this is the familiar simple oscillator response expression (2.21).

The alternative arrangement is one we have not met before. It corresponds to an oscillatory motion applied to the end of a spring, the other end of which is connected to a mass that is free to move. The circuit is now a little more complicated, with the inductor and the capacitor both connected across a current source. We have added the resistive block to the inductor, corresponding to some sort of viscous damping on the motion of the mass, but it could have been added in series with the capacitor, representing losses in the spring. We shall look at a general method for solving more complicated circuits such as this in Section 8.8, but this particular circuit is quite easy to treat. If we denote the current from the generator by i, the current through the inductor by i_1, and the current through the capacitor by i_2, then clearly $i = i_1 + i_2$. Also, since the capacitive and inductive branches of the circuit are connected together at both ends, the voltages across them must be equal, so that

$$i = i_1 + i_2 \qquad i_1(j\omega L + R) = \frac{i_2}{j\omega C}. \qquad (8.8)$$

These two equations allow us to write

$$i_1 = \frac{i}{(1 - LC\omega^2) + j\omega RC} \qquad i_2 = \frac{j\omega C(R + j\omega L)i}{(1 - LC\omega^2) + j\omega RC}. \qquad (8.9)$$

Clearly these two currents are both greatest when $\omega^2 = 1/LC$, and the magnitude of this resonance current varies inversely with the damping resistance R. The response curve for i_1 is just a simple resonance curve of the type shown in Fig. 2.5, and the shape of i_2 is not very different.

This analysis is not just an empty exercise, for a vibration detector based on this principle does exist in some insects. The mass represented by the inductor is simply the mass of the body of the animal, and the spring is provided by the elastic properties of the animal's legs. At one of the joints of the leg is a detector, called the subgenual ("below the knee") organ, which responds to the elastic flexing of the joint, represented by the current i_2. The vibratory signal is provided by waves propagating in the ground, or in the substrate on which the animal rests, and this is so massive relative to the animal that its motion is virtually unaffected by the presence of the animal, thus constituting a velocity generator. The response of the sub-genual organ has a simple resonance shape, so that it can be optimized by evolutionary processes for the detection of predators or prey.

8.3 Levers

It is useful here to introduce briefly the analysis of a lever, such as might be formed by the ossicles of a mammalian ear, connecting the tympanum to the neural transducers in the cochlea. A lever behaves very much like an ideal electrical transformer, as we see in a moment, though a real lever differs in behavior from a real transformer. Consider first an ideal lever with perfectly light and stiff arms of lengths l_1 and l_2 as shown in Fig. 8.3. Let the velocity of motion of the end of lever arm 1 be v_1 and that of arm 2 be v_2, in the directions shown. Then the perfect stiffness guarantees that $v_2/v_1 = -l_2/l_1$. Equally, if F_1 and F_2 are the forces on the arms, then the balance of turning moments requires that $F_2/F_1 = l_1/l_2$. Such a mechanical device has an electrical analog in a transformer with turns ratio $n_2/n_1 = l_1/l_2$ (the same way round as the force ratio). In order that the lever transmit static forces, we must require that the transformer be ideal in the sense that the actual number of turns is infinite, so that the inductance of either winding with the other open circuited is also infinite. When analyzing the circuit, as we discuss in Section 8.8, we simply require that the currents or voltages on the two sides of the transformer have the specified ratios

LOW-FREQUENCY NETWORK ANALOGS

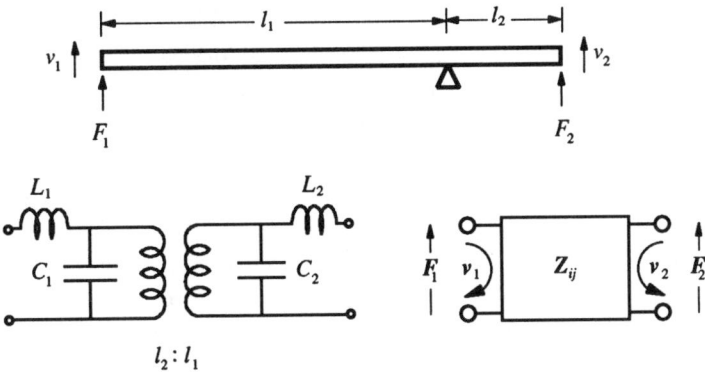

Figure 8.3 A simple lever and its electrical analog. If the lever is ideal, having completely stiff arms of zero mass, then the analog is an ideal transformer with appropriate turns ratio. For a real lever this model must be supplemented by elements representing the mass and compliance of the lever arms. The network analog is a two-port element characterized by the impedances Z_{11}, Z_{22} and $Z_{21} = Z_{12}$.

$$\frac{V_2}{V_1} = \frac{l_1}{l_2} \qquad \frac{i_2}{i_1} = -\frac{l_2}{l_1}. \tag{8.10}$$

If we consider an ideal lever with a load $Z_2^{(M)}$ attached to arm 2, then the impedance seen at the end of arm 1 is

$$Z_1^{(M)} = \frac{V_1}{i_1} = \left(\frac{l_2}{l_1}\right)^2 \frac{V_2}{i_2} = \left(\frac{l_2}{l_1}\right)^2 Z_2^{(M)}. \tag{8.11}$$

Such a lever therefore acts as an ideal impedance transformer.

Figure 8.3 illustrates such a transformer analog and shows also how it should be supplemented to allow for the non-zero mass and finite stiffness of the lever arms. In each branch the elements L and C must be chosen to represent the point impedance of the lever at the point of application of the force when the lever is held fixed at its pivot point. Usually in our applications, only the first vibration mode of the lever beam needs to be considered, and the analog inductances L_n will be approximately one-third of the mass of the relevant lever arm. The analog capacitances C_n depend upon the cross-sectional area S, radius of gyration κ (see Fig 3.6), and Young's modulus E of the lever material. To an adequate approximation, based on the discussion in Section 3.9, we can show that

$$L_n = \frac{\rho_S l_n S_n}{3} \qquad C_n = \frac{l_n^3}{4ES\kappa^2} \tag{8.12}$$

where ρ_S is the density of the lever material. These relations are derived in discussion example 5 at the end of this chapter.

Even without solving a network including a real lever, we can draw some conclusions about its effect on total system behavior. An ideal transformer operates down to arbitrarily low frequencies, so that a lever can transmit static forces. At non-zero frequencies, however, the shunting capacitors, which represent the elastic compliances of the lever arms, are able to carry some of the current, so that there is a loss of velocity in going through the lever, even when account is taken of the lever-arm ratio. There is also a loss in transmitted force because of the inductive effect of the mass of the lever arms. Both these losses become increasingly severe at higher frequencies, though complicated by lever arm resonances. In brief then, a real lever, in addition to its impedance transforming function, tends to have a low-pass characteristic and to suppress high frequency coupling between the two circuits that it links.

We can go one step further than this and replace the real lever by a "black box", as shown in Fig. 8.3, the behavior of which in an analog circuit is defined completely by the relation between the voltages and currents (or forces and velocities) at its two ports. Because all the individual components of the lever network are linear in behavior, implying that doubling the forces simply doubles the velocities, the behavior of the black box is also linear, and can be represented by the equations

$$F_1 = Z_{11}^{(M)} v_1 + Z_{12}^{(M)} v_2$$

$$F_2 = Z_{21}^{(M)} v_1 + Z_{22}^{(M)} v_2. \tag{8.13}$$

(Here we have written the equations in terms of forces and velocities for clarity, but it is only a matter of changing the symbols to write them in terms of electrical quantities.) The impedance coefficients $Z_{ij}^{(M)}$ are generally complex, because of the effect of the capacitance and inductance elements representing the mass and compliance of the lever arms.

It is straightforward, but a little tedious, to solve the network according to the methods to be discussed in Section 8.8 so as to find explicit expressions for the impedances $Z_{ij}^{(M)}$. The only subtle point is that we use the relations (8.10) to describe the behavior of the ideal transformer embedded in the network. The results are

$$Z_{11}^{(M)} = j\left[\omega L_1 - \frac{1}{\omega C_1} + \frac{l_1^2 C_2}{\omega C_1 (l_1^2 C_2 + l_2^2 C_1)}\right]$$

$$Z_{22}^{(M)} = j\left[\omega L_2 - \frac{1}{\omega C_2} + \frac{l_2^2 C_1}{\omega C_2 (l_1^2 C_2 + l_2^2 C_1)}\right]$$

$$Z_{12}^{(M)} = Z_{21}^{(M)} = -j\left[\frac{l_1 l_2}{\omega (l_1^2 C_2 + l_2^2 C_1)}\right] \tag{8.14}$$

It is obvious that the expression for $Z_{12}^{(M)}$ is symmetric under interchange of the subscripts 1 and 2, so that necessarily $Z_{12}^{(M)} = Z_{21}^{(M)}$. At first sight this may be a little surprising, since the lever itself is not physically symmetric. The result is, however, completely general for all linear systems that do not involve magnetic fields. This is a particular example of the reciprocity theorem, which we have already met in relation to transfer impedance, and which we shall meet several times again.

It is worthwhile to remark that, if the arms of the lever become ideally stiff, then $C_1 = C_2 = 0$, and all the $Z_{ij}^{(M)}$ become infinite. If we wish to treat a system under this approximation because we do not know the arm stiffnesses, then it is generally easier to revert to the simpler equations (8.13).

With the internal behavior of the lever now conveniently represented by a black-box component, we can now simply use this as a circuit element. The only point to watch is that the forces and velocities have to appear in a mirror-symmetric way at the two ends of the black box, as indeed they are shown for the physical lever. If we reverse the directions of all the forces and velocities, then the equations (8.13) still apply, but if the directions of only some of these quantities are reversed, then they should be replaced by negative currents or negative forces in the symmetric direction. Thus, for example, if the direction of the current v_2 is reversed in the circuit, then it should be treated as $-v_2$ in the equations.

We shall return to a further brief consideration of the role of the lever in auditory systems when we consider the middle ear in Chapter 12. The peripheral auditory system is usually taken to terminate at the tympanum, and the inner ear to be the neural transducer organ. The middle ear, when the distinction is made, is the link between these two and, in vertebrates, takes the form of some sort of lever system. We shall see that the impedance-matching function of a lever is important to the functioning of the total auditory system.

8.4 Analogs for Acoustic Components

Acoustic systems can also be examined by first constructing electric circuit analogs. Indeed, because acoustic systems often consist of interconnected tubes and cavities, the analogy is generally rather more directly obvious than for mechanical systems. It is convenient in our discussion to consider acoustic systems in two stages. Here we take up the simpler stage in which it is assumed that all the components in the system are small compared with the sound wavelength involved. This leads to great simplification, and gives good results up to system dimensions of about one-tenth of a wavelength. The method is adequate for analyzing the acoustic behavior of quite a number of animals, from frogs to cicadas, and tells us a great deal about their auditory and sound-producing systems. In Chapter 10 we shall relax this low-frequency assumption and derive methods for analyzing more general systems.

It is possible to use circuit analogs for acoustic systems only in cases in which we can think of acoustic flow as a one-dimensional quantity. Systems made up of

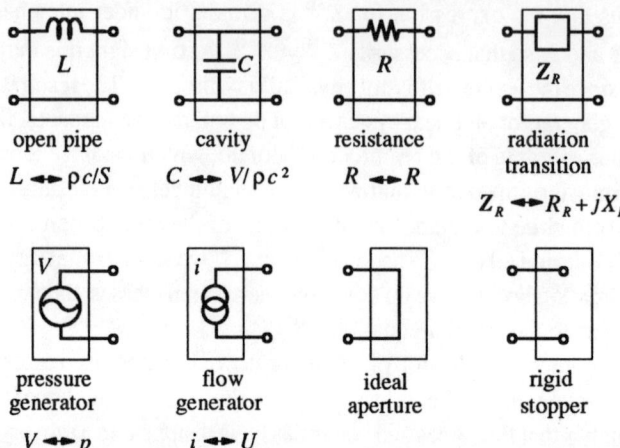

Figure 8.4 Electric analogs for acoustic elements. In the case of the pressure generator and the flow generator, the sign is taken to be that of the upper terminal. All other elements are non-polar.

tubes, small volumes, membranes, and apertures generally satisfy the necessary conditions. We cannot, however, do the same thing for wave propagation in volumes with appreciable extent in two or three dimensions.

In analyzing acoustic systems, it is natural that we deal with the acoustic quantities pressure p and volume flow U, and that we carry out the analysis in terms of the acoustic impedance $Z = dp/dU = p/U$. The equality given here applies since both p and U are very small quantities. In constructing electrical analogs, the natural analogy is between acoustic pressure and electric potential and between acoustic volume flow and electric current. Once again we have three basic acoustic components, as illustrated in Fig. 8.4. The first is a short open pipe. If we apply an alternating pressure difference between its two ends, the air in the pipe accelerates backwards and forwards like a simple mass. The fact that the air is compressible has rather little effect, for we have assumed the pipe to be short compared with the sound wavelength. If the cross-sectional area of the pipe is S and its length l, then the enclosed mass of air is $\rho l S$, where ρ is the density of air, and the force acting upon it is pS. The equation of motion is therefore

$$pS = \rho l S \frac{dv}{dt} \quad \text{or} \quad p = \frac{\rho l}{S}\frac{dU}{dt} = j\omega\frac{\rho l}{S}U. \tag{8.15}$$

Comparing this with the electrical equation (8.2), we see that the pipe behaves like an analog inductance with $L = \rho l/S$.

In a real pipe there will also be a small resistive component to this inductance, caused by viscous forces in the moving air near the walls of the pipe. This can be represented by a small resistor in series with the inductor representing the mass of

LOW-FREQUENCY NETWORK ANALOGS

air in the pipe. We discuss the magnitude of this viscous force in more detail in Chapter 10, where we see that it depends upon frequency and upon the radius of the pipe. Clearly a narrow pipe has a greater surface-to-volume ratio and therefore a more significant loss component. For the present it is adequate to use the approximation

$$R_V \approx 1.2 \times 10^{-5} \left(\frac{\rho c}{S}\right) \frac{\omega^{1/2} l}{a} \approx 2 \times 10^{-3} \frac{\omega^{1/2} l}{a^3} \tag{8.16}$$

where the dimensions must all be in SI units to give the correct numerical value in acoustic ohms. Expressions giving numerical values for all the impedances are summarized in Table 8.1.

An extreme case of viscous losses occurs if the pipe is packed with material such as wool, so that the air must flow through narrow constricted passages and past many obstacles. Effectively we have many very small capillaries, each of radius a, in parallel to produce a plug of radius r. Because by (8.16) the resistance of each capillary is inversely proportional to the cube of its radius, a^{-3}, and the number of capillaries is proportional to $(r/a)^2$, the total resistance varies as $1/ar^2$. In this case the flow is limited by the viscous resistance rather than the achievable acceleration, and the flow rate is simply proportional to the pressure difference. This component is then an acoustic resistance, and its electric analog is an electric resistance of appropriate magnitude. An approximate expression for the numerical

Table 8.1 Analogs for acoustic components

	R	L	C
Open tube, radius a, length l	$1.2 \times 10^{-5} \dfrac{\rho c \omega^{1/2} l}{\pi a^3}$	$\dfrac{\rho l}{\pi a^2}$	—
Closed cavity, volume V	—	—	$\dfrac{V}{\rho c^2}$
Fine-tube bundle, radius r, length l, tubule radius a	$\approx 2 \times 10^{-3} \dfrac{\omega^{1/2} l}{ar^2}$	—	—
Radiation impedance at aperture, radius a	$\dfrac{\rho \omega^2}{4\pi c}$	$0.2 \dfrac{\rho}{a}$	—
Irregular aperture, area S	—	$0.7 \dfrac{\rho}{S^{1/2}}$	—
Tympanum, area S, thickness d, resonance frequency ω_T, quality Q	$\dfrac{\omega_T L_T}{Q}$	$\dfrac{1.4 \rho_s d}{S} = L_T$	$\dfrac{1}{\omega_T^2 L_T}$

value is given in Table 8.1, but this neglects the volume occupied by the pipe walls and typically underestimates the resistance by a factor of at least 3.

The third basic component is a simple closed cavity with volume V and a single inlet. An applied acoustic pressure p causes air to flow into the cavity until the pressure is equalized. Again, we assume the cavity to be small compared to the sound wavelength so that mass acceleration effects can be neglected. The equation of state for the air in the cavity is the simple adiabatic law $(p_0 + p)V^\gamma = $ constant, where p_0 is the normal atmospheric pressure and γ is the ratio of the specific heats of air. The discussion in Section 6.2 then shows that the bulk modulus is $K = \gamma p_0$, which can be written in the form $K = \rho c^2$, where c is the velocity of sound in air. If a volume dV of air flows into the cavity, the original contents must be compressed an amount dV and this leads to a change in pressure that must just balance the applied acoustic pressure. Thus

$$p = -K\frac{dV}{V} = \frac{\rho c^2}{V}dV = \frac{\rho c^2}{V}\int U dt. \qquad (8.17)$$

If we compare this with the electrical equation (8.5) for the behavior of a capacitor, we see that the analog capacitance for the volume V is just $V/\rho c^2$. The impedances of all these elements are measured in acoustic ohms, as we discussed in Chapter 7, so that the analogy with electric ohms is clear.

As with mechanical systems, we need to define two terminating elements and two acoustic generators. The terminations, as shown in Fig. 8.4, are a solid stopper and an aperture. The solid stopper presents no difficulties, except that we must remember that, if we block off a previously open pipe, we must take account of the small volume of contained air and replace it by an analog capacitance. This could have been included in the circuit analog, but adds undue complication at this stage. The acoustic impedance of an aperture is something that we treated in some detail in Section 7.7. As illustrated in Fig. 7.5, the motion of air in the end of an open tube, or at any aperture, is not unrestricted, for it must drive the external air in the fashion of a vibrating piston. The air exerts a radiation impedance with both resistive and reactive parts, as given explicitly by (7.17), and its magnitude depends upon the area of the aperture—normally the area of the tube itself. At low frequencies, for a circular aperture of radius a and area S, (7.17) gives

$$R_R \approx \left(\frac{ka}{2}\right)^2 \left(\frac{\rho c}{S}\right) = \frac{\rho \omega^2}{4\pi c} \qquad X_R \approx 0.6ka\left(\frac{\rho c}{S}\right) = \omega\frac{0.6\rho a}{S}. \qquad (8.18)$$

Because the radiation impedance is associated with the transition from a plane wave in the pipe to a spherical wave in the outside world, we have shown its analog in Fig. 8.4 as a four-terminal network that must then be terminated by an ideal opening. This convention helps when we have to consider the effect of an incident sound wave.

LOW-FREQUENCY NETWORK ANALOGS

Once again we have two types of acoustic sources to consider. The first is a pressure source, exemplified by the effect of an incident sound wave or the action of a sound generator with a large coupling volume between it and the system we are studying. Ideally such a system has zero impedance, but this is not fully realized in acoustic systems. However, the acoustic source impedance at low frequencies is sufficiently low in the two cases mentioned that it can usually be neglected. We return to this point later. The second type of generator is an acoustic current source of the type we discussed in detail in Chapter 7. It is exemplified by a vibrating piston set into the side wall of an enclosure, or by a pulsating flow of air entering from a narrow pipe.

8.5 End Correction for a Pipe and an Aperture

As a brief but important application of the analogs so far developed, let us look at the transition between the end of a short open pipe and the free air. Let us compare the behavior of a pipe with a hypothetical ideally open end exhibiting zero acoustic impedance with that of a pipe with a realistic radiation impedance. If the pipe length is l, its radius a, and its cross-sectional area S, then the impedance at the input end of an ideally open pipe is just $j\omega\rho l/S$. For a realistically terminated pipe, the impedance at the input end is equal to the radiation impedance in series with the pipe impedance, so that

$$Z_{IN} \approx \frac{\omega^2 \rho}{4\pi c} + j\frac{0.6\omega\rho a}{S} + j\frac{\omega\rho l}{S} \rightarrow j\frac{\omega\rho(l+0.6a)}{S}. \tag{8.19}$$

The last form of writing shows that, since from Fig. 7.5 the real part of the radiation impedance is very small compared to the imaginary part at low frequencies, the essential effect of the open end is simply to increase the acoustic length of the pipe by 0.6 times its radius. We can often simply make this correction instead of concerning ourselves with radiation impedance, unless the radiation damping term is also important.

A similar thing happens if we have a simple aperture of area S in an otherwise solid plane baffle. There is effectively a transition from spherical waves to plane waves on each side of the baffle, and therefore two aperture impedances in series. If we neglect the thickness of the baffle and assume the aperture to be circular, this gives a total effective tube length for the aperture of $1.2a \approx 0.7S^{1/2}$ and an acoustic impedance of $0.7j\omega\rho S^{-1/2}$. This result applies quite well for apertures of other shapes.

In this chapter we carefully keep the notation for acoustic systems and electric analogs separate, by always using the appropriate electrical symbols in the network. In the next chapter it will be more convenient to use a rather different notation in which we take the electric analogs for granted and write the currents in their acoustic form as U and the voltage generators as pressures p. This will keep the real system more firmly in mind and, since the impedances are numerically identical in either system, it will save one transformation step. Where we have to consider a mechanical system, we shall similarly use v and F.

8.6 The Helmholtz Resonator

We defer consideration of biological systems until Chapter 9, but one simple and important example will serve to illustrate the use of the electric circuit analogs. This example, called a Helmholtz resonator after its originator, the great German physicist and physiologist Hermann Helmholtz (1821–1894), consists simply of a flask or other rigid container with a neck open to the air, as shown in Fig. 8.5. We explain its use after we have completed the analysis.

The analog circuit is simple to construct, since one part leads directly to the next in a straightforward way. The incident sound wave, essentially a pressure generator, acts upon the aperture at the end of the neck, represented by its radiation impedance, then on the air in the neck tube, and finally on the volume of air in the flask. Any other outlet from the flask is closed by a rigid stopper. We see immediately that the circuit is that of a simple resonator of the type discussed for mechanical systems in Fig. 8.2. The resonance frequency $\omega_H = (LC)^{-1/2}$ is given in terms of the physical dimensions of the components as

$$\omega_H = c\left[\frac{S}{(l+0.6a)V}\right]^{1/2}. \tag{8.20}$$

If the enclosed volume joins the neck in a nearly plane surface, then it might be better to take an end correction $0.6a$ at this aperture as well. At the resonance, the

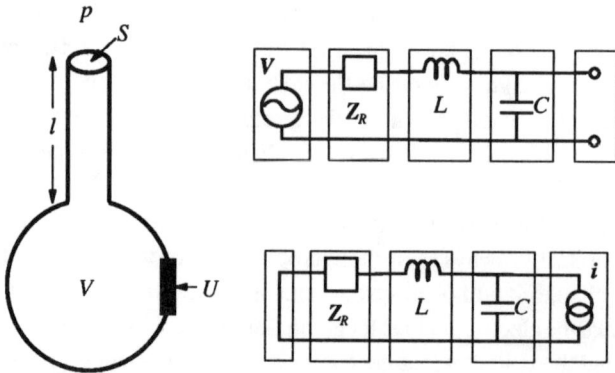

Figure 8.5 A Helmholtz resonator consists of an enclosed volume with an open neck. If the resonator is driven by an external plane-wave acoustic signal, then the analog circuit is as shown in the upper diagram. If instead the resonator is driven by a small internal piston source, then the analog network is as shown in the lower diagram.

LOW-FREQUENCY NETWORK ANALOGS

acoustic flow through the neck of the flask and the pressure variation within it are both maximal. For a wide range of convenient dimensions this resonance frequency is a few hundred hertz. Note that this calculation is self-checking, for if the resonance frequency turns out to be high enough that the dimensions of the resonator exceed about $\lambda/10$, then the analysis is no longer quantitatively valid.

It is now appropriate to mention the purpose for which Helmholtz built and used these resonators. In his case the flask had a long wide neck and also another short and very narrow neck that could be fitted into his ear. The ear presents a high impedance, relative to that of the open neck, and adds very little volume, so that the behavior of the resonator is unaffected. It therefore amplifies the sound pressure level at the ear for a narrow range of frequencies around ω_H, and a set of graduated resonators can be used as a primitive but effective sound spectrograph.

In the second arrangement shown in Fig. 8.5, we suppose a small piston set in the interior wall of the flask to act as an acoustic volume source. This is in some ways analogous to the sound-producing apparatus in some animals, the source being provided by the buckling of a shell-like ribbed tymbal. The analog circuit now has the form shown in the lower part of Fig. 8.5, which is just the same as the second circuit of Fig. 8.2. We have already analyzed the behavior of this circuit, and equation (8.9) shows that the current i_1 through the inductor, which now represents the acoustic flow out of the neck of the resonator, has a simple resonance behavior with a maximum at the frequency ω_H given by (8.20). If the frequency at which the tymbal source is driven matches this resonance, then the acoustic output of the system is greatly enhanced. In fact some insects have ribbed tymbals that buckle progressively in a series of short pulses, as we discussed in Section 5.10. The muscles driving the tymbals operate at a relatively low frequency, but the rate of buckling is adjusted so that the train of pulses produced by each buckling event is at a frequency closely matched to the Helmholtz frequency of the resonator system.

8.7 The Tympanum

A very important component of many auditory systems is a tympanic membrane or eardrum, stretched across the auditory canal. An incident sound wave causes the tympanum to vibrate, and these vibrations are communicated by means of a mechanical attachment at the center of the tympanum to some sort of auditory capsule containing the neural transduction device. Let us examine this in a little more detail from an acoustic, rather than a mechanical point of view.

The fundamental vibration of the tympanic membrane, which is the mode that primarily concerns us, can be represented in shape by a normalized eigenfunction ψ_1. Higher modes do exist but, as discussed in Section 5.5, they are not driven very efficiently by the acoustic pressure and so can generally be neglected. We can include them, as shown below, if we think them to be important in a specific case. Suppose the tympanum is driven by an acoustic pressure p at frequency ω.

Then the displacement response $z(x, y)$ at a point (x, y) on the tympanum is given by (5.19), with m, n chosen to represent the fundamental mode and designated here simply by 1. We take this notation to include the case of a circular or elliptic membrane with (x, y) replaced by (r, ϕ). The displacement velocity is found by taking $\partial z/\partial t$, and the acoustic volume velocity U is then given by the integral of this velocity over the surface of the membrane,

$$U = \left(\frac{j\omega}{\rho_s d}\right) \frac{\left[\int_S \psi_1(x, y) dS\right]^2 p\, e^{j\omega t}}{(\omega_1^2 - \omega^2) + 2j\omega\alpha}. \tag{8.21}$$

This is just the resonance behavior of a simple vibrator with resonance frequency ω_1, damping α and acoustic inertance (or acoustic mass)

$$L_T = \frac{\rho_s d}{\left[\int_S \psi_1(x, y) dS\right]^2}. \tag{8.22}$$

The integral in this expression depends only a little upon the shape of the tympanum, and is approximately equal to $0.7 S^{1/2}$, where S is the area of the membrane. Thus

$$L_T \approx 2\rho_s d/S. \tag{8.23}$$

If the tympanum were able to move like a simple piston, rather than being attached at its edges, the integrals in (8.22) would each have been equal to $S^{1/2}$, eliminating the factor 2 in (8.23).

When we construct an acoustic analog circuit for a tympanum, we can therefore take the inductor to have the numerical value L_T given by (8.23). The numerical value for the capacitor can be fixed by the requirement that the resonance frequency of the membrane with no air loading is $\omega_T = (L_T C)^{-1/2}$, and the resistor can be fixed to give the required Q value, again in the absence of air loading. The acoustic impedance of the tympanum is thus

$$Z_T = j\left(\omega - \frac{\omega_T^2}{\omega}\right) L_T + \frac{\omega_T}{Q} L_T. \tag{8.24}$$

We can treat higher modes of the tympanum in just the same way, by substituting the appropriate mode eigenfunctions into (8.22). If the membrane has a symmetrical shape, rectangular or elliptical, then the integrals for all the antisymmetric modes—those with an odd number of nodal lines in either direction for the rectangular case or with any nodal diameters at all in the elliptical case—vanish exactly. The pressure signal cannot drive these modes, and even if excited they would yield no acoustic flow. The first significant higher mode is thus the one denoted by (3,1) for the rectangular case or (0,2) for the elliptical case.

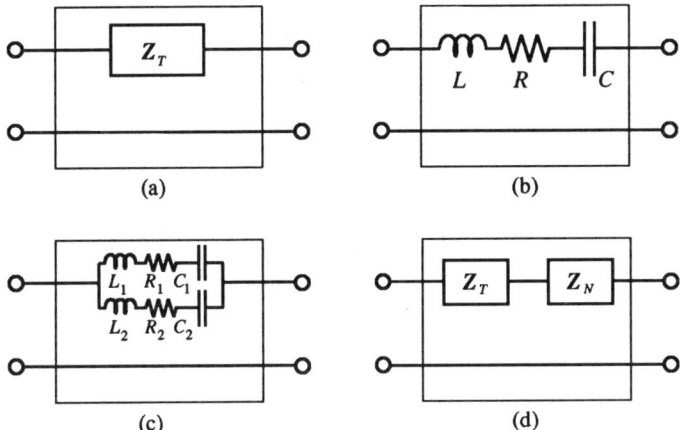

Figure 8.6 Electric analog for a tympanum (a) in a general symbolic form, (b) showing the circuit elements for just the fundamental mode, (c) allowing for a higher mode as well as the fundamental, and (d) allowing for an additional mechanical connection to a neural transducer with mechanical impedance $Z_N^{(M)}$ and acoustic impedance $Z_N = Z_N^{(M)}/S^2$.

For the elliptical case, each of the surface integrals in (8.22) has a value near $0.4S^{1/2}$, so that the factor in front of the acoustic inertance in (8.23) is not 2 but rather about 7. The factor is similar for the case of a rectangular membrane. The acoustic pressure couples rather poorly to drive these higher modes, and when driven they give a very small acoustic flow. Of course biological membranes are never exactly symmetrical, tending to oval rather than elliptical shapes or having non-uniformities of thickness. These features could enhance the coupling of at least some of the higher modes, and may allow the otherwise excluded antisymmetric modes to contribute, but the fundamental remains the most important.

When construct the circuit analog for a tympanum as an acoustic element, we must recognize that the two pairs of terminals of the element represent the two sides of the membrane, rather than the top and bottom of a spring. The resulting analog is shown in Fig. 8.6. If we wished to include the possibility of contribution from a higher mode of the tympanum, then, since membrane admittances from different modes are simply additive as shown by (5.19), the impedances of the separate modes are connected in parallel inside the element as shown in the figure. As we have indicated, however, the acoustic impedances of the higher modes are at least four times larger than that of the fundamental, except very near to their resonances, so that this refinement is generally not required. From our discussion in Chapter 5 we also recall that the second resonance of a membrane is at a frequency about three times that of the fundamental for all approximately equiaxial membranes. If we do include a higher mode in this way, then we find that, at a frequency just below its resonance, it makes a contribution to the acoustic flow that is comparable to and almost out of phase with the flow associated with the fundamental. These two contributions nearly cancel at this particular frequency, giving a maximum in the membrane impedance.

A more significant refinement is to include the loading effect of the connection of the membrane to the neural transducer capsule, generally effected by means of a link to the center of the membrane. The general case can be complicated, as we discussed in Sections 5.6 and 5.7, but evolutionary optimization has ensured that in most animals the impedance of the connection is sufficiently small that it does not greatly upset the shape of the fundamental membrane mode. If the mechanical impedance of the neural transducer is $Z_N^{(M)}$, and if it is connected near to the center of the tympanum, then it makes an additive contribution to the acoustic impedance of the tympanum of approximately $Z_N = Z_N^{(M)}/S^2$, where S is the membrane area. There should be a small correction to this, reflecting the difference between the average motion of the membrane and its motion at the point of connection—the difference between a point impedance and a transfer impedance. This correction factor is as large as 2 for connection at the center of the membrane and becomes significantly less than unity if the connection is made close to the membrane edge, as we discuss below. This geometrical correction allows some improvement to be made in the mechanical impedance matching between membrane and transducer capsule. For illustrative calculations we take the factor to be unity.

When we have solved our network problem, as we discuss in detail below, we will generally end up with an expression for the acoustic current U through the tympanum. We will usually be concerned to convert this to a vibration velocity amplitude communicated to the connection to the neural transducer mechanism. We first convert the acoustic flow U to a mean deflection velocity $\bar{v}$ for the membrane by the simple relation $\bar{v} = U/S$. We then seek a relation between $\bar{v}$ and the velocity $v(r)$ at the point of connection of the transducer, a distance r from the center of the membrane. If the membrane is roughly circular, then it will vibrate in its fundamental mode with a pattern closely similar to the Bessel function $J_0(k_1 r)$ where k_1 is required to satisfy the equation $J_0(k_1 a) = 0$, a being the radius of the membrane. From the graph of the Bessel functions in Fig. 5.3, we see that $k_1 \approx 2.4/a$. Since the relative amplitude at radius r is just $J_0(k_1 r)$, the necessary conversion is

$$v(r) = \frac{J_0(k_1 r)}{\int_0^a J_0(k_1 r) 2\pi r\, dr} U \approx \frac{2.3 J_0(2.4 r/a)}{S} U \qquad (8.25)$$

since the value of the integral in the denominator is $S/2.3$. For attachment at the center of the membrane $J_0(0) = 1$, so that $v(0) \approx 2.3 U/S$. This means that the displacement at the center of the membrane is 2.3 times the average membrane displacement in the case of the fundamental mode.

If we apply a similar discussion to higher modes of the membrane, then the numerical coefficient is larger than the value 2.3 found for the fundamental, because the integral in the denominator of (8.21) is smaller if k_1 is replaced by a value k_n appropriate to a higher mode. This increased factor essentially balances one of the integrals in the denominator of the acoustic inertance (8.18), but still leaves higher modes less effective than the fundamental by at least a factor 2 in amplitude, corresponding to -6 dB.

8.8 Solution of Networks

The simple electric analog circuits we have so far encountered have been easy to solve so as to find the currents resulting from given applied voltages. When we come to consider more complex acoustic systems, however, we shall require a more systematic approach to solving the network analogs. We shall outline such an approach here, leaving its application to the next chapter, where we deal with real model systems.

A general network can be connected in a very complex manner, but always consists of branches joined together at nodes. An arbitrary example is shown in Fig. 8.7. Most of the branches will simply contain impedances, but a few may contain voltage generators or current generators. To solve the network we need to be able to calculate the currents in all the branches, and for this we require a set of equations equal in number to the number of branch currents. These equations derive from two simple rules, known as Kirchhoff's laws. The first states that the sum of all the currents flowing into any node must equal the sum of the currents

Figure 8.7 As shown in the upper part of the figure, a general network can be analyzed either in terms of currents in each of its branches or in terms of circulating currents. The lower part of the diagram illustrates the analysis of a circuit representing a real lever, driven by a force and connected to a mechanical load.

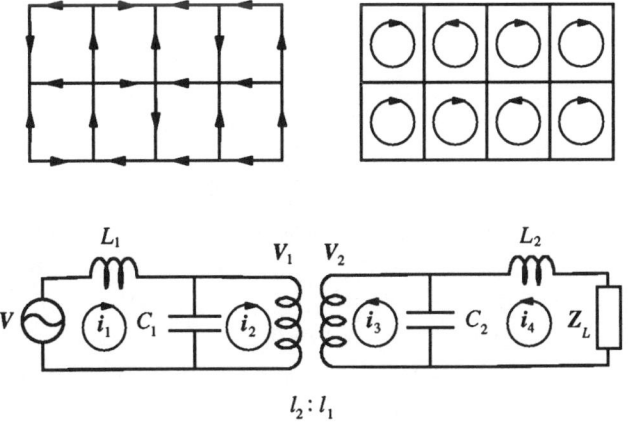

flowing out, so that no charge accumulates. The second states that, when we go around any closed circuit in the network, the sum of all the voltages contributed by the voltage generators in the circuit must equal the sum of all the voltage drops across the impedances. Current generators simply specify the value of the current in a particular branch.

The simplest way to solve a network is not to do so in terms of the currents in its branches, but rather in terms of the currents in its circuits, as shown in Fig. 8.7. In this way we end up with many fewer equations, for the circuit currents automatically satisfy Kirchhoff's first law because they flow into and out of any node they encounter. The circuits in the network can often be chosen in several different ways, all of which lead to identical results. The only requirement is that we choose enough circuits so that the number of currents passing through any node is at least equal to one less than the number of branches linked at that node. This will generally happen automatically as we draw the circuit currents. We then go around each circuit and write down the sum of the voltage drops across each impedance, noting that the current flowing through that impedance may involve contributions of appropriate sign from several circuit currents. We set this sum equal to the sum of all the voltage generators in the circuit. If we encounter a current generator, then its value simply fixes the total current flowing through it. In this way we have a set of linear equations equal in number to the number of circuit currents, and these are sufficient to determine them all.

We make this procedure clear by setting up the equations for the example in Fig. 8.7, which actually represents a real lever with slightly flexible arms, as discussed in Section 8.3, excited by a force at one end and connected to a mechanical load $Z_L^{(M)}$ at the other. The current loops could have all been taken as running clockwise, but it is sometimes convenient in networks that look symmetrical to use this same symmetry in the currents. We are then less likely to make errors of sign in the equations.

The loops for circuits 1 and 4 are straightforward, and give the equations

$$j\omega L_1 i_1 + \frac{1}{j\omega C_1}(i_1 - i_2) = V \qquad (8.26)$$

$$j\omega L_2 i_4 + Z_L i_4 + \frac{1}{j\omega C_2}(i_4 - i_3) = 0. \qquad (8.27)$$

The generator voltage V appears with a positive sign because the current emerges from its upper (positive) terminal and enters through its lower (negative) terminal. The two inner loops are a little more complicated in this case because of the transformer. We do not know, or care, what the actual voltages at the transformer are, but there are certainly voltages across it that must be taken into account. Let us call these V_1 and V_2 as in the diagram. The Kirchhoff loop law applied to these two circuits then leads to the equations

LOW-FREQUENCY NETWORK ANALOGS

$$\frac{1}{j\omega C_1}(i_2 - i_1) = -V_1 \qquad \frac{1}{j\omega C_2}(i_3 - i_4) = -V_2. \tag{8.28}$$

Both V_1 and V_2 appear with negative signs because the current enters the upper terminal instead of leaving it. We can now write down the equations describing the ideal transformer which, from (8.10) are simply

$$V_2 = \frac{l_1}{l_2} V_1 \qquad i_3 = \frac{l_2}{l_1} i_2. \tag{8.29}$$

Notice how the symmetry helps us in checking all these equations against each other. We can now use (8.29) in (8.28) to eliminate V_1, V_2 and one of the currents i_3 or i_2. After a little simplification, the result is

$$\frac{l_2}{C_2}\left(\frac{l_2}{l_1}i_2 - i_4\right) = \frac{l_1}{C_1}(i_2 - i_1). \tag{8.30}$$

We can also use the second equation of (8.29) to eliminate i_3 from (8.27). Equations (8.26), (8.30) and this modified (8.27) can now be solved to find each of the remaining independent currents i_1, i_2 and i_4. We are particularly interested in i_1 and i_4, since these are the analogs of the velocities of motion of the two ends of the lever. We might also be interested in the force transmitted to the load, which is just $Z_L i_4$. We leave the actual solution as an exercise, since the result is not of immediate importance.

Any network can be solved in just this way. Indeed most of the cases we shall meet are rather simpler, since they do not involve transformers. Once we have written down the initial equations, we can either solve them algebraically, which has advantages if the system is not too large because it gives us an explicit answer, or else numerically using a computer. This second approach is important, and we look at it briefly in the next section.

8.9 Computer Solutions

The network equations written down in the previous section as (8.26) through (8.29) constitute a total of six equations in the six complex unknowns i_1, i_2, i_3, i_4, V_1 and V_2. There is just one independent variable V, which occurs in the first equation. If we were to write the unknown variables as $x_1, x_2, ..., x_6$ and the independent variables for each equation as $y_1, y_2, ..., y_6$ (as we noted, all are zero in our particular case except for y_1), then the whole set of equations has the form

$$\begin{aligned}
a_{11}x_1 + a_{12}x_2 + \ldots + a_{16}x_6 &= y_1 \\
a_{21}x_1 + a_{22}x_2 + \ldots + a_{26}x_6 &= y_2 \\
\vdots \\
a_{61}x_1 + a_{62}x_2 + \ldots + a_{66}x_6 &= y_6
\end{aligned} \tag{8.31}$$

where the coefficients a_{ij} are generally frequency-dependent complex impedances. We could have simplified the equations algebraically, as we did, to eliminate some of the unimportant variables, resulting in a smaller number of equations and independent variables. The three final equations in three unknowns at which we arrived could have been written in just the same block form. In a large network, we might easily have 20 equations in 20 unknowns and with several of the independent variables y_i, corresponding to voltage sources, non-zero. In fact, since all the variables are complex, we need to treat their real and imaginary parts separately. This doubles the number of variables and, by taking the real and imaginary parts of each equation, we also have twice as many equations.

Fortunately there are very efficient computer algorithms for solving sets of linear equations of this form with constant coefficients. A particular computer program to perform this solution is given in the book *Numerical Recipes* listed in the Bibliography. What we have to do to make use of such a program is first to fix the frequency ω, so that all the coefficients a_{ij} have definite numerical values. We then solve the equations to find all the variables x_j and in particular the one in which we are interested. The whole process is then repeated for a different frequency and so the behavior of the system is explored.

We shall need to resort to such numerical solution of some of our more complex acoustic systems, but many of the simpler ones can be solved algebraically to give an explicit solution, and we shall do this wherever it is conveniently possible.

References

Electric analogs: [6] Ch 3; [7] Ch 1; [8] Ch 8; [9] Ch 4,5
Electric networks: [6] Ch 3; [7] Ch 1
Acoustic components: [6] Ch 5; [9] Ch 5

Discussion Examples

1. Calculate the acoustic impedance of the following elements at 1000 Hz: (a) an open tube of length 10 mm and diameter 4 mm, (b) a cavity of volume 1 cm^3, (c) a slack diaphragm of diameter 4 mm and thickness 100 μm.
2. An open tube consists of a short length l_1 with radius a_1 connected to another short section of length l_2 and radius a_2. Show that it behaves acoustically in the same way as a tube of radius a_1 with total length $l_1[1 + (l_2 a_1^2/l_1 a_2^2)]$ if end corrections are neglected.
3. A tympanum has area 60 mm^2 and thickness 20 microns. Its resonance frequency is 2000 Hz and its effective Q value is 2.0. Calculate its approximate acoustic impedance (a) at its resonance frequency, (b) at a frequency of 5 kHz.
4. A human mouth has a volume of about 100 cm^3, and the open lips, about 10 mm thick, form an aperture of area about 2 cm^2 for a particular sound made by "popping" the lip opening with a finger tip, the throat passage remaining closed. Estimate the Helmholtz resonance frequency. What will be the new resonance frequency if the lips are opened to an aperture of area 5 cm^2?

5. Using the discussion of Chapter 3 and the analog circuit of Fig. 8.3, derive the expressions given in (8.12) for the lumped impedances describing a real lever.

Solutions

1. (a) $6 \times 10^6 j$ (b) $-2.2 \times 10^7 j$ (c) $1.5 \times 10^6 j$ acoustic ohms.
3. (a) Taking $\rho_s \approx 1000$, $L \approx 330$ and $R = L\omega_T/Q$ where ω_T is the resonance frequency. Thus at resonance $Z_T = R \approx 2 \times 10^6$ acoustic ohms. (b) Off resonance, $Z = L(\omega - \omega_T^2/\omega)j + R$, so at 5 kHz $Z \approx 2 \times 10^6 + 1 \times 10^7 j$.
4. Allowing for end correction both internally and externally on lip opening, effective $l \approx 20$ mm. For (a) this gives $f_H \approx 600$ Hz and for (b) $f_H \approx 900$ Hz.
5. (a) Consider a lever with just one arm. For pivoted motion, by (3.40), the lowest eigenfunction is $\psi = (3/l^3)^{1/2} x$ where l is the lever arm length and x is the distance from the pivot. The resonance frequency of this pivoted motion is zero. By (3.26), with $n = 1$ and $\omega_1 = 0$, the mechanical impedance at the end of the arm is

$$Z^{(M)} \approx j(\pi a^2 \rho_s l/3)\omega$$

which is equivalent to a simple mass equal to one-third that of the lever arm. This verifies the first result of (8.12).

(b) Suppose we clamp the pivot of the lever in Fig. 8.3. The transformer then reflects an infinite impedance at its input, and the primary circuit is a simple LC resonant circuit which has resonance frequency $\omega_1 = (LC)^{-1/2}$. We can identify this physically with the first resonance of a bar clamped at one end, as given by (3.36) and (3.37). Insertion of C from (a) then gives L as in (8.12).

9 LOW-FREQUENCY AUDITORY MODELS

SYNOPSIS. Animals generally have a plane of symmetry, and therefore two ears. There is usually an acoustic connection between the two ears through some sort of tube or cavity, but in animals such as mammals the connecting Eustachian tubes are so long and narrow that the ears are essentially independent. For sound at frequencies sufficiently low that the distance between the ears is much less than the sound wavelength, the sound pressure is the same at the two ears, but there is a difference in phase that depends upon the separation of the ears and the direction from which the sound is coming, as illustrated in Fig. 9.1. Auditory systems in which the two ears are acoustically coupled have generally developed in such a way as to respond to this phase difference directly, but animals with independent ears have to rely upon signal processing at a higher neural level in order to derive information about the direction of the acoustic source.

An unusual auditory detector, such as might be found in the wing of an insect, is illustrated in Fig. 9.2. It consists simply of a taut diaphragm mounted in a frame. It has a simple peaked frequency response, as shown in Fig. 9.3 and a figure-eight directional pattern.

A simple model for a mammalian ear is just a taut membrane, the tympanum, stretched over the entry to a cavity and loaded by the impedance of the neural transducer mechanism, as shown in Fig. 9.4. Such an ear is omnidirectional at low frequencies and has a frequency response similar to that of a simple resonant vibrator. The resonance frequency of the tympanic membrane is lowered by the loading of the neural transducer, but raised by the resilience of the air confined in the middle-ear cavity. The motional response of the tympanum at the point of attachment of the neural transducer is shown in Fig. 9.5, for the physical parameters given in Table 9.1. In order that such an ear transfer as much mechanical energy as possible to the neural transducer, it is desirable that the resistive component of the transducer impedance be equal in magnitude to the resistive component of the tympanum impedance itself. The transducer thus lowers the Q value of the tympanum by a factor 2, and Q values in the range 1 to 5 are probably the rule.

In a simple coupled auditory system we might find two tympana coupled together through a body cavity as shown in Fig. 9.6. If the impedance of the tympana, their resonance frequencies, and the volume of the cavity are properly chosen as shown in Table 9.2, as might have occurred through evolutionary processes, then such an auditory system has a broad resonant response and a considerable difference in response level between ipsilateral and contralateral ears, as is shown in Fig. 9.7. The directional response is cardioid in shape in the important frequency region, so that the system has useful directionality. There is also a significant phase difference between the response of the two ears, depending upon the incidence direction of the sound, as shown in Fig. 9.8. A sophisticated neural processing system could make use of this information.

For an animal such as a frog, with ears opening directly into the mouth cavity, we might wish to construct a rather more detailed system model taking into account the short Eustachian tubes and the nares, as shown in Fig. 9.9. The calculated acoustic response of this system for the physical parameters given in Table 9.3 is shown in Fig. 9.10. Once again there is a useful difference in response amplitude between ipsilateral and contralateral ears, and a cardioid response pattern, but the axis of the cardioid is shifted towards the rearward direction.

Calculations based on system models such as these, as well as providing a guide to the behavior of typical auditory systems, can also serve to define the non-invasive measurements that must be made on a living animal in order to characterize its auditory system in terms of acoustic impedances.

9.1 Constructing Models

We now have all the apparatus necessary to model the behavior of passive acoustic systems, such as auditory systems, in the low-frequency domain in which their dimensions are small compared with the wavelength of the sound waves involved. What we shall do is to look at a number of representative systems, corresponding broadly to the auditory systems of certain insects and vertebrates, and see what we can discover about the mechanical signals fed to the neural transducer in sound fields with different frequencies and incidence directions. These mechanical signals provide the stimuli upon which any analysis at a higher neural level must be based. In accord with the philosophy of this book we adopt a general approach that is potentially applicable to a wide variety of different animals, rather than entering into a detailed analysis of any particular cases.

Nearly all terrestrial animals have bilateral symmetry—the notable exceptions being certain bottom-dwelling fishes, with distorted asymmetric forms, and certain shell-fish with chiral symmetry. This means that most auditory systems contain at least two symmetrically related acoustic receivers, usually physically connected by means of some sort of cavity or tube. In many animals however, for example mammals, the acoustic connection is almost nonexistent. It therefore makes sense to examine some cases of pairs of isolated single ears, in which the correlation between their responses in a sound field is made at a neural level. In most animals, however, the two ears are coupled acoustically, and must be treated in a combined acoustic model. Analysis of such systems will provide many of our most interesting examples.

In analyzing these systems we must first make a mental picture of the physical model, with all the components and their interconnections carefully defined. As outlined in our general philosophy in Chapter 1, this model should initially include only the minimum number of essential elements, without neglecting anything required by the physics; embellishments can always be added at a later stage. We then draw an electrical analog circuit for the system, as outlined in Chapter 8, to aid in the formulation of the equations describing system performance. We then solve those equations, usually in the present chapter simply by algebraic manipulation, but if necessary by numerical methods, and examine their predictions. For simplicity of notation we denote voltage (pressure) sources by the acoustic quantities p_n and currents (acoustic volume flows) by U_n. Our network diagrams will generally show just blocks representing impedances Z_i with the subscript i chosen to denote the particular acoustic component.

9.2 The Incident Sound Field—Diffraction

The acoustic fields encountered by animals in the environment are often complex, involving the direct waves radiated by one or more sound sources together with waves reflected from objects in the immediate neighborhood. It requires much sophisticated signal processing at a neural level to derive information from such fields and, while higher animals such as humans certainly carry out this analysis at a subconscious level, simpler animals respond only to simpler cues, represented primarily by the direct field from the nearest or strongest source. We shall therefore make all our analysis in terms of the simplest such case—an incident plane wave with a single frequency. Such a wave is specified simply by its frequency, intensity or sound pressure, and propagation direction. In Chapter 15, which deals with signal processing, we shall see how the response of a linear system to such a simple and well defined stimulus can be used by superposition to find what happens in more complex cases.

Such a free-field stimulus can be readily generated in the laboratory using a loudspeaker in an anechoic room, provided that the distance of the loudspeaker from the animal is large compared with the size of the animal and compared with the acoustic wavelength involved. In other laboratory experiments it may be useful to use a different arrangement in which sound is fed to the ear from a headphone source, consisting essentially of a small loudspeaker in a small enclosure that is sealed to the ear. This situation requires further analysis, as we see later.

The free-field situation is illustrated in Fig. 9.1(a). It is helpful to choose a coordinate system relative to the symmetry plane of the animal, as shown, so that the angle θ gives the direction of incidence of the sound wave relative to this plane. For the more complex auditory systems we may need to use full three-dimensional coordinates, and thus two angles (θ, ϕ), but for most of our examples this complication is unnecessary.

Analysis of the sound field around even a small object is rather complicated. If the object is small compared with the wavelength, which is our assumption in this chapter, then it turns out that the sound pressure amplitude is the same over its whole surface; all that differs is the phase. If we neglect diffraction effects around the animal, then the phase difference between the acoustic signals at the two ears can be found simply from the path difference Δ shown in the Fig. 9.1(a). The pressure wave has the form $pe^{j(\omega t - kx)}$, where $k = \omega/c$, so that the phase difference between the waves at the two ears is $k\Delta = kD \sin\theta$, where D is the distance between the two ears. We can therefore write

$$p_1 = pe^{jk\Delta/2}e^{j\omega t} \qquad p_2 = pe^{-jk\Delta/2}e^{j\omega t} \qquad \Delta = D\sin\theta. \qquad (9.1)$$

Note that we have chosen to write the pressures in a symmetrical way to emphasize the symmetry of the system. To be more realistic we must allow for the effects of diffraction around the head or body of the animal, as shown in Fig. 9.1(b). It turns

LOW-FREQUENCY AUDITORY MODELS

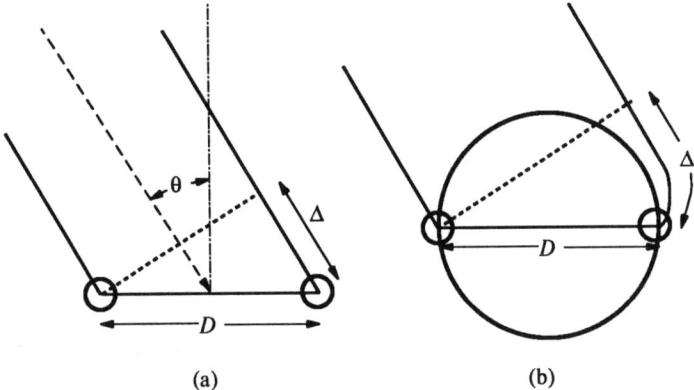

Figure 9.1 (a) Free-field incidence of a plane wave upon two ears separated by a distance D. The path difference Δ depends on the angle of incidence θ. (b) A more realistic situation, taking into account diffraction around the head of the animal. For a spherical head, Δ is increased by a factor 1.5 over the free-field case.

out that, if we can regard the part of the animal on which the ears are mounted as a long cylinder, then diffraction effects double the phase shift from the simple result calculated above, at least to a first approximation. If it is a better approximation to regard the ears as mounted on a solid sphere, then the phase shift is increased by a factor close to 1.5. Thus

$$\Delta_{cylinder} \approx 2D \sin \theta \qquad \Delta_{sphere} \approx 1.5D \sin \theta. \qquad (9.2)$$

The head or body of the animal thus has the effect of increasing the acoustic distance between the two ears, a result that is not really surprising, since the acoustic wave must effectively travel around the outside of the head to reach the ear remote (contralateral) from the sound source and thus covers a distance rather greater than the direct separation between the ears.

At high frequencies, as discussed in Section 6.9, the head may have a more complex effect, essentially because there can be interference between waves traveling to the contralateral ear by a variety of paths. In addition to a phase delay, this interaction gives an amplitude difference between the two ears, interpreted essentially as a shadowing of the contralateral ear by the bulk of the head. Formulae giving the amplitudes of sound incident on each ear are given explicitly in (6.42) and (6.43), and depend upon the general shape of the head as well as upon the angle of incidence. It is straightforward to include this additional complication in all the results to be derived in this and later chapters, but we shall omit it in the interests of simplicity of exposition.

If we are concerned to analyze a laboratory situation in which the sound is applied to the ears by some form of headphone, then it is important that the exact acoustic arrangement is specified and analyzed, if necessary by including the sound system as an integral part of the model for analysis. The possibilities are so numerous that there is little to be gained by general discussion here.

9.3 Response and Sensory Threshold

When we analyze an auditory system, the initial quantity calculated will be the acoustic current through some component, nearly always a tympanum, for a given value of applied acoustic pressure. Following the methods discussed in Section 8.7 we can relate this acoustic current to the mechanical motion of the tympanum, essentially by dividing by the area of the tympanum. A further numerical adjustment, related to its point of attachment, then gives the motion of the connection to the neural transducer. In this and subsequent chapters we shall present response curves in a uniform manner by plotting the r.m.s. displacement response and the r.m.s. velocity response at the link to the neural transducer as functions of frequency for both ipsilateral and contralateral presentation of the free-field sound at a level of 74 dB re 20 µPa, equivalent to 0.1 Pa r.m.s., a typical level for human conversation. We use logarithmic scales on both axes. Because the response is linear, in the mathematical sense, it can be simply scaled for other applied acoustic pressures.

There are several reasons for displaying the results in this way, rather than as sensory threshold curves. The first reason is that we are able to plot absolute values for displacements and velocities, and these can be compared with the results of measurements using, for example, laser vibrometers. The second, and more basic, reason is that to plot a sensory threshold curve implies an assumption about the transduction mechanism. It is most likely that mechanical velocity is usually the quantity transduced into nerve impulses, but we are not justified in making this assumption. Even if the assumption is basically correct, it is most unlikely that the transduction mechanism responds equally well at all frequencies. In any case, if we wish to convert the figures into a sort of primitive sensory threshold curve, all that is necessary is to reflect the curve in the frequency axis (invert it), and note that a factor of 10 in the response corresponds to a threshold shift of 20 dB.

In plotting the directional response of each system we use a polar plot, as is natural, and display the response at angle θ, as defined in Fig. 9.1, in decibels relative to the peak response. We generally do this at several frequencies. The use of a decibel scale is common in such plots, but requires interpretation. Our scale will usually run from 0 to −30 dB, with the central circle of the plot left blank. This is to emphasize that the central origin does not correspond to a null response, as it would on a linear scale; a null response is minus infinity on a decibel scale, and cannot be accommodated on the figure. With this caveat, a polar plot with a decibel scale gives a more useful representation of directional discrimination than would a similar plot with a linear scale.

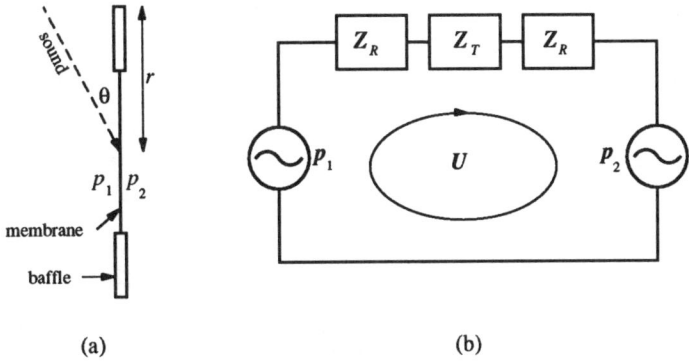

Figure 9.2 (a) A simple ear consisting of a taut membrane surrounded by a rigid baffle. (b) The analog circuit from which the response of the system can be calculated. Z_T is the acoustic impedance of the membrane and Z_R that of the radiation load upon it.

9.4 Baffled Diaphragm

The first type of ear we mention stands in a class by itself and warrants only brief discussion. This is an acoustic detector consisting of a flexible membrane or thin shell supported on a rigid frame and perhaps surrounded by some form of more rigid baffle panel, as shown in Fig. 9.2(a). One could imagine such a detector as a part or whole of an insect wing. Fairly clearly the motion of the membrane responds to the difference in acoustic pressure between its two sides, which is itself determined by the phase shift in propagation to these two sides. This is an extreme case of the correction to the phase shift along the direct propagation path discussed in Section 9.2. It is not easy to derive an exact answer, but we expect the path difference to be about equal to the distance from the center of the membrane to the edge of the surrounding baffle, or to the edge of the membrane if the baffle is simply its thickened edge, multiplied by the angular factor $\sin\theta$, where the axis from which θ is measured is taken here as lying in the membrane plane. Denoting this distance by r, then the mean pressures acting on the two sides of the membrane are given by (9.1) with $\Delta \approx r\sin\theta$. The diaphragm itself is loaded on each side by its radiation impedance, so that the analog circuit appears as in Fig. 9.2(b).

Analysis of the circuit is particularly easy, since it has only one loop. The two pressure sources simply subtract, since they are connected into the circuit in opposite phase. The formal result is

$$U = \frac{p_1 - p_2}{Z_T + 2Z_R} \approx \frac{jp(\omega/c)\Delta \sin\theta}{(R_T + R_R) + j\omega(L_T + 2L_R) + 1/j\omega C_T} \tag{9.3}$$

where in the second form of writing we have taken advantage of the fact that $k\Delta \ll 1$

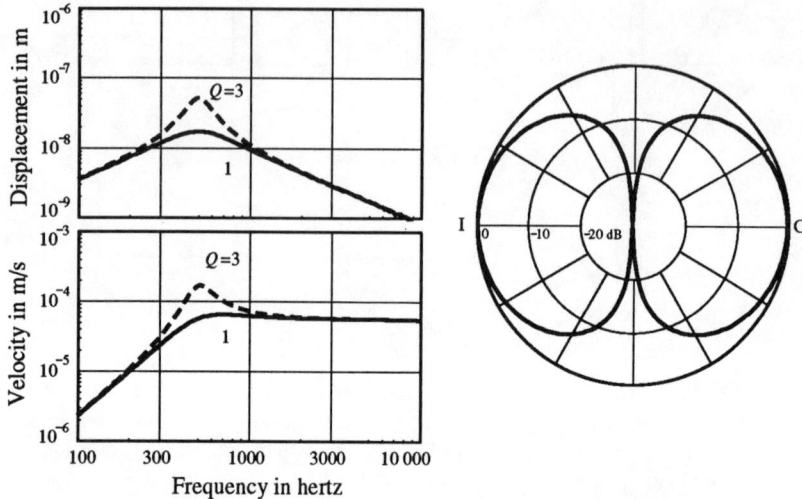

Figure 9.3 Frequency response and directionality for a simple insect ear, consisting of a small tympanum of area 20 mm² and loaded resonance frequency 500 Hz, set in a baffle of diameter 10 mm. The membrane of the tympanum has thickness 10μm, and is connected to an auditory capsule of effective mass 1 mg. The displacement and velocity responses are calculated for two different Q values as shown. Note that the assumptions of the model are not valid above about 3000 Hz.

and have written $k = \omega/c$. We see that there is a resonance in the response at the frequency ω^*, given by

$$\omega^* = [(L_T + 2L_R)C_T]^{-1/2} \qquad (9.4)$$

where the denominator of (9.3) has its minimum value.

The frequency dependence of the response is shown in Fig. 9.3, as also is the directional response, which is the same at all frequencies provided $kr \ll 1$. We have converted the acoustic volume flow U to a mechanical velocity v by the simple relation $v = U/S$, since we have no information about neural transducer attachment. This system is known as a pressure-gradient transducer, for obvious reasons. The figure-eight directional response in terms of power, which is proportional to U^2 and so has the form $\sin^2\theta$, is typical of such systems, as is the decline in response at low frequencies. It clearly gives useful directional information.

Since we shall not return to consider this system again later, it is worthwhile to make a few comments about its high-frequency behavior. This is complicated by higher resonances of the diaphragm but, as we discussed in Section 8.7, these generally couple rather inefficiently to a simple acoustic field. We can gain an idea of high-frequency performance by using the reciprocity theorem, which states broadly that the behavior of the system used as a receiver will be similar to that when it is used as a transmitter, and thus similar to the behavior of a vibrating disc as discussed in Section 7.5. From that discussion we see that the angular response becomes increasingly narrow as the wavelength decreases below the diaphragm diameter, as illustrated in Fig. 7.4 (rotated through 90° to agree with our present coordinate system).

The behavior of the response also changes for $kr > 3$, as expected from the sharp change in behavior of the radiation resistance shown in Fig. 7.3. For $kr \ll 3$, as we have discussed, the diaphragm is driven by nearly equal pressures on its two sides, with a phase difference between them. This phase difference, and thus initially the driving force, increases linearly with frequency. For $kr > 3$, in contrast, the remote side of the diaphragm is essentially in a sound shadow, so that the driving force is simply proportional to the incident sound pressure. Both the displacement and velocity responses therefore have a further drop of 6 dB/octave, additional to that shown in Fig 9.3, above the frequency for which $kr = 3$ (~30 kHz for the case calculated in the figure).

9.5 Simple Cavity-Backed Ear

In many mammals, as remarked above, the two ears are essentially separate acoustically, because the Eustachian tubes are very long and narrow. We can therefore treat each ear as acoustically independent. The external pinna and meatus represent complications that we add later; for the present let us just consider an ear consisting of a simple tympanum (with attached connections to the neural transducer), backed by a cavity of volume V. The physical model is shown in Fig. 9.4, together with its analog circuit. The figure actually shows the Eustachian tube as well, but we ignore this initially. Again we have only one current loop to consider and the network is readily solved to give the acoustic current

$$U = \frac{p}{Z_R + Z_T + Z_N + Z_V} = \frac{p}{R_T' + j\omega L_T' + (1/j\omega)(C_T'^{-1} + C_V^{-1})} \quad (9.5)$$

where $Z_T' = Z_T + Z_N + Z_R$ includes the loadings Z_N contributed by the neural transducer and Z_R from radiation. The response is that of a simple resonant circuit with mass L_T', stiffness $C_T'^{-1} + C_V^{-1} = (C_T' + C_V)/C_T' C_V$, and damping resistance R_T'. The resonance frequency ω^* is given by

$$\omega^* = \left(\frac{C_T' + C_V}{L_T' C_T' C_V}\right)^{1/2} = \omega_T \left(1 + \frac{C_T'}{C_V}\right)^{1/2} \quad (9.6)$$

where ω_T is the resonance frequency of the tympanum loaded by the neural transducer and the radiation impedance. For a small backing cavity, $C_V = V/\rho c^2$ may be comparable with or even smaller than the membrane compliance C_T, so

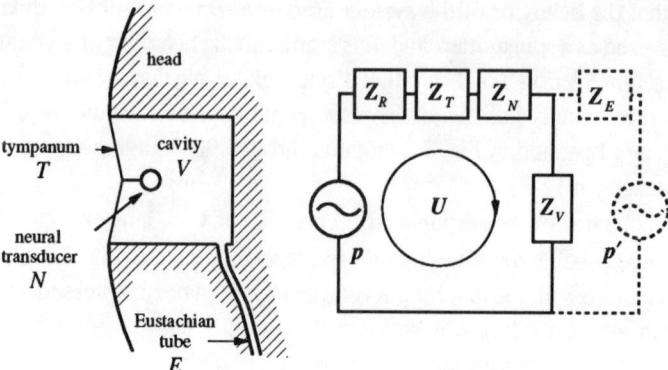

Figure 9.4 A simple ear, such as found in many animals, consisting of a taut membrane with acoustic impedance Z_T and radiation impedance Z_R, connected to a neural transducer capsule with effective acoustic impedance Z_N, and backed by a cavity with acoustic impedance Z_V. In reality the cavity will be vented to the exterior through a narrow Eustachian tube with acoustic impedance Z_E. The analog circuit, with or without the Eustachian tube branch, allows calculation of system behavior.

that usually ω^* will be significantly greater than ω_T. In a typical ear of this general pattern, the linkage to the neural transducer adds significant mass and dissipation to the tympanum itself, as we discuss in the next section, and must always be considered when evaluating the acoustic impedance and resonance frequency of the tympanum. The radiation load is much less important.

When we have calculated the acoustic flow U through the tympanum, we need to convert this to a mechanical motion v of the point of attachment of the neural transducer. From the discussion in Section 8.7 we see that we first divide U by the area S of the tympanum to convert the volume flow to an average velocity over the tympanum surface, and then multiply by a factor equal to the ratio of the tympanum deflection at the point of attachment of the neural transducer to the average deflection in the fundamental mode. If the attachment point is near the center of the tympanum, this factor is approximately 2.3, so that $v \approx 2.3 U/S$. Response curves for displacement and velocity at the neural transducer attachment for a typical ear with the parameters given in Table 9.1 are shown in Fig. 9.5. The response is the same in form as that of a simple vibrator, as discussed in Chapter 2, and the ear is omnidirectional at these relatively low frequencies.

Real ears of this pattern necessarily have some sort of Eustachian tube, partly to prevent the cavity filling with fluid, and partly to provide a means for compensating for changing atmospheric pressure. This additional component is shown in Fig. 9.4, both in the model and in the analog circuit. In the type of ear we are

LOW-FREQUENCY AUDITORY MODELS

Table 9.1 Assumed physical parameters for a simple ear

Tympanum area	$S_T = 20$ mm^2
Tympanum thickness	$d = 10$ μm
Tympanum resonance	$\omega_T/2\pi = 500$ Hz
Effective transducer mass	$m_N = 1$ mg
Q of loaded tympanum	$Q = 1$
Cavity volume	$V = 1$ cm^3
Length of Eustachian tube	$l_E = 50$ mm
Diameter of Eustachian tube	$2a_E = 0, 0.5, 1$ mm

discussing, it is a long narrow tube, and its impedance Z_E is large and essentially resistive, with resistance R_E given in Table 8.1. The tube typically terminates in a nasal passage that is open to acoustic pressure p' through the nostrils. At ordinary frequencies $Z_E \gg Z_V$ so that the Eustachian-tube branch of the circuit can be ignored. At very low frequencies, however, Z_V increases greatly and Z_E decreases, and we must solve the complete network. If we write U' for the acoustic current through the Eustachian tube, then the circuit equations are

$$U(Z_T + Z_V) + U'Z_T = p$$
$$UZ_V + (Z_E + Z_C)U' = p' \tag{9.7}$$

and, if we make the appropriate low-frequency approximation $p' = p$, the solution for the acoustic current through the tympanum is

$$U = \frac{p}{Z_T' + Z_V + Z_T'Z_V/Z_E}. \tag{9.8}$$

This is a simple generalization of the result (9.5), with an added term $Z_T'Z_V/Z_E$ in the denominator that clearly vanishes if Z_E is very large. At very low frequencies the presence of this term reduces the response of the system, as shown in Fig. 9.5 for several different Eustachian tube diameters. This reduction of sensitivity may be an advantage to certain animals, since it reduces the interfering effects of unavoidable internal noises such as blood flow and internal resonances of the body during motion.

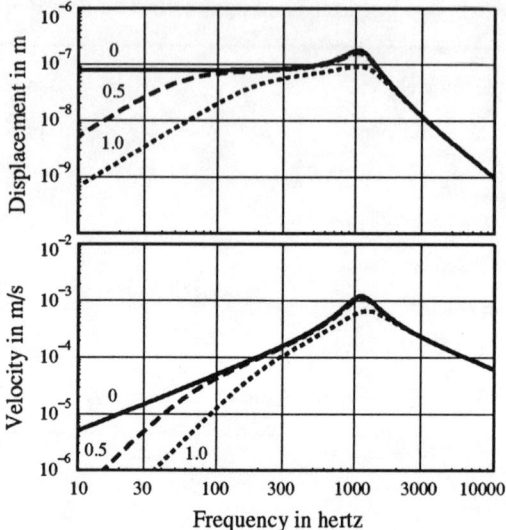

Figure 9.5 Frequency response of a simple cavity-backed ear with the parameter values shown in Table 9.1, for three different diameters of the Eustachian tube, given in millimeters as a parameter on the curves. The assumed acoustic pressure is 0.1 Pa r.m.s., equivalent to 74 dB. Note that the assumptions of the model are not valid above about 3000 Hz.

9.6 Neural Transducer Matching

So far we have considered the neural transducer as an essential but rather arbitrary component of the ear, but this is by no means true in reality. Evolutionary forces have ensured that the transducer is nearly optimally matched to the tympanum to which it is connected so that it can function in the most efficient manner possible. It is time for us to examine what this means from a physical point of view, and it is because this can be done only in relation to a complete ear that we have postponed it until now.

The object of the auditory system is to convert acoustic stimuli to neural signals. Detailed criteria for efficiency may vary from one case to another, but a common requirement is that as much mechanical energy as possible, within the chosen frequency band, should be transferred to the neural transduction apparatus. This transducer can vary over a large range in its size and mechanical properties, but it is the dissipative part of its impedance that is responsible for the transduction process; the reactive part is reversible and takes no energy from the stimulus. Exactly how the transduction process works need not concern us, though we note that it actually involves the distortion of cell membranes, which opens ion channels and initiates electrical activity. From a physical point of view, the question is how we can best match the mechanical impedance of the transducer to that of the rest of the ear in order to extract the greatest amount of mechanical power.

Referring to the simple circuit of Fig. 9.4, we see that the power dissipated in the transducer is $P = \frac{1}{2} R_N U^2$, where R_N is the resistive part of the impedance Z_N of the transducer capsule and U is the magnitude of the acoustic current U in the network, given by (9.5). We have specified P in terms of acoustic quantities for convenience, since we arrive at an exactly equivalent expression if we convert everything to mechanical quantities. If we write each impedance in the form $R + jX$ with a suitable subscript, then the power transferred to the neural transducer becomes

$$P = \tfrac{1}{2} R_N |U|^2 = \frac{R_N p^2/2}{(R_R + R_T + R_N)^2 + (X_R + X_T + X_V + X_N)^2} \tag{9.9}$$

and we wish to choose values for R_N and X_N so as to make this quantity a maximum. The choice of X_N is simple—we just choose it so that

$$X_N = -(X_R + X_T + X_V) \tag{9.10}$$

which brings us to the resonance frequency of the complete system with the transducer load. We can then differentiate (9.9) with respect to R_N and set the result equal to zero to find the appropriate value for R_N. The result is

$$R_N = R_R + R_T. \tag{9.11}$$

We must therefore arrange for the acoustic resistance of the transducer, as seen at the tympanum, to be equal to the total resistance in the rest of the main acoustic circuit. This result is so general that it is called a theorem—the Maximum Power Transfer Theorem.

The first part of the result, as expressed in (9.10), is fairly obvious; we expect power transfer to be highest at the resonance frequency of the system. The second part is much less obvious. The transducer must load the tympanum to an amount sufficient to reduce its Q value by a factor 2 from what it would be when simply loaded by the purely mechanical parts of the system. This is a very important result, since evolutionary processes may reasonably be assumed to have led to approximate optimization of the system subject to certain rather general anatomical constraints. From the point of view of analysis, it allows us to estimate some of the system parameters about which we would otherwise have little information. The unloaded Q of the tympanum is not likely to be high, since biological material generally has rather high internal viscous losses, and an upper limit of about 10 seems reasonable. We expect the loading of the transduction apparatus to reduce this value to a maximum of about 5. In an actual auditory system, evolution may have favored a trade-off between sensitivity and bandwidth in the direction of a lower initial Q value, so that a final Q close to unity may be expected in higher animals.

We should perhaps add a minor qualification to these statements. They apply in detail to mechanically passive systems only, and there is increasing evidence that mammalian auditory systems have neural transducers that are, or may become, mechanically active. What is thought to happen is that mechanical stimulation of a neural transducer cell causes electrical activity which itself produces a change in the dimensions of the cell, involving the consumption of metabolic energy. This mechanical disturbance may then propagate back through the mechanical linkage of the transducer to affect the tympanum. Such an active system may still be linear, simply reducing the tympanum damping or even going into oscillation under unfavorable conditions, leading to oto-acoustic emission, but is more likely to be nonlinear, as is easily possible at an electro-biochemical level. Detailed analysis clearly involves discussion of the magnitude and phase of any such cellularly generated mechanical effects, and would take us outside the scope of this book.

This is as far as we need to take consideration of the operation of the neural transducer organ at the present stage, since we are dealing for the moment only with the peripheral auditory system, which is conventionally taken to terminate at the tympanum. In Chapter 12 we shall examine in a little more detail the mechanics of the transducer organ, derive an expression for its actual input impedance in typical cases, and discuss the role of the middle ear as an impedance-matching device.

9.7 Cavity-Coupled Ears

In many animals, in contradistinction to mammals, the two ears are quite closely coupled acoustically by means of an auditory canal with diameter comparable to that of the tympana themselves. We reserve discussion of those cases in which the canal is essentially a pipe until Chapter 11, though our development here can be applied to their behavior at low frequencies. Our present concern will be rather with the case of frogs and other anurans, in which the ears connect almost directly with the mouth cavity, and the case of certain insects such as cicadas, in which two tympana are set into opposite sides of a body cavity. The model for this system is shown in Fig. 9.6, along with the electrical analog network. For notational convenience we have combined the impedance of the radiation load and of the neural transducer with that of the tympanum and written $Z_T' = Z_T + Z_R + Z_N$. The two loop equations are

$$Z_T'U_1 + Z_V(U_1 + U_2) = p_1$$

$$Z_T'U_2 + Z_V(U_2 + U_1) = p_2 \qquad (9.12)$$

where p_1 and p_2 are given by (9.1). The solution for U_1 is

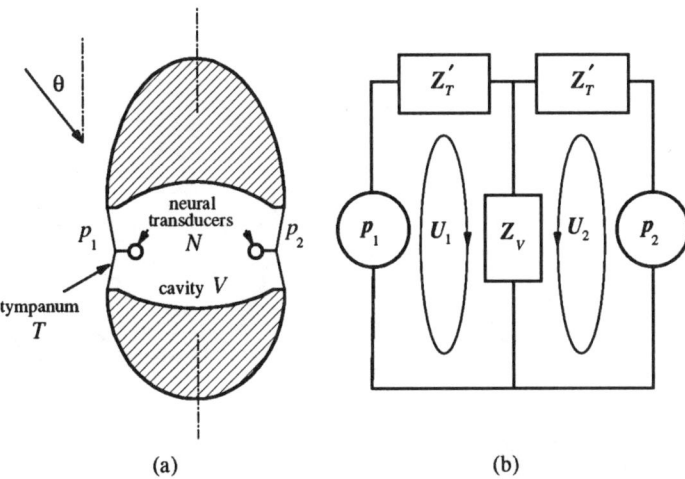

Figure 9.6 An auditory system consisting of two simple ears coupled acoustically through a closed cavity. The acoustic impedance Z_T' includes contributions from the neural transducer and the radiation load, as well as from the tympanum itself.

$$U_1 = \frac{p_1(Z_T' + Z_V) - p_2 Z_V}{Z_T'(Z_T' + 2Z_V)} \tag{9.13}$$

and U_2 can be written down simply by interchanging p_1 and p_2, since the circuit is symmetric.

The response (9.13) has two maxima, one at the frequency for which Z_T' is a minimum and one at the frequency for which $Z_T' + 2Z_V$ is a minimum. These correspond respectively to a mode in which the tympana move antisymmetrically (one going in and one going out, giving no compression of the air in the cavity) and a mode of higher frequency in which they move symmetrically. The first mode is driven by the pressure gradient across the head of the animal and the second by the average pressure. In practice, because of the low Q value of the loaded tympana, the two modes may overlap substantially.

It is useful to look at the directional behavior of the system, as expressed by the numerator of (9.13), which can be written, for $k\Delta \ll 1$, as

$$\{ [R_T' - (X_V + \tfrac{1}{2} X_T' k\Delta \sin\theta)] + j[X_T' + \tfrac{1}{2} R_T' k\Delta \sin\theta] \} p . \tag{9.14}$$

If we choose the parameters of the system appropriately, then both brackets can be made to vanish at some particular frequency for $\theta = -90°$, corresponding to a null response for the contralateral ear. The ipsilateral ear, with $\theta = 90°$ in this expression, has a large response under these same conditions. Such directionality would clearly be of evolutionary advantage to the animal concerned, and we might

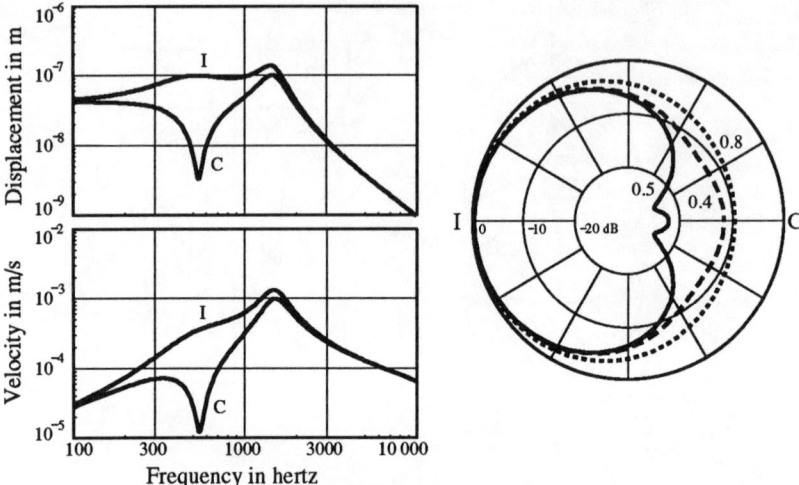

Figure 9.7 Frequency response, given as the displacement and velocity at the center of the tympanum for an acoustic pressure of 0.1 Pa r.m.s. (74 dB), for the cavity-coupled auditory system of Fig. 9.6, with the parameter values given in Table 9.2. The two curves in each diagram refer to ipsilateral (I) and contralateral (C) incidence of the sound respectively. The polar diagram shows the angular response of the left ear at frequencies of 0.4, 0.5, and 0.8 kHz, in decibels relative to the maximum response at each frequency. Note that the assumptions of the model are not valid above about 3000 Hz.

expect that the system should have developed in this direction, provided that the necessary magnitudes of the parameters can readily be achieved physically. We should not expect optimization to be perfect in a real animal, of course, since auditory behavior needs to be robust as the animal grows and ages. We should also look for robustness in general behavior as the physical parameters are varied over small ranges.

It is quite easy to write a computer program to calculate the magnitude and phase of the acoustic flow U_1 in ear 1, given the values of the other quantities involved in (9.13). The subroutines written need to be embedded in a program giving a graphical display of the frequency response for ipsilateral and contralateral ears. We can then vary the parameters one by one in an interactive manner to optimize the behavior of the acoustic system according to whatever criteria we may choose. In this case we have chosen to optimize so as to give maximum directional discrimination near the frequency of peak sensitivity, since this appears to be the likely evolutionary direction.

Figure 9.7 shows the calculated behavior of an ear optimized in this way, the relevant physical parameters having the anatomically reasonable values given in Table 9.2, which might apply to a large cicada or lizard or to a small frog. The

LOW-FREQUENCY AUDITORY MODELS

Table 9.2 Assumed physical parameters for a cavity-coupled auditory system

Separation of ears (sphere)	$D = 20$ mm
Tympanum area	$S_T = 20$ mm^2
Tympanum thickness	$d = 10$ μm
Effective transducer mass	$m_N = 1$ mg
Loaded Tympanum resonance	$\omega_T/2\pi = 500$ Hz
Q of loaded tympanum	$Q = 1$
Cavity volume	$V = 1$ cm^3

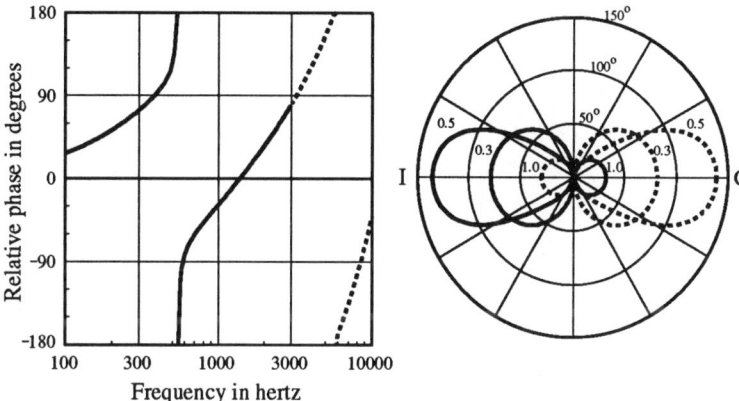

Figure 9.8 Phase difference between motion of left and right tympana for the simple cavity-coupled auditory system of Fig. 9.6 and Table 9.2. The left part of the figure shows the phase difference as a function of frequency for lateral presentation, $\theta = 90°$, the assumptions of the model being no longer valid above about 3000 Hz. The polar plot gives phase difference as a function of angle of incidence for 300 Hz, 500 Hz, and 1 kHz sound. The full line corresponds to a positive and the broken line to a negative relative phase, and the pattern at each frequency is antisymmetric about the symmetry plane of the animal.

response has been plotted out to a frequency of 10 kHz for comparison with later more detailed calculations, though our assumption that the system is small compared with the sound wavelength ceases to hold above about 3 kHz. Clearly the model auditory system has a quite realistic frequency response, particularly if the neural transducer is assumed to respond to velocity rather than displacement. The system also has a good measure of directionality over its most sensitive frequency range and, for the particular parameter values chosen, a very deep contralateral null at a frequency a little below that of maximal sensitivity. This null actually occurs near the frequency of the lower resonance, for which the two tympana move in antiphase, while the peak sensitivity occurs at the frequency for which they move in phase. The depth of the contralateral minimum is a sensitive function of the parameters, but it still remains usefully deep—at least -10 dB relative to the ipsilateral response—for variations of $\pm 30\%$ in most of the parameters, and over a 2:1 frequency range. The whole acoustic behavior of the auditory system is thus acceptably robust.

Finally, if the animal is able to use more sophisticated neural processing to extract relative phase information from its two ears, the separate signals do in fact differ in this respect too. The phase difference between the motion of the tympana in the two ears can also be calculated from (9.11). The calculated result, again for the parameter values given in Table 9.2, is shown in Fig. 9.8.

9.8 A More Elaborate Model

The cavity-coupled auditory system described in the previous section is a good physical model for the actual auditory system of an insect such as a cicada, but we may wish to elaborate it somewhat when it is considered as a representation of the auditory system of a frog. The additional features to be added are the short wide Eustachian tubes connecting the ears to the mouth cavity, and the nostrils or nares

Figure 9.9 A more complex auditory system model, corresponding to an animal such as a frog. The ears communicate with the mouth cavity through short broad Eustachian tubes, and the mouth is also vented by short narrow nostrils or nares. In the analog network, which has three loops, the tympanum impedances Z_T' include the loads due to radiation, to the neural transducers, and to the Eustachian tubes.

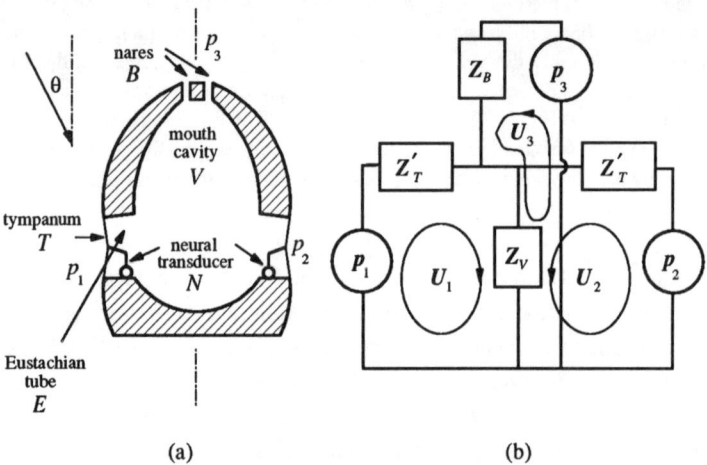

LOW-FREQUENCY AUDITORY MODELS

which provide an additional vent to the mouth. We could of course go further and allow the possibility of vibration of the skin of the throat, or even acoustic coupling through the trachea to the lungs, but this can be deferred to a still more complex model. As discussed in Chapter 1, it is best to add complications one at a time so that their importance can be assessed.

Figure 9.9 shows a simplified model incorporating these extra features, together with the analog electrical network. The Eustachian tubes appear simply as small inductive impedances Z_E in series with the impedance of the respective tympana. The two nostrils can be treated as a single impedance Z_B, essentially resistive since they are narrow, and close enough together that their acoustic admittances can be simply added. The extra circuit they contribute to the network simply connects across the impedance Z_V representing the mouth cavity, since they inject acoustic flow into that cavity. This extra circuit contains an acoustic pressure generator p_3 of appropriate phase to represent the acoustic field immediately outside the nares. Other possible circuits involving acoustic flow into the mouth through flexible throat walls or through the lungs would be treated similarly.

The three circuit equations from which the behavior of this network is to be derived can be written down very simply, following our general rules. They are

$$Z_T'U_1 + Z_V(U_1 + U_2 + U_3) = p_1$$
$$Z_T'U_2 + Z_V(U_1 + U_2 + U_3) = p_2 \qquad (9.15)$$
$$Z_B U_3 + Z_V(U_1 + U_2 + U_3) = p_3 .$$

where here $Z_T' = Z_T + Z_R + Z_N + Z_E$. The number of these equations is small enough that one can once again derive an explicit solution for U_1 algebraically, the result being

$$U_1 = \frac{p_1(Z_T' + Z_V) - p_2 Z_V + (p_1 + p_3) Z_T' Z_V / Z_B}{Z_T'(Z_T' + 2Z_V + Z_T' Z_V / Z_B)} . \qquad (9.16)$$

The result (9.16) is clearly a generalization of (9.13), and reduces to it when the nares are closed and $Z_B \to \infty$. It would equally have been possible to solve the three complex equations (9.15) for particular cases by the general numerical method given in the book *Numerical Recipes* listed in the Bibliography.

As an investigation of the effect of adding the nares and Eustachian tubes to our simpler cavity-coupled model of the previous section, we have kept the anatomical parameters as nearly unaltered as possible at the values given in Table 9.2. Clearly additional anatomical parameters are required, as listed in Table 9.3, and it is found that the system performs better, from the point of view of directional discrimination, if the Q value of the tympana is reduced from 1.0 to 0.5. As shown in Fig. 9.10, the overall acoustic behavior is not greatly changed by the anatomical modifications, though several features are worthy of comment. The peak response

Table 9.3 Assumed physical parameters for a frog-like auditory system (additional to Table 9.1)

Q of loaded tympanum	$Q = 0.5$
Diameter of nares (each)	$2a_B = 0.5$ mm
Length of nares	$l_B = 3$ mm
Forward distance of nares	$d_B = 15$ mm
Area of Eustachian tubes	$S_E = 20$ mm^2
Length of Eustachian tubes	$l_E = 3$ mm

has been reduced by about a factor 3, or 10 dB, and declines further at low frequencies because of back venting through the nares, as we discussed in Section 9.5 for a simple ear. (Note that the axis scale for the velocity response differs by a factor 10 from that in Fig. 9.7.) As a compensation for slightly reduced sensitivity, the auditory system now shows directional discrimination exceeding 20 dB at all frequencies below about 2 kHz, as well as a deeper null near 500 Hz. The axis of maximum sensitivity for each ear is shifted backwards through an angle of about 45° for the particular parameters used. The extent of this shift is proportional to the amount by which the entry to the nares lies forward of the line joining the two ears.

It is interesting to note that the presence of the nares, which are necessary for breathing in higher animals, actually improves the overall performance of the auditory system, if high weighting is given to directional discrimination rather than simple sensitivity. They do, however, significantly reduce response at very low frequencies.

In the particular examples calculated in this and previous sections, many of the anatomical parameters were chosen in a rather arbitrary manner, representing some particular fictitious animal. Because of the way in which almost all the acoustic quantities scale, it is possible to translate the calculated response curves simply to larger or smaller animals of the same type. For example, if all the dimensions are doubled, keeping things such as elastic stress in the membrane constant, then the whole frequency response pattern simply shifts downwards by a factor 2. The one exception to this statement relates to the acoustic resistance in narrow tubes such as the nares, which scales in a rather more complicated way because of the factor $\omega^{1/2}$ in the governing equation (8.12). This can be corrected by minor rescaling of the tube diameter. In making this comment on scaling, however, it is important to note that it is a convenience in analysis but not an acoustic necessity. It is possible to change the frequency behavior of a system

LOW-FREQUENCY AUDITORY MODELS

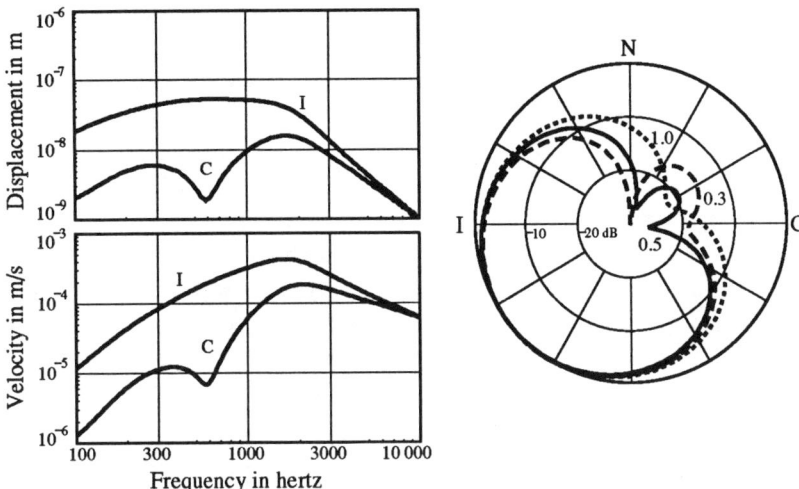

Figure 9.10 Frequency response, given as the displacement and velocity at the center of the tympanum for an acoustic pressure of 0.1 Pa r.m.s. (74 dB), for the frog-like auditory system of Fig. 9.9 and the parameter values given in Table 9.3. The two curves in each diagram refer to ipsilateral (I) and contralateral (C) incidence of the sound respectively. The polar diagram shows the angular response of the left ear at frequencies of 0.3, 0.5, and 1.0 kHz, in decibels relative to the maximum response at each frequency. The nares are in the forward direction N. Note that the assumptions of the model are not valid above about 3000 Hz.

without changing its overall size, or conversely to maintain a general frequency response despite dimensional changes, by appropriate adjustments of more subtle parameters such as tympanum thickness and tension. The precise behavior of the system cannot be maintained against overall size changes, but most of the important features can be preserved.

9.9 Laboratory Studies

It would be valuable confirmation of the theoretical treatment to know the values of all the physical parameters, such as tympanum thickness, tension, and damping, and neural transducer impedance, for a particular case, and to insert them in a model to see how well it is able to reproduce the observed performance of the auditory system. There are obvious difficulties in obtaining reliable anatomical and physiological data from dead animals, but most of the quantities can be measured or estimated with reasonable confidence. There are then two possible approaches.

In the physical domain, if the tympana are visible from outside the animal, and if we have access to a laser vibrometer, then tympanal motion can be measured to high precision. The animal can be placed in a free sound field, and the measurements compared directly with the predictions of the model.

In the biological domain there are two approaches to the measurement, one behavioral and one neurophysiological. The behavioral method essentially determines perception threshold by observing the reactions of a conscious animal in a natural situation. We might hope that the threshold determined in this way correlates well with calculated response curves, such as those of Fig. 9.7, when they are inverted. In the neurophysiological approach, it is usual to record the response of a single nerve fiber in the pathway between the neural transducer and the brain, using an anesthetized animal. Interpretation is then complicated by the fact that the transducer cell to which the fiber is connected may itself have a characteristic frequency response, different for each cell. Indeed in such experiments it is probably appropriate to accept the predictions of the model as mechanical input to the transducer, and to treat any modification of this observed in the measurements as characteristic of the transducer cell.

The model analysis can also be used in an inverse way to estimate the internal parameters of an auditory system from non-invasive external acoustic measurements. Direct measurements of membrane thicknesses, cavity volumes, and other physical dimensions are also possible and give incontrovertible data, but generally only upon non-living specimens.

The first thing we must do in this sort of experimental analysis is construct and analyze a conceptual model, as we have done in the preceding sections of this chapter. From this model we know the number of acoustic impedance elements believed necessary to describe the system, and hence, considering the real and imaginary parts of each impedance, the total number of unknown parameters. This will not be just twice the number of impedance elements, since some, such as cavities, require only one parameter while others, such as tympana, require three. We can then devise a measurement strategy that is adequate to allow us to determine these parameters.

The measurement of acoustic pressure in the immediate vicinity of the ports to the system, or in accessible cavities such as the mouth, can be achieved readily by means of a small calibrated microphone, but we also need some means to determine acoustic flow. The most straightforward way is by use of a laser vibrometer, which gives a spot measurement of mechanical velocity at a point on a vibrating object such as a tympanum. We can either make a set of measurements across the tympanum and integrate to find the acoustic flow, or simply make a single measurement near the center and divide by a factor of about 2.3 to convert to mean velocity. A third method is to deliver a known acoustic flow to the ear, for example by generating and measuring a large pressure amplitude in an enclosure and leading this to the ear through a short tube of known high acoustic resistance, either an annular capillary or a tube packed with wool. Essentially what is measured in each case is the acoustic input impedance at the ear. Such measurements are used in simpler form in otological diagnosis, and have also been extensively used in the study of musical instruments. It is important for our present analysis that both the magnitudes and the phases of the pressure and volume flow be measured.

The exact measurement strategy will depend upon the acoustic system being studied, but will generally involve several measurement situations devised to be as different as possible, and a range of frequencies for each situation. Solution is straightforward only if we can determine a sufficient number of acoustic flows and pressures that the network equations describing the system have only the impedances Z_i as unknowns. The system of equations is then linear and easily solved, either algebraically or by use of a computer algorithm. We illustrate this below for a simple system. If it is impossible to make an adequate number of measurements, then the system of equations is no longer linear, since it retains some of the currents U_j as unknowns in combination with unknown impedances as $Z_i U_j$. In such cases it is often best simply to model the measurement situations and adjust the system parameters heuristically, using physically reasonable values, until an acceptable fit is obtained. The set of physical parameters obtained in this way will generally be appropriate, but in some cases more than one set may appear equally possible.

To return to the simpler situation, suppose we are investigating a cavity-coupled auditory system of the type illustrated in Fig. 9.6, along with its analog network, and described by the equations (9.12). It is our aim to determine the acoustic impedances Z_T', describing the tympanum with associated loadings, and Z_V describing the coupling cavity. It is not possible to separate the radiation, Eustachian tube, and transducer components of Z_T' from the component intrinsic to the tympanum, since they simply add in series. We already know from simple measurement the distance between the ears, and also the area of the tympana.

For the first measurement we block the tympanum of ear 2, for example by flooding it with water, thus setting $U_2 = 0$ in (9.12) and giving

$$p_1 = (Z_T' + Z_V)U_1 \quad \text{or}$$
$$Z_T' + Z_V = U_1/p_1 = M_1(\omega) \tag{9.17}$$

where $M_1(\omega)$ can be defined to be the result of this first measurement of U_1 and p_1 at frequency ω. This requires either a laser vibrometer or a calibrated acoustic flow source and a probe microphone, as discussed before.

For the second measurement we might supply sound just to ear 1 and measure the motion of the tympana of both ears, with no tympanal blocking. This requires a laser vibrometer, since one free-field situation is involved. In the network equations (9.12) we simply set $p_2 = 0$, and the response equation for the right-hand loop of the network becomes

$$Z_V U_1 + (Z_T' + Z_V)U_2 = 0 \quad \text{or}$$
$$(Z_V + Z_T')/Z_V = -U_1/U_2 = M_2(\omega) \tag{9.18}$$

where $M_2(\omega)$ defines the result of this second measurement. From (9.17) and (9.18), the required impedances are given by

$$Z_V(\omega) = \frac{M_1(\omega)}{M_2(\omega)}; \qquad Z_T'(\omega) = M_1(\omega) - \frac{M_1(\omega)}{M_2(\omega)}. \qquad (9.19)$$

Z_V is given in explicit form in Table 8.1, and from this we can deduce the cavity volume. Similarly we know the functional form of $Z_T'(\omega)$ from Table 8.1, and its components can be defined by measurements at two frequencies respectively above and below the resonance or, more accurately, by measurements over a range of frequencies covering the resonance. The explicit expressions in Table 8.1, together with knowledge of the area of the tympanum, then yield its total effective mass, including that of the neural transducer, Eustachian tube, and radiation load, together with its resonance frequency, and its Q value.

If the measurement must be made by supplying a known acoustic flow U_1 to ear 1 and measuring pressure, rather than with the help of a laser vibrometer, then it may be necessary to make measurements on one ear only, for we have no simple way of measuring U_2.. Using the arrangement specified above for the second measurement, we can write the additional equation for the left-hand loop as

$$p_1 = (Z_T' + Z_V)U_1 + Z_V U_2 \qquad (9.20)$$

and we can eliminate U_2 between this equation and (9.18) to give

$$\frac{(Z_T' + Z_V)^2 - Z_V^2}{Z_T' + Z_V} = \frac{p_1}{U_1} = M_3(\omega) \qquad (9.21)$$

where $M_3(\omega)$ is the result of the measurement operation. Combining this measurement with $M_1(\omega)$, defined by (9.17), we find

$$Z_V(\omega) = (M_1^2 - M_1 M_3)^{1/2} \qquad Z_T' = M_1 - (M_1^2 - M_1 M_3)^{1/2}. \qquad (9.22)$$

These impedances can be associated with physical quantities as discussed above.

More complex systems necessarily require more extensive measurements to define their parameters, but this analysis illustrates the procedure. It is also possible to use the basic model analysis to predict quantities such as the acoustic pressure in the ear cavity, given by $Z_V(U_1 + U_2)$, for comparison with more invasive measurements if these are contemplated.

9.10 An Aquatic Auditory System

Relatively little is known about the auditory systems of fishes, compared with the extensive information now available on terrestrial animals. The auditory detection problem is very different in the two cases. In terrestrial animals the medium is air, with an acoustic wave impedance ρc of about 400 rayls, while the body of the animal has a very large wave impedance and essentially reflects all incident sound energy. Detection is accomplished in most cases through the development of very light membranes, backed by air-filled cavities, as discussed in the preceding sections. In aquatic animals the situation is quite different, since the wave impedance of the body of the animal is very similar to that of the surrounding medium, and sound waves penetrate the animal and displace it bodily in the same way as the water. Otolith detectors, as discussed in Chapter 4, are able to respond to acceleration signals because the density of the otolith is about three times that of the surrounding tissue. The other common acoustic detection device is associated with the swim bladder, in those fishes that have one, and again relies upon the difference in density, and so in acoustic wave impedance, between the air in the bladder and the surrounding water and tissue.

Because the sound speed in water is so high, about 1480 m s^{-1}, the wavelengths are correspondingly long, and most potential auditory system components are small in size compared with the wavelength. This simplifies the analysis, as we found above for terrestrial systems. Our immediate need is to know, under these conditions, how a flexible air-filled cavity responds to the incidence of an acoustic wave. We can idealize this by considering the scattering of sound by a spherical bubble in water. We expect two distinct effects. In the first place the acoustic pressure exerted on the sphere oscillates with time, so that we expect a corresponding oscillation in its radius. Secondly, the phase of this pressure changes over the surface of the sphere, depending upon the direction of incidence of the sound wave, so that there should be a bodily displacement of the sphere relative to the surrounding liquid. We need to know the magnitudes of these two effects.

Suppose the sphere has radius a, then it acts as an enclosed volume and can be modeled by a capacitor of magnitude $C = V/\rho c^2 = {}^4/_3 \pi a^3/\gamma p_W$ where γ is the ratio of the specific heats of air and p_W is the pressure in the sphere, equal to that of the surrounding water. As far as oscillation is concerned, the mass of the air in the sphere is negligible, and what matters is the mass of water moving with it. This is given by the inertive part of the radiation load, and by (7.8) and (7.9) has the magnitude $j\omega L = j(\rho_W c_W/S) k_W a = j\omega \rho_W/4\pi a$ where the subscript W refers to water. The resonance frequency of the sphere for radial oscillations is therefore

$$\omega_0 = \frac{1}{\sqrt{LC}} = \frac{1}{a}\left(\frac{3\gamma p_W}{\rho_W}\right)^{1/2}. \tag{9.23}$$

For a sphere diameter of 10 mm in shallow water this leads to a resonance frequency of about 600 Hz, which is in a useful region of the audio spectrum.

This resonance is damped by both radiation and viscous losses. From (7.8) and (7.9) applied to water as a medium, the radiation resistance is $R_R = (\rho_W c_W/S)(k_W a)^2$, and it turns out that the internal damping is adequately represented by a term of the form $R_V \approx 1.6 \times 10^{-4} \rho_W \omega^{3/2}/4\pi a$ for $\omega \approx \omega_0$. The Q value of the resonance is therefore approximately

$$Q \approx \frac{L\omega_0}{(R_R + R_V)} \approx (k_W a + 1.6 \times 10^{-4} \omega_0^{1/2})^{-1}. \tag{9.24}$$

For the 10 mm spherical bubble considered above, each term in the denominator of this expression is about 0.01 at the resonance, so that $Q \approx 50$. We thus expect considerable amplification of the motion near the resonance, the amplitude of the radial vibrations being calculated from a simple resonant circuit to give the acoustic current $U = 4\pi a^2 v$. An acoustic sensor based upon this radial vibration is, of course, non-directional.

Calculation of the displacement behavior is more difficult, but the result is simple. Provided $k_W a \ll 1$, the air-filled sphere moves back and forth in the direction of the sound wave with an amplitude three times that of the wave itself. This means that its motion relative to the surrounding fluid is twice the amplitude of the wave. The direction of motion is the same as the propagation direction of the wave so that, if the displacement could be distinguished against the radial oscillations, the motion could serve as a direction indicator.

When we apply these ideas to make quantitative estimates for a fish or other aquatic animal, we note that the swim bladder is not spherical, and that the surrounding tissue has appreciable shear strength and viscous loss. The shape of the bladder is not immensely important for estimating magnitudes, but the properties of the surrounding tissue may be. Elastic attachments to the bony structure of the fish may serve as neural transducers when the volume of the bladder oscillates, or when it undergoes the much smaller translation motion, while losses in the biological tissue will greatly increase the damping of the bladder over the value estimated above. Even with a much lower Q value however, or substantially away from the resonance frequency, the radial oscillations of the bladder are several orders of magnitude larger than its translational motion, so that its response is probably essentially non-directional.

Systems such as this, based upon the swim bladder, are not the only acoustic organs of fish. We have already mentioned otoliths, and brief mention should also be made of the lateral-line organ, which consists of a line of sensory hair detectors spaced along each side of the fish. These sensors presumably respond to flow of water in their immediate vicinity, but are not likely to be efficient acoustic detectors since there is little relative motion between the fish and its environment in an acoustic field. The fact that this organ is extended along the body of the fish suggests that its directivity properties are important to the animal, since the angular resolution is about $\delta\theta = \lambda/l$ where λ is the wavelength involved, l is the total length of the organ, and $\delta\theta$ is given in radians.

9.11 Conclusions

This brief survey of simple auditory systems at low frequencies shows that it is possible to calculate optimization strategies that will produce ears or pairs of ears with a reasonably wide hearing range localized about a preferred frequency and, in the case of coupled ears, with quite good intensity and phase discrimination for ipsilateral and contralateral sound presentation. Our modeling technique mimics, in fact, the path taken by evolutionary processes in refining auditory anatomy and physiology.

The simple approach set out in this chapter serves also as a tool for determining the properties of real auditory systems in the living state by means of non-invasive measurements. This ensures that living rather than dead systems are studied, and also defines the set of measurements required to characterize the system, a point often missed in the absence of a guiding theory.

References

System modeling: [21]
Analogous electro-acoustic systems: [6] Ch 6; [8] Ch 11; [9] Ch 8
Bubbles: [8] pp. 228–231

Discussion Examples

1. Write a computer program to evaluate $|U_1|$ from (9.13) for arbitrary incidence direction. (Hint: find the absolute values of numerator and denominator and divide.) Explore the behavior of the model as physical quantities are varied, starting from those in Table 9.2.
2. Do the same for a cavity-backed ear with Eustachian tube, starting from (9.8) and the parameter values in Table 9.1.
3. Do the same for the frog model of (9.16), starting from the parameter values in Table 9.3.
4. A spherical bubble of diameter 10 mm in shallow water has $Q = 50$ for radial oscillations. Calculate the amplitude of these radial oscillations in an acoustic field of level 20 dB re 0.1 Pa at the resonance frequency of the bubble.
5. Compare the result of example 4 above with the translational displacement amplitude of the bubble in the same field.

Solutions

4. $p = 1$ Pa. At resonance, $Y \approx 50/L\omega_0 \approx 10^{-6}$. Thus $U \approx 10^{-6}$ m^{3}s^{-1} and $v \approx 0.003$ m s^{-1} at 600 Hz. So $a \approx 1$ μm.
5. Water displacement $= p/z_W\omega \approx 2 \times 10^{-10}$ m. So bubble displacement amplitude $\approx 6 \times 10^{-10}$ m.

10 PIPES AND HORNS

SYNOPSIS. Pipes and horns are important components of many auditory and vocal systems, and their dimensions are often comparable with or larger than the wavelength of the sounds with which they interact. It is wrong to simply apply naive ideas of "open" or "stopped" pipes when analyzing such systems, for only occasionally are these approximations valid. More generally we must consider the waves flowing in both directions in the pipe or horn and use a set of four impedance coefficients Z_{ij}, as shown in Fig. 10.1, to represent the element by an electric analog network. The values of these impedance coefficients can be readily calculated in terms of the dimensions of the pipe or horn and the operating frequency. In narrow pipes, as shown in Fig. 10.2, the speed of sound may be somewhat reduced, and the attenuation α can become quite high.

To illustrate the behavior of pipes in acoustic systems, Fig. 10.4 shows the results of a calculation of the resonance frequencies of a pipe terminated by a cavity, as shown in Fig. 10.3, as the cavity volume is increased. The lowest resonance frequency moves from the stopped-pipe value to a much lower Helmholtz frequency, while each upper resonance shifts from the stopped-pipe to the next-lower open-pipe value. In biological systems, pipes are rarely either rigidly stopped or fully open.

Horns are symmetrical flaring structures as shown in Fig. 10.5. For the rather short horns found in biological systems, the exact horn profile has only a minor influence on acoustic behavior, and we must analyze the whole system to determine its response. It is instructive, however, to calculate the pressure gain in the throat of a horn, blocked by a very high impedance, for a plane wave falling normally on its mouth. The results, for three particular horns having the same length and the same throat and mouth diameters but flare profiles of parabolic, conical, or exponential shape, are given in Fig. 10.6. All horns have high pressure gain in a limited pass-band between upper and lower cutoff frequencies. The lower cutoff frequency is nearly the same for all horns and depends on the mouth and throat diameters and the length. The upper cutoff frequency varies more and is influenced by the horn profile. The width of the pass band increases as the length of the horn increases. The effects of terminating the horn with a matching load or with a resonant diaphragm are shown in Fig. 10.7. In both cases the gain is severely degraded, showing that the termination must be taken into account when considering total system behavior.

It is interesting to consider the behavior of an obliquely truncated horn, as shown in Fig. 10.8, since this bears some resemblance to an animal pinna. The direction of maximum acoustic sensitivity varies somewhat with frequency, as shown in Fig. 10.9, and the long oblique flap adds appreciably to the gain.

At very high frequencies, the behavior of a horn becomes complicated, and higher modes, such as illustrated in Fig. 10.10, can propagate, though they are quickly attenuated when the horn diameter narrows to less than the sound wavelength. A horn the shape of a shallow cylinder with an off-axis coupling to the auditory system, as illustrated in Fig. 10.11(a), constitutes a simple model for the human pinna. Obliquely incident sound can excite both the plane-wave mode and the first antisymmetric mode, at appropriate resonance frequencies for each, and the auditory system will be driven by the sum of these two pressures, while normally incident sound can excite only the plane-wave mode. The horn gain will thus have a different frequency dependence for different angles of incidence, as illustrated in Fig. 10.11(b).

In the region where the horn diameter is much greater than the sound wavelength, it

PIPES AND HORNS 179

is appropriate to use an optical analog for wave propagation, with reflection from the sides of the horn. Because such a reflector cannot focus the sound into a region smaller than a wavelength in diameter, it is advantageous to use the reflector to focus sound onto the mouth of a smaller horn, as shown in Fig. 10.12, the restriction being that the horn mouth must be at least one wavelength in diameter for optimum efficiency. This appears to be an appropriate model for the pinnae of animals such as bats.

10.1 Acoustic Elements

Two or the most important elements of acoustic systems in biology are pipes and horns, as exemplified by the trachea in the vocal system, and by the meatus and the external pinna in auditory systems. In nearly all cases in which they occur, the dimensions of these structures are comparable with the wavelength of sound, at least for the higher frequencies with which the system must deal. Their exact shapes and dimensions therefore become significant, and we need a detailed discussion to elucidate their behavior. This more detailed treatment must, of course, reduce to the simpler version presented in Chapter 8 at frequencies low enough that the wavelength greatly exceeds the component dimensions.

The analysis of these extended systems can be made arbitrarily complex, by including refinements of geometry and material properties, but we shall be content with a simple treatment because we are concerned with generalities of behavior, rather than with fine details. The principal extension necessary to the electric analog theory developed in Chapter 8 is the recognition that pipes and horns have two ends, and that the acoustic flow in at one end is not necessarily equal to the flow out at the other. We therefore need a more general circuit element to describe their behavior and to serve as an analog in our electrical networks. This will emerge naturally as we proceed.

10.2 Pipes and Tubes

For our purposes there is no real distinction between a pipe and a tube, so we shall generally use the term pipe to mean any long duct with constant cross section. It turns out that the behavior of such a pipe does not depend greatly upon the shape of its cross-section, and is not much affected if it is bent, provided that the bend is not too sharp on the scale of the pipe diameter. An initial treatment in terms of long straight pipes can therefore be readily generalized.

It would be possible to develop our discussion in terms of the normal modes of pipes, in much the same way as we did for strings, but there is a major disadvantage to this approach. It works well for strings because they are generally anchored to a structure of relatively high rigidity, so that the boundary conditions at the two ends of the string are well defined. Simple treatments of pipes do the same thing, and discuss "open" and "stopped" pipes and their normal modes. In biological systems, however, the boundary conditions are almost never as simple as this. An "open" end is often either partly obstructed by a protective flap, or terminates in a horn, while a "stopped" end is usually blocked by a tympanum or similar structure that is far from rigid, or terminates in a cavity. Many discussions of auditory and vocal systems are marred by simplistic application of these concepts. Instead, therefore, we develop our discussion in terms of waves in the pipe. This immediately gives greater generality, but allows recovery of the simple ideas when the boundary conditions are appropriate.

Suppose we have a pipe of cross-section S extending in the x direction from $x = 0$ to $x = l$. If we confine our attention to the behavior of the system at a fixed frequency ω, then the most general possible situation reduces to a wave with wave number $k = \omega/c$ and amplitude A (the complex quantity including a phase factor) traveling in the $+x$ direction, together with a similar wave of amplitude B traveling in the $-x$ direction. The acoustic pressure $p(x)$ at any point x is then simply the sum of the contributions from these two waves, so that

$$p(x) = Ae^{-jkx} + Be^{jkx} \tag{10.1}$$

where we have dropped the time factor $e^{j\omega t}$ for convenience. Note that the minus sign in the exponent refers to the wave traveling in the $+x$ direction, as discussed in Section 6.2. Now a wave in a pipe propagates in just the same way as a plane wave in free space, if we neglect the very small viscous friction at the walls, so that the acoustic particle velocity in the direction of propagation is $v = p/\rho c$ as we saw in (6.17). Because the cross-section area is S, this means that the acoustic volume flow associated with a wave in the pipe is $U = pS/\rho c$. Adding the contributions of the two waves in (10.1) and taking note of their opposite propagation directions then gives

$$U(x) = (S/\rho c)(Ae^{-jkx} - Be^{jkx}). \tag{10.2}$$

It is convenient to define the quantity

$$Z_0 = \rho c/S \tag{10.3}$$

known as the characteristic impedance of the pipe. From (10.1) and (10.2) it is seen to be the input impedance p/U for an infinitely long pipe with no returning wave.

To describe the acoustic volume flow in the pipe in a symmetrical manner, we write the flow into the pipe at $x = 0$ as U_1 and the flow into the pipe at the other end $x = l$ as U_2, with the pressures similarly labeled as in Fig. 10.1. Then we can write the relation between these quantities as

$$p_1 = Z_{11}U_1 + Z_{12}U_2$$
$$p_2 = Z_{21}U_1 + Z_{22}U_2 \tag{10.4}$$

where the complex quantities Z_{ij} are impedances of some sort. Note that the flows and pressures are arranged symmetrically in both acoustic and electric analogs. If,

PIPES AND HORNS

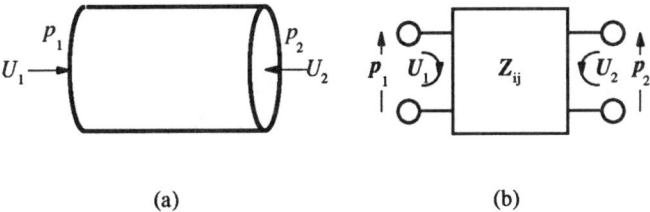

Figure 10.1 (a) Acoustic flows U_i and pressures p_i for a length of pipe; (b) the corresponding electric analog element.

for some reason, the direction of one flow or the sense of one pressure or voltage is opposite to that shown, then it should carry a minus sign with it. Of course, if we measure everything in the reverse sense, corresponding to turning the diagram upside-down, then all the minus signs cancel.

A little algebraic manipulation of (10.4) using (10.1) and (10.2) then leads to the results

$$Z_{11} = Z_{22} = -jZ_0 \cot kl$$
$$Z_{12} = Z_{21} = -jZ_0 \operatorname{cosec} kl \ . \tag{10.5}$$

The equalities $Z_{11} = Z_{22}$ and $Z_{21} = Z_{12}$ clearly follow from the symmetry of the problem, since the pipe looks exactly the same from both ends. The fact that $Z_{21} = Z_{12}$ is an expression of the reciprocity theorem, which we have mentioned several times before, and is actually true even for systems, such as horns, that are not symmetrical.

It is easy to see how we can deduce the normal mode frequencies for ideally open and stopped pipes from these results. For a pipe ideally open at both ends, $p_1 = p_2 = 0$. A little algebra then shows, from (10.4) and (10.5), that this requires that $\cos^2 kl = 1$, which means that $kl = n\pi$ or $\omega_n = n\pi c/l$, with $n = 1, 2, 3, \ldots$. For a pipe open at end 1 and stopped at end 2, $p_1 = 0$ and $U_2 = 0$ so that, from the first equation of (10.4) and (10.5), $\cot kl = 0$ or $kl = (n - \frac{1}{2})\pi$ and $\omega_n = (n - \frac{1}{2})\pi c/l$.

Of rather more importance is the input impedance $Z_{IN} = p_1/U_1$ for open and stopped pipes. A little algebra immediately gives the results

$$Z_{IN}^{open} = jZ_0 \tan kl \to \frac{j\omega l}{S} \tag{10.6}$$

$$Z_{IN}^{stopped} = -jZ_0 \cot kl \to \frac{\rho c^2}{j\omega l S} \ . \tag{10.7}$$

The final form of writing in each case is for the low-frequency limit in which $kl \ll 1$. Clearly the results are equivalent to those given in Table 8.1 for the lumped-component low-frequency approximation, the stopped pipe behaving like an enclosure of volume $V = lS$.

From the trigonometric forms of the impedance coefficients Z_{ij} given in (10.5), it is clear that their behavior with frequency is complicated. Their values go through a sequence of infinities and zeros as the frequency is increased, and these are related to potential resonances of the system. The actual resonances may not occur at these singular points, but may be displaced because of the impedance of other acoustic components connected to the two ends of the pipe. We look at an example of this in Section 10.4.

10.3 Wall Losses

We have already noted in Section 8.4 that we should make allowance for viscous and thermal losses to the walls of narrow pipes. These corrections become relatively less important at high frequencies, but they can have very significant influence on the behavior of an acoustic system when the pipe involved is very narrow, as may be the case in insect auditory systems. The exact results are a little complicated, but essentially what happens is that the wave number k should be modified from its simple value ω/c to become a complex quantity

$$k = \frac{\omega}{c'} - j\alpha. \tag{10.8}$$

The propagation speed c' of the wave in the pipe is reduced somewhat below its open-air value c, and the wave amplitude attenuates with distance as $e^{-\alpha x}$. The behavior of c' and α as functions of tube radius and frequency is shown in Fig. 10.2. We can usually neglect the change in wave speed, except in the very narrow pipes of insect systems, but the attenuation losses may be important in other systems. For pipes more than about 1 mm in diameter it is an adequate approximation to take

$$\alpha \approx 1.2 \times 10^{-5} \omega^{1/2} a^{-1}. \tag{10.9}$$

For very narrow pipes, the numerical coefficient in this equation is approximately doubled.

When it comes to taking account of these wall losses in the behavior of the pipe, the formal procedure is simply to substitute the complex value of k into the relations (10.5). As discussed in Appendix A, trigonometric functions with complex arguments can be simply expanded in terms of the hyperbolic functions sinh and cosh, which are readily evaluated. The formal generalizations of the expressions given in (10.5) are set out in Appendix B. In analyzing the behavior of a system, as we shall see in the next chapter, we do not need to do this until we write the final computer program, and then it is just a case of being careful with the algebra.

An extreme case of wall interaction can occur if the wall is actually porous,

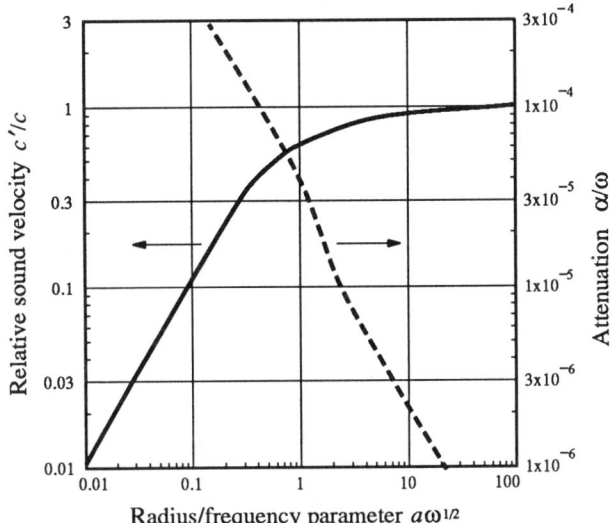

Figure 10.2 Sound speed c' and attenuation coefficient α in a pipe of radius a at angular frequency ω. The parameter $a\omega^{1/2}$ is 1.7 times the ratio of the pipe radius to the viscous boundary layer thickness.

as may be the case in some avian auditory systems. We shall not examine this in great detail, but a quick calculation shows the possible magnitude of the effect. If L is the acoustic inertance per unit length of pipe and C the acoustic compliance per unit length, then the sound propagation speed in the pipe is $c' \approx (LC)^{-1/2}$ and the characteristic impedance of the pipe is $Z_0' = (L/C)^{1/2}$. The inertance is contributed only by the mass of air in the free part of the pipe, of cross section S, while the compliance is contributed by the whole cross section, including the air permeating the porous walls. Suppose that the effective cross-section of pipe including the entire porous area is S'. Simple substitution of $L = \rho/S$ and $C = S'/\rho c^2$ then shows that the propagation speed and characteristic impedance in such a pipe are

$$c' \approx c(S/S')^{1/2} \qquad Z_0' \approx Z_0(S/S')^{1/2}. \tag{10.10}$$

For a relatively narrow auditory canal connecting two simple ears, it is not hard to contemplate a situation in which S' is several times S, so that the effective speed of sound within the auditory canal is perhaps only half that in the free air. We should also expect significant attenuation due to motion of the air into the pores of the wall.

10.4 Helmholtz Resonator

In Section 8.6 we discussed the simple Helmholtz resonator, which consists of an enclosed volume vented through a pipe, using low-frequency approximations. Let

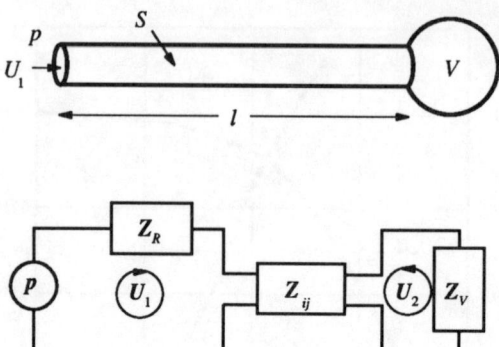

Figure 10.3 A Helmholtz resonator, together with its electric analog circuit. The directions of the circulating currents U_1 and U_2 are chosen, for convenience, so that they enter the two-port pipe analog symmetrically.

us now repeat the analysis with a better treatment of the pipe. There are two reasons for doing this. The first is to illustrate how we should use the impedance coefficients Z_{ij} when solving analog networks, and the second is to show the fallacy of treating pipes as "open" or "stopped" in real acoustic systems.

Figure 10.3 shows the Helmholtz resonator together with its analog network. We have assumed that the enclosed cavity of the resonator is small enough that it can be treated as a simple compliance, though a more complex treatment is, of course, possible. In drawing the currents in the network, we have to be careful that they enter the pipe element Z_{ij} in the same sense at each end, otherwise some of the signs will need to be changed. The network equations are simply

$$(Z_R + Z_{11})U_1 + Z_{12}U_2 = p \tag{10.11}$$

$$Z_{21}U_1 + (Z_V + Z_{22})U_2 = 0 \tag{10.12}$$

and these can be solved to give the flow U_1 through the neck as

$$U_1 = \frac{(Z_V + Z_{22})p}{(Z_R + Z_{11})(Z_V + Z_{22}) - Z_{12}Z_{21}}. \tag{10.13}$$

The resonance frequencies are given by the minima of the denominator.

For the simple problem of resonance frequencies we are examining here, we can neglect Z_R in comparison with Z_{11} and ignore the resistive parts of all the other impedance coefficients. The denominator is then entirely real, and it vanishes at the resonance frequencies. If we take V to be the volume of the cavity so that $Z_V = -j\rho c^2/V\omega$, and use the explicit expressions for the Z_{ij} given in (10.5), we find the resonance condition to be

$$\frac{\omega V}{Sc}\tan\frac{\omega l}{c}=1. \tag{10.14}$$

If the length of the tube is much less than the sound wavelength, so that $\omega l/c \ll 1$, we can take the tan function to be approximately equal to its argument, so that (10.14) reduces to the value found before for the Helmholtz frequency, $\omega_H = c(S/lV)^{1/2}$. More generally, however, (10.14) has many possible solutions, and it is informative to see how these resonance frequencies vary as we change the volume of the terminating cavity. This is illustrated in Fig. 10.4. When the cavity volume is very small, the resonance frequencies of the system are $\omega_n = (n - \frac{1}{2})\pi c/l$, which are the resonance frequencies of a stopped pipe of length l. As the cavity volume increases, all the resonance frequencies fall progressively. The lowest resonance, for $n = 1$, tends to the Helmholtz resonance frequency for the system, while the higher resonances with $n > 1$ tend to the resonances of an open pipe, though with the index n reduced by 1, so that $\omega_n = (n - 1)\pi l/c$.

The lesson to be learned from this analysis applies to many biological systems, whether the pipe terminates in a cavity, a tympanum, or some other kind of impedance. That lesson is that the nature and magnitude of this terminating impedance has a very significant effect on the resonance frequencies of the pipe,

Figure 10.4 Resonance frequencies for the first five modes of the Helmholtz resonator of Fig. 10.3 as a function of the cavity volume V. The pipe length l is 10 cm and the pipe diameter 10 mm. For very small volumes, the resonance frequencies are those of a stopped pipe, while for large volumes they are those of an open pipe, supplemented by a very low-frequency Helmholtz resonance. The assumptions underlying the model cease to be valid for very large cavities and high frequencies.

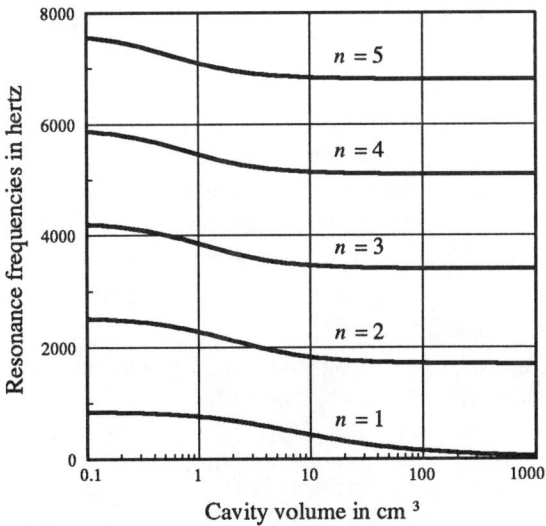

particularly those of the lower resonances. A very similar conclusion comes from a study of the resonance frequencies of a pipe closed with a light slack diaphragm, as set out in the discussion examples.

Before leaving the subject of this section we must insert a cautionary remark. It might seem that our discussion of the Helmholtz resonator is now exact, and that all higher modes have been properly included, but this is not true. We have progressed from the simple single-mode model of Section 8.6 to a much more sophisticated model, it is true, but our treatment of the behavior of the cavity is still oversimplified. It is valid to treat the cavity as a lumped acoustic compliance, as we have done here, only so long as its dimensions are small compared with the sound wavelength involved. This is appropriate for the first few modes while the cavity remains small compared with the pipe length, but is certainly not true for higher modes. Our next model in the series must therefore include the modes of the cavity itself. We do not need to do this for any of our biological problems, but it is a good example of the use of an appropriately complex model at each stage of the discussion.

10.5 Simple Horns

A striking feature of many mammalian auditory systems is the pair of large horn-like pinnae mounted on the sides of the head. It is qualitatively clear that they act in some way to collect sound energy and funnel it to the tympanum, but we require some analysis to evaluate just how this occurs and to explore the frequency response and directionality of the horn. The horns of typical pinnae are not truncated at right angles to their axis, but rather at an oblique angle. We shall return to consider this complication in the next section after we have understood the behavior of normally truncated horns.

It is traditional in acoustics texts to investigate the behavior of horns of infinite length and having one of a number of idealized geometric profiles—conical, exponential, "hypex" (Salmon), "Bessel", etc. The results are elegant, but are of little use in the present context because biological horns are necessarily finite in length and often quite short, and their geometry rarely conforms closely to any of these idealized types. We shall therefore consider only horns of finite length, and examine the behavior of just three types—a conical horn, a horn with a more rapid and quasi-exponential flare, and a horn with a slower and quasi-parabolic flare—as illustrated in Fig. 10.5. Between them they serve as models for all the horns encountered in biological systems.

The wave equation for propagation in a horn with cross section $S(x)$ is the so-called horn equation or Webster equation

$$\frac{\partial^2 p}{\partial t^2} = \frac{c^2}{S}\frac{\partial}{\partial x}\left(S\frac{\partial p}{\partial x}\right). \tag{10.15}$$

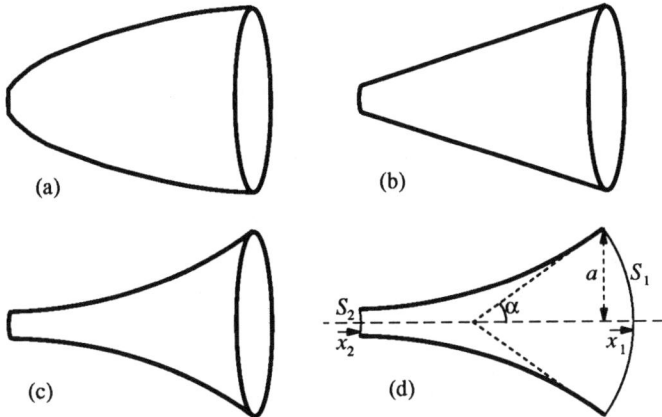

Figure 10.5 (a)–(c) Three horn shapes, all with the same mouth and throat dimensions and the same length. The horn profiles are (a) parabolic, (b) conical, and (c) exponential. Diagram (d) defines the coordinates and illustrates the curvature of the wavefront at the mouth of the horn, together with the associated tangent cone.

Clearly it reduces to the ordinary wave equation (6.8) when S is constant, as in a cylindrical pipe. It is not an exact equation, particularly at high frequencies and for large flare rates on the horn, but is a useful approximation. The explicit forms for the profiles of the horns we shall consider are

$$S(x) = S_0 x^2 \quad \text{(conical)} \tag{10.16}$$

$$S(x) = S_0 e^{2mx} \quad \text{(exponential)} \tag{10.17}$$

$$S(x) = S_0 x \quad \text{(parabolic)} \tag{10.18}$$

The parabolic and conical horns are both members of a more general family with $S(x) = S_0 x^m$. These are called Bessel horns because the propagating waves within them can be written quite generally in terms of Bessel functions.

Evaluation of the coefficients Z_{ij} for a finite horn follows just the same lines as the method used in the previous section for a cylindrical pipe. The one difference is that, instead of simple exponential behavior e^{-jkx} describing the propagating waves, we must use rather more complicated functions. For the conical horn, for example, the propagating wave has simple spherical form, so that its pressure and velocity components have the spherical-wave forms given in (6.33) and (6.35). For parabolic and exponential horns the behavior is a little more complicated. We need not go into details of the algebra involved, and it will suffice to quote the final results, which are collected in Appendix B for convenience. In each case we take the radius of the large end, or mouth, to be a, so that the mouth area is $S_1 = \pi a^2$, and the radius of the small end, or throat, to be b, so that the throat area is $S_2 = \pi b^2$. The length of the horn between these ends is taken to be l.

The coefficients Z_{11} and Z_{12} thus apply to the horn viewed from its mouth, and the coefficients Z_{22} and Z_{21} to the same horn viewed from its throat. Because the horn looks different from its two ends, it is no surprise that $Z_{22} \neq Z_{11}$. However, when we work out expressions for Z_{12} and Z_{21}, we find that $Z_{21} = Z_{12}$. We have remarked on this somewhat unexpected result before, in Section 8.3, in relation to levers, which are similarly asymmetric. It is a general property of linear systems without steady magnetic fields, and is called the reciprocity theorem.

The expressions for the Z_{ij} are, in each case, moderately complicated, but straightforward to evaluate. For all horns, the impedance coefficients are pure imaginary, as for the cylindrical pipe, if wall losses are neglected. For the exponential horn, and indeed for hypex horns more generally, there is a discontinuity in the mathematical form of the impedance coefficients at the frequency for which $k = m$, where m is the flare constant defined by (10.17). Below this frequency, called the flare cutoff frequency, we have only an exponentially attenuated disturbance, in-phase at all points, instead of propagating waves. This behavior blocks propagation completely in horns of infinite length, but its effect is much less severe in short horns. Indeed, the behavior of short exponential horns in this respect is very little different from that of horns of other profiles.

When we come to look more carefully at the propagation of waves in a horn, we find that the wavefronts are not plane, but rather curved, as indeed we have recognized in the case of the conical horn by using the spherical wave propagation equation. This means that the cross-section area $S(x)$ in the horn equation (10.15) is actually the area of the curved wavefront meeting the axis at position x, as shown in Fig. 10.5(d), rather than the geometric cross-section of the horn. Provided the horn does not flare too widely—in which case some of the other assumptions underlying the horn equation (10.15) also cease to hold—this does not make a great deal of difference to propagation within the horn. It does, however, have a significant effect at the mouth of the horn, where an incoming plane wave must make the transition to a curved wavefront in the horn. The incoming plane wave in fact drives different parts of the curved wavefront in different phase, so that the transformation of energy is incomplete. We need not go into the analysis of this effect in detail, and the result is relatively straightforward. Suppose that the radius of the horn mouth is a and the flare angle at the mouth—the semi-angle of a cone that would be tangent to the horn surface at its mouth, as shown in Fig. 10.5(d)—is α. Then the curvature mismatch reduces the coupling between a wave in the horn and an incident or outgoing plane wave by a factor

$$F_\alpha(ka) = \frac{\sin[(ka/2)\tan(\alpha/2)]}{(ka/2)\tan(\alpha/2)}. \tag{10.19}$$

PIPES AND HORNS

This factor is unity at low frequencies, but reduces the efficiency of the horn at high frequencies. The coupling becomes zero when $ka = 2n\pi(\tan \alpha/2)^{-1}$.

In a complete treatment of horns it would be appropriate to make allowance for wall losses as we did for cylindrical pipes. The principle is the same, but exact calculation is difficult because the imaginary part of the wave number k varies along the horn as its radius changes. We could reasonably average the value of the reciprocal of the radius along the horn, since losses are proportional to this quantity, and then use the results (10.8) and (10.9) to assign an appropriate imaginary part to k. We shall not usually bother to do this, since in most auditory systems most of the damping is contributed by the tympanum and by radiation losses at the wide horn mouth. Only for long narrow horns, such as the trachea in the vertebrate vocal system or horns in certain insect auditory systems, are wall losses significant.

There is another effect that must be taken into account at high frequencies, and this arises from the fact that we are trying to build a simple one-dimensional model with a single acoustic current for a situation that is really three-dimensional. When we model a very small component in a sound field, there is no ambiguity about the sound pressure at the entry to the component; it is just equal to the sound pressure in the field. For an object of size comparable with or larger than the sound wavelength, however, the situation is more complex. We already know, for example, that the sound pressure close to a large plane baffle is twice the free-field sound pressure, because of the effect of the reflected wave.

Although it is difficult to analyze this situation in detail, we can arrive at the final result very simply. By a fundamental theorem in electric networks, Thévenin's theorem, any real generator can be replaced by an ideal generator in series with some appropriate impedance. Applying this to the pressure acting on the mouth of the horn, or of a pipe of cross-section S_1, we suppose the effective generator pressure to be p_E and the series impedance Z_E. There are two steps in the calculation.

If we look at the case of an infinitely long pipe, parallel to the propagation direction and with walls that are infinitely thin and smooth, then the presence of this pipe makes no difference to the wave propagation. An analog network built up using the components of Fig. 8.4 shows a pressure generator p' feeding through a radiation transition Z_R to the input impedance $Z_0 = \rho c/S_1$ of the infinite pipe. If the flow into the pipe is to be the same as that in a plane wave, so that $U = p/Z_0$, then we must have an effective generator magnitude

$$p' \approx p(1 + Z_R S_1/\rho c) \qquad (10.20)$$

where Z_R is the radiation impedance of an aperture of area S_1 as given by (7.17) and Fig. 7.5. To a good approximation, if $S_1 = \pi a^2$, then $Z_R S_1/\rho c \approx 0.6jka$ for $ka < 1.7$ and ≈ 1 for $ka > 1.7$. For small apertures, therefore, $p' \approx p$, while for large apertures $p' \approx 2p$. The source impedance Z_E is just the radiation impedance Z_R, which is, of course, routinely included in our analog network anyway, so that we need only replace the free-field pressure p by the equivalent pressure p'.

In the case of a horn, we must also make allowance for the curvature of the wavefront at the horn mouth, as expressed by the function $F_\alpha(ka)$ of (10.19). This can be done simply by generalizing (10.20) to make the effective pressure source strength

$$p_E = p'F_\alpha(ka) = p(1 + Z_R S_1/\rho c)F_\alpha(ka) . \qquad (10.21)$$

This expression is what we must use in our analog network for a horn excited by a plane wave parallel to its axis. We see in Section 10.6 what modification we should make when the sound is incident at an angle to the horn axis.

Although it is of only limited use to calculate the behavior of a horn as a separate component, since the acoustic impedance of whatever is connected to its throat has a large influence, nevertheless such a simple calculation is instructive. We therefore calculate the acoustic pressure p_T generated in the throat of a horn by a plane wave incident along the axis, for three simple cases. The first case is that of a stopped horn, in which the throat is rigidly blocked. A small probe microphone of very high acoustic impedance might then be inserted to measure the pressure. The second case is that of a matched horn, in which we suppose the throat to be connected to an infinite pipe of matching diameter. Again a small probe microphone could be used to measure the throat pressure. Thirdly, we suppose the throat to be terminated by a moderately thick and only lightly damped resonant membrane.

The network equations for a horn with an acoustic load Z_L at its throat are easily written down with the help of the analog network of Fig. 10.6. Note that the driving pressure is taken as p_E, in accord with our discussion above, and the internal impedance of the generator is represented by the radiation impedance Z_R at the mouth of the horn, which we always include anyway. These equations are

$$p_E = (Z_R + Z_{11})U_1 + Z_{12}U_2$$
$$0 = Z_{21}U_1 + (Z_L + Z_{22})U_2$$
$$p_T = Z_{21}U_1 + Z_{22}U_2 = -Z_L U_2. \qquad (10.22)$$

The third equation is not usually one of the set, and is very nearly the same as the second, but is divided up so as to give two equivalent expressions for the pressure p_T in the throat of the horn.

For the case of a horn with a rigidly stopped throat, $Z_L = \infty$, which means that $U_2 = 0$, either from physical considerations or from the second of equations (10.22). The first of (10.22) then gives U_1 and the first form of the third equation gives the throat pressure as

PIPES AND HORNS

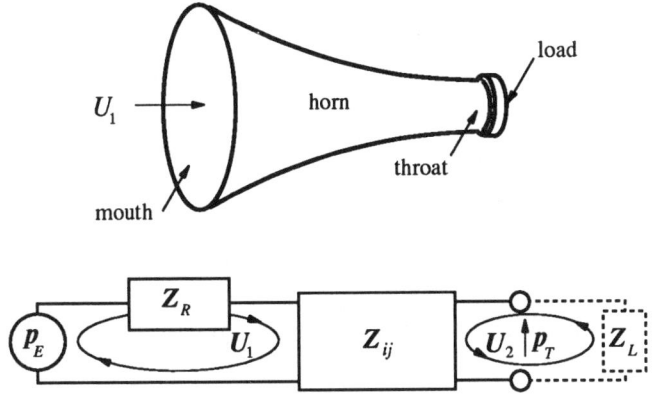

Figure 10.6 Electric network analog for a horn, loaded at its throat by an acoustic impedance Z_L. p_E is the analog driving pressure and Z_R the radiation impedance at the horn mouth.

$$p_T = \left[\frac{Z_{12}}{Z_R + Z_{11}}\right] p_E \qquad (10.23)$$

with p_E given by (10.21). For the matched horn, we define the matching load at the throat to be $Z_2 = \rho c/S_2$, and the network equations are then

$$p_E = (Z_{11} + Z_R)U_1 + Z_{12}U_2$$
$$0 = Z_{12}U_1 + (Z_{22} + Z_2)U_2. \qquad (10.24)$$

Again, these are easily solved to obtain $p_T = -Z_2 U_2$ in the form

$$p_T = \left[\frac{Z_2 Z_{12}}{(Z_{11} + Z_R)(Z_{22} + Z_2) - Z_{12}^2}\right] p_E. \qquad (10.25)$$

The solution when the horn is terminated by a resonant diaphragm is formally the same as (10.25) if we replace the terminating impedance Z_2 by the impedance Z_T appropriate for a tympanum, as given in Table 8.1.

It is easy to calculate these responses as functions of frequency, using the explicit expressions given in Appendix B for the Z_{ij} for different types of horns. The calculation is simplified since we are interested only in the absolute value of the pressure, not its phase. The first and simplest case is that in which the throat of the horn is assumed to be rigidly stopped. Figure 10.7 shows the calculated gain $G = 20\log_{10}(p_T/p)$ in decibels for three horns, each having the same length, mouth and throat diameters, but being of parabolic, conical, or exponential profile respectively. The average flare angle is about 27° in each case, but the flare is differently distributed along the length of the horn. Physically these horns are

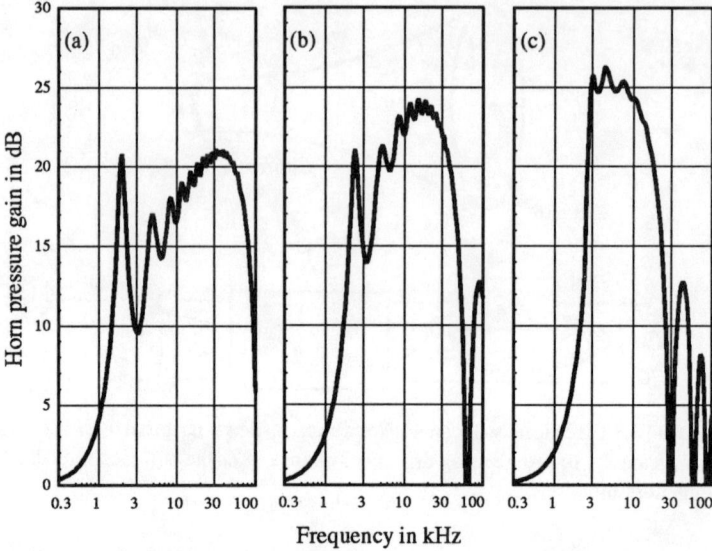

Figure 10.7 Response of (a) parabolic, (b) conical, and (c) exponential horns of the same length (50 mm), mouth diameter (50 mm), and throat diameter (5 mm), with the throat rigidly stopped.

about the size of the pinnae of a large mammal. The curves can be scaled to any absolute size of horn of the same shape, an increase in dimensions by a factor K changing the frequency scale by a factor $1/K$. These curves have been verified, at least for the case of a conical horn, up to a scaled frequency that includes the first subsidiary maximum above the cutoff. The fact that the simple theory works so well, even at high frequencies, arises from the fact that higher modes generated at the mouth, which we discuss in Section 10.8, are unable to propagate to the narrow throat. There is some doubt, however, about the calculation for the parabolic horn at high frequencies, because of focusing effects to be discussed later.

It is notable that the performance of all three horns is broadly similar. The gain falls to unity, or 0 dB, as the frequency falls below a lower cutoff which, for these particular horns, is at about 2 kHz. This corresponds to the frequency of the actual flare cutoff (2.5 kHz here) in the case of an exponential horn, but we can see that such behavior is quite general. There is then a pass band in which the gain is quite high, extending to an upper cutoff above which the gain once more becomes small, and may even be less than 0 dB in certain frequency intervals. The upper cutoff frequency varies somewhat with horn profile.

It is useful to have approximate formulae for estimating the frequencies of the lower and upper cutoff in an arbitrary horn, and these can be provided if the flare angle at the mouth is not too large. The lower cutoff frequency is given approximately by

$$\omega_{\text{lower}} \approx \frac{c}{l} \log_{10}\left(\frac{S_1}{S_2}\right) \qquad (10.26)$$

and the upper cutoff approximately by

$$\omega_{\text{upper}} \approx \frac{2n\pi c}{a \tan \alpha/2}. \qquad (10.27)$$

These results are accurate to within about ±30% for ordinary horn shapes. Clearly a more careful calculation is necessary if the result is important. Note that the pass band of a horn with given mouth and throat diameters can be extended at both low and high frequencies by making it longer, while its performance at high frequencies alone can be improved somewhat, at the expense of the lower parts of the passband, by modifying the profile towards parabolic. There is, however, a measure of overestimation in the parabolic calculation at high frequencies because of the neglect of reflection effects, which we discuss below.

Within the pass band, the gain reaches a peak value of about

$$G_{\text{stopped}} \approx 10 \log_{10}(S_1/S_2) + A \quad \text{dB} \qquad (10.28)$$

for the stopped exponential horn. The constant A comes from (10.20), for we wish to compare the pressure in the blocked throat of the horn with the pressure we would measure with the same microphone set up in front of the throat-blocking plate alone, or equivalently with the pressure measured by a microphone of the same diameter as the throat. $A \approx 6$ dB if the mouth diameter is large compared with the sound wavelength and the throat diameter small, but $A \approx 0$ dB if both are large or both small compared with the sound wavelength. Conical and parabolic horns do not achieve a gain quite as high as this.

In addition to the general shape of the stopped-throat horn gain characteristic, it is worthwhile to note the presence of resonance peaks, the first, at about 2 kHz for the particular horns in Fig. 10.7, being the most prominent. The extent to which the resonances stand out above the general curve depends upon the ratio of the length of the horn to its mouth diameter. If the horn is short and wide, then the radiation resistance at the mouth at the frequency of the first resonance is nearly $\rho c/S_1$ and the resonance Q is very low. This is usually the case for auditory pinnae, which have about the shape for which the figures were calculated. For long narrow horns, as found in musical instruments or in the mammalian vocal system, the first few resonances are much more pronounced. Note, incidentally, that the frequency of the first resonance, for a given horn length, depends appreciably upon the profile of the horn, as illustrated in Fig. 10.7.

Figure 10.8 shows the effect of the other two model terminations on the pressure gain of the conical horn. The effect on the gain of the other profiles is similar. When the horn throat is matched to an infinite-pipe load, the shape of the

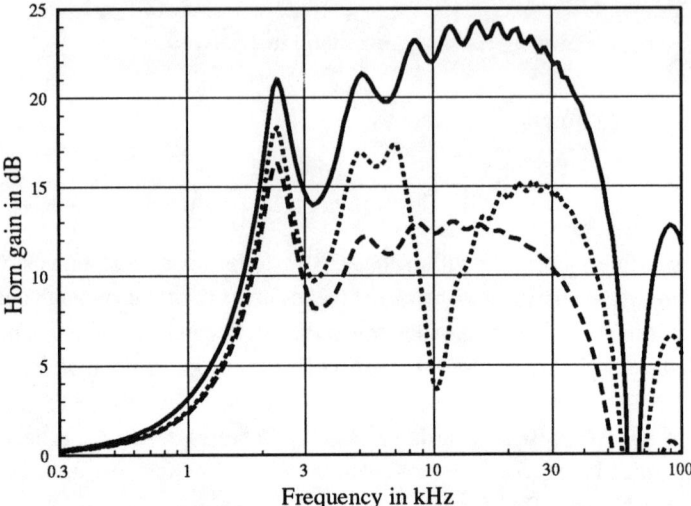

Figure 10.8 Response of the conical horn of Fig. 10.6 with the throat rigidly blocked (full curve), matched to an infinite-pipe load (broken curve), and terminated by a resonant diaphragm (dotted curve). The diaphragm has a thickness of 65 μm, a resonance frequency of 10 kHz, and a Q of 5.

gain curve is very little altered, but the actual pressure gain, measured in decibels, is approximately halved, representing about a 10 dB loss in performance. In the case of the resonant-membrane termination, there is a very marked decrease in pressure gain near the frequency of the diaphragm resonance, but a less severe effect at other frequencies. These calculations emphasize that we need to consider the performance of the whole system, and cannot simply combine the performance of isolated components measured under different conditions.

10.6 Directionality of a Horn

In the frequency range below that corresponding to the upper cutoff frequency of the horn, the wavefront at the mouth is curved with a dome height that is less than about half a wavelength, so that it is a reasonable approximation to regard it as plane for the purposes of calculation. The angular distribution of radiation when the horn is used as a transmitter, or the angular distribution of sensitivity when it is used as a receiver, is therefore nearly the same as that of an open pipe of the same diameter, as shown in Fig. 7.6. The angular response narrows as the frequency increases, the angular displacement of the −3 dB points being about

$$\theta_{-3\,\mathrm{dB}} \approx \pm \sin^{-1}(1.6/ka) \tag{10.29}$$

at least in the range $1 < ka < 10$. For $ka > 10$ wavefront curvature can no longer be neglected and the beam broadens and develops side lobes.

PIPES AND HORNS

We can include this directional effect in the effective strength p_E of the network pressure generator, by generalizing (10.21) to include a directional factor $D_{ka}(\theta)$. The complete result for a wave of free-field pressure p incident at an angle θ to the axis of a horn with mouth area $S_1 = \pi a^2$ is then

$$p_E \approx p\left(1 + \frac{Z_R S_1}{\rho c}\right) F_\alpha(ka) D_{ka}(\theta) . \tag{10.30}$$

There is no simple expression for $D_{ka}(\theta)$, which is given by the curves of Fig. 7.6, but we note that $D_{ka}(0) = 1$. In the forward direction it is approximately equal to the corresponding expression for a baffled pipe,

$$D_{ka}(\theta) \approx 2J_1(ka \sin \theta)/ka \sin \theta \qquad \theta < \pi/2. \tag{10.31}$$

10.7 Obliquely Truncated Horns

It is a good first approximation to consider the external part of the pinna of many animals to have the form of an obliquely truncated simple horn, as shown in Fig. 10.9. The actual profile is often not far from conical, though our discussion in Section 10.5 leads us to expect that the exact flare shape is not critical.

Any moderately accurate treatment of such an obliquely truncated horn is very difficult, and even an approximate calculation such as the one referred to in the Bibliography [25] is quite complex and uncertain. The horn commences its flare with complete sides and then tapers away to an extended flap. We expect the normal part of the horn to behave simply, while the extended flap presumably both changes the direction of greatest sensitivity and adds somewhat to the acoustic gain.

Calculations confirm these expectations. For a typical upright pinna shape, as in a rabbit or a marsupial, the angle of greatest sensitivity, which we might call the acoustic axis, is shifted away from the horn axis in the direction of the open face of the pinna. The shift of the acoustic axis is frequency dependent, as shown

Figure 10.9 An obliquely truncated conical horn. The axis of maximum acoustic gain makes an angle θ with the geometric horn axis. The mouth of the "complete" part of the horn is shown dotted.

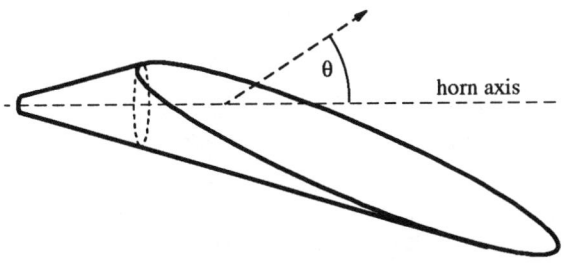

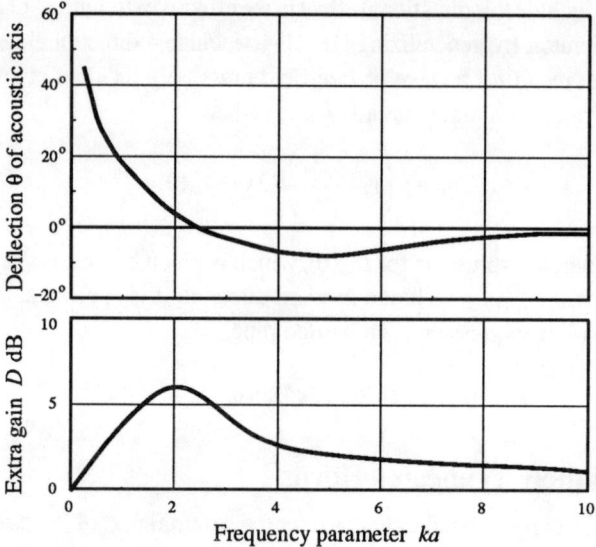

Figure 10.10 Angular displacement θ of the acoustic axis of an obliquely truncated conical horn with mouth radius a at the end of the complete part, and apex semi-angle 20°, truncated at an angle of 30°. The shape is about as shown in Fig. 10.9. Also plotted is the extra gain relative to a normally truncated horn the size of the complete part. (From [25].)

in Fig. 10.10, and it actually moves surprisingly far from the horn axis at very low frequencies. The importance of this shift at very low frequencies is, however, illusory, since the angular response is then extremely wide and the position of the acoustic axis hardly detectable. Of greater significance is the more moderate displacement of the axis at intermediate frequencies where the angular response is much sharper. We must be careful, however, about applying these ideas simplistically to real pinnae, in which the horn axis is often curved. This introduces reflection effects at high frequencies, as discussed in Section 10.9.

The extended curved flap of the pinna also adds to the gain of the horn, but only to a modest extent, the peak addition being about 6 dB for the case calculated. This too is frequency dependent, and the extra gain is added in a useful part of the pass band of the normal horn.

10.8 Higher Modes in Horns

As the frequency is increased, for a horn or pipe of given diameter, we ultimately reach a point where it is possible to have standing waves across the diameter. It is probably only near the mouth of auditory pinnae that this effect is significant, but it is worthy of a little comment, if not much detailed analysis.

Suppose we look carefully at the propagation of an acoustic wave of frequency

PIPES AND HORNS

ω along a rather wide pipe of radius a. To analyze this properly we need to write down a full three-dimensional wave equation, since it is not clear, in a pictorial sense, that the wave will necessarily travel straight down the pipe, rather than moving at an angle and reflecting off the sides. Cylindrical polar coordinates (r, ϕ, z) are appropriate for this problem and, after some algebra, we find that the pressure in a general wave can be written as a sum of functions of the type

$$p_n(r, \phi, z) = A e^{-jkz} J_n(\kappa r) \sin n\phi \; e^{j\omega t} \qquad (10.32)$$

where J_n is a Bessel function of order n. Bessel functions almost always turn up when we consider problems with circular symmetry, as we found also for the circular membrane. The axial wave number k is related to the frequency by

$$k^2 + \kappa^2 = (\omega/c)^2 \quad \text{or} \quad k = \sqrt{(\omega/c)^2 - \kappa^2} \qquad (10.33)$$

and the allowed values of κ are those corresponding to standing waves across the pipe, for which $dJ_n(\kappa r)/dr = 0$ at the pipe wall $r = a$. If $n = 0$, the waves are axially symmetric and ϕ does not enter. The lowest mode then has $\kappa = 0$ and is the familiar plane wave. The next axially symmetric mode has $\kappa = 3.83/a$ and the cross-section shown in Fig. 10.11(b), with the central part moving in the opposite direction to the edge. If the frequency is low so that $\omega/c < \kappa$ then, from (10.33), k becomes

Figure 10.11 Patterns of acoustic pressure for the higher modes in a circular pipe or horn: (a) the plane-wave mode, (b) the second axially-symmetric mode, (c) the first antisymmetric mode, (d) a higher mode. Axial acoustic flow has opposite directions in regions marked + and −, and there is radial acoustic flow across the boundaries between these regions.

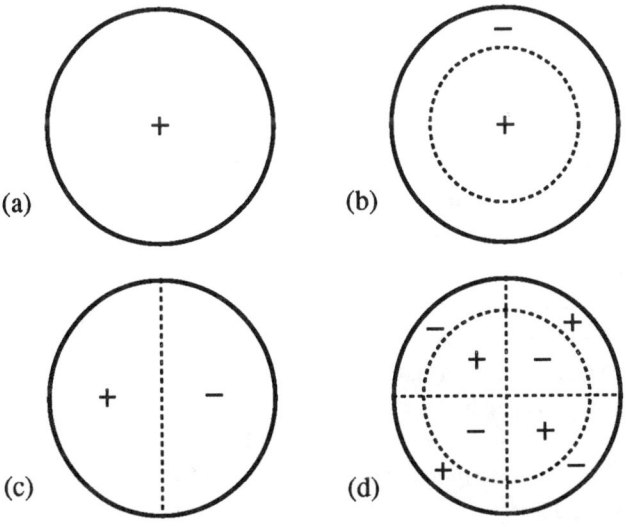

imaginary and the mode is exponentially damped with distance along the pipe rather than propagating. In wide pipes, however, the mode can propagate, its group velocity—the speed of propagation of a tone burst—being rather less than that of a plane wave, though the phase velocity ω/k is actually higher. Modes of this type are generated by a plane wave falling normally on the mouth of a horn, through the effects of wavefront curvature, but cannot propagate far towards its throat unless the frequency is very high. At very high frequencies additional modes of this type, with more annular regions of opposite motion, can be generated and propagate.

Perhaps more important are the modes with $n = 1$, which are antisymmetric across the pipe, one half of the air moving in antiphase with the other as shown in Fig. 10.11(c). Such waves are excited when a plane wave falls obliquely on the end of a horn, or quite generally in an obliquely truncated horn or pinna. The cutoff frequency for propagation of this mode is lower than for the first higher axial mode, because for it $\kappa = 1.84/a$. Again, the wave is sharply attenuated in the narrow part of a horn but, as we shall discuss in more detail in the next section, modes of this antisymmetric type may play a part in shallow pinnae such as found in primates and in humans. At much higher frequencies, modes with higher angular dependence $\sin n\phi$ and with several nodal circles, as shown in Fig. 10.11(d), can propagate.

It is useful to have an estimate of the rate of attenuation with distance for higher modes that cannot propagate. If the frequency is well below the propagation cutoff for the mode involved then, from (10.33), $k \approx j\kappa$ and the attenuation behavior is as $e^{-\kappa x}$. For the lowest asymmetric mode with $n = 1$, we have seen that $\kappa a = 1.84$, so that the amplitude of the mode is reduced by $e^{-3.7}$, or about 30 dB, over a distance equal to the diameter of the pipe. For higher modes the attenuation is even more rapid, so that only modes that are close to or above their propagation cutoff frequency have appreciable effect on the behavior of the system.

10.9 Shallow Asymmetric Horns

As an example of the importance of such higher horn modes in some auditory systems, consider the simple model shown in Fig. 10.12. This model has been used as a basis for analysis of the human external ear by Shaw (see the Bibliography). In this model the horn has become a simple short cylinder, with the throat a small aperture in its inner face, and we are concerned with the pressure driving flow into this throat. A plane wave incident at an angle to the axis of the cylinder can generate both a plane-wave mode and, if its frequency is high enough, modes with angular dependence $\sin(2n - 1)\phi$, the first and most important of which is the antisymmetric mode with $n = 1$. Since the throat is placed off-center, it is affected by the pressure in this mode as well as by the pressure in the symmetric plane-wave mode.

PIPES AND HORNS

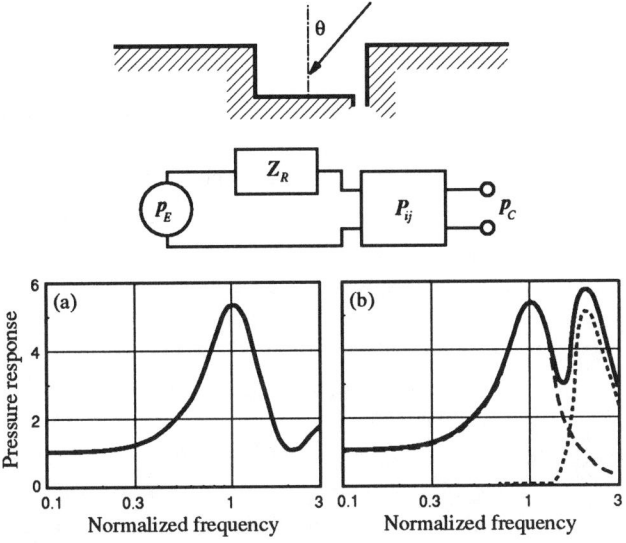

Figure 10.12 A simple model for an ear with an off-set auditory canal as examined experimentally by Shaw. The behavior can be analyzed with the help of the analog circuit shown, but a separate analog is required for the plane-wave and the antisymmetric modes. For sound incident at $\theta = 0$, only the plane-wave mode is excited, and the response at the entry to the auditory canal is as in (a). For oblique incidence ($\theta = 45°$) as in (b), the plane mode response is shown dashed, the antisymmetric mode dotted, and the total response as a full curve. (Adapted from [23].)

It is possible to calculate the response of the simple system shown in Fig. 10.12(a), measured at the auditory canal aperture, for sound incident at an arbitrary angle θ, but we shall not go through this exercise in detail. An outline shows the basis of the analysis and indicates the results in at least semiquantitative fashion. Shaw's investigation, incidentally, was experimental rather than theoretical. For simplicity we assume the impedance presented by the canal to be very high, so that it is effectively stopped, but we must repeat the warnings in Section 10.5 about interpretation of the calculated results!

If a is the radius of the cylindrical cavity and l its depth, then the plane-wave mode has a first resonance when the acoustic length $l' \approx l + 0.8a = \lambda/4$, the end correction being a little larger than usual because of the flange. If $l \approx a$, then this resonance occurs for $ka \approx 0.9$ or $\omega_+ \approx 0.9c/a$. The pressure will be a maximum at the inlet to the auditory canal at this frequency, and indeed since Fig. 7.5 shows that $R_R \approx 0.2Z_0$ for $ka = 0.9$, the pressure amplification will be about a factor 5, or 13 dB, at resonance.

It is easy to make this discussion quantitative, using the analog circuit shown in Fig. 10.12. Since there is no flow into the auditory canal, the pressure p_C at its entry is readily seen to be

$$p_C = \frac{P_{12} p_E}{Z_R + P_{11}} \tag{10.34}$$

where the generator pressure p_E is given by (10.30) and includes a factor $D_{ka}(\theta)$ taking account of the direction of incidence of the plane wave. The response for normal incidence ($\theta = 0$) is shown semiquantitatively in the graph of Fig. 10.12(a). For oblique incidence with $\theta = 45°$, the plane-wave mode gives the pressure contribution shown as a dashed curve in Fig. 10.12(b).

For the first resonance of the antisymmetric mode, we have already seen that the transverse wave number is $\kappa = 1.84/a$, and the axial wave number must still satisfy the requirement $ka \approx 0.9$, if we neglect possible modification of the end correction. From (10.33) we then find the resonance frequency $\omega_- \approx 2c/a$. Thus for the particular geometry considered, $\omega_- \approx 2\omega_+$. This antisymmetric resonance is rather less damped than the plane-wave resonance, because of the dipole nature of the motion, so the response peak is comparable in height to that of the plane-wave resonance. A significant difference arises, however, from the fact that the antisymmetric mode makes almost no contribution to the response below its cut-off frequency.

It is possible to formulate an expression very much like (10.34) for the pressure contribution of the antisymmetric wave, but this would take us beyond the scope of our discussion since we have not derived the appropriate pipe impedance coefficients for anything but the plane-wave mode. We can note, however, from (10.33), that P_{12} will be very small below the cut-off frequency of the cylindrical duct, so that the antisymmetric mode makes no contribution to the canal pressure below this frequency. We can also note that the expression in p_E taking account of incidence direction has a quite different angular dependence from the plane-wave case. Indeed for an antisymmetric wave $p_E = 0$ for normal incidence, and it goes through a maximum at an angle that depends on the system dimensions. The pressure at the auditory canal coupled through the antisymmetric mode therefore has the form shown in the dotted curve of Fig. 10.12(b) when the sound is incident from an angle $\theta = 45°$.

When the ear model is stimulated by a sound wave coming from along its axis, then only the plane-wave mode is excited, and the frequency response at the entry to the auditory canal is of the form shown in Fig. 10.12(a). For a sound wave incident at $\theta = 45°$, however, both modes are excited to comparable amplitudes, and their pressures add at the entry to the auditory canal, giving the frequency response shown as a full curve in Fig. 10.12(b). If the diameter of the cavity is taken to be about 20 mm, then the resonance frequency for the plane-wave mode is about 5 kHz and for the antisymmetric mode about 10 kHz, so that this mechanism may provide useful directional clues for animals such as primates whose external pinnae have something of this shape and size.

10.10 Hybrid Reflector Horns

At the limit of very high frequencies and very wide horns, say with $ka > 100$, a large number of higher modes can propagate in the horn if they are generated by unsymmetrical excitation. Instead of the analysis becoming impossibly complex in this limit, it becomes simple again, and we can think of sound rays propagating in straight lines in the same way as light, with reflection when they strike the walls of the horn. An appropriately shaped pinna may then, to some extent, focus the incident sound to the inlet meatus of the auditory system, while a badly shaped pinna, such as a deep paraboloid, may reflect it uselessly. To gain some feeling for the range of applicability of such an optical ray model, we should recall that it is impossible for an optical system to focus incident rays into an area smaller than the wavelength in diameter, and the same is true of sound. The ray approach has approximate validity, therefore, only in that part of a horn that is significantly greater than one wavelength in diameter. As the pinna narrows, it behaves first as a multi-mode and then as a single-mode horn.

Figure 10.13(a) shows the optical focusing effect achieved by a paraboloidal

Figure 10.13 (a) Focusing behavior of a paraboloidal horn when the horn diameter is much greater than the sound wavelength. The region of focus, shown dotted, is about a wavelength in diameter. (b) In the case of a very shallow primary horn, addition of a small reflector, appropriately positioned, can focus the reflected waves into an aperture. Both reflector and aperture must be larger than about one wavelength in diameter. (c) Addition of a subsidiary horn near the position of focus allows further gain. The mouth of the subsidiary horn must be about one wavelength in diameter.

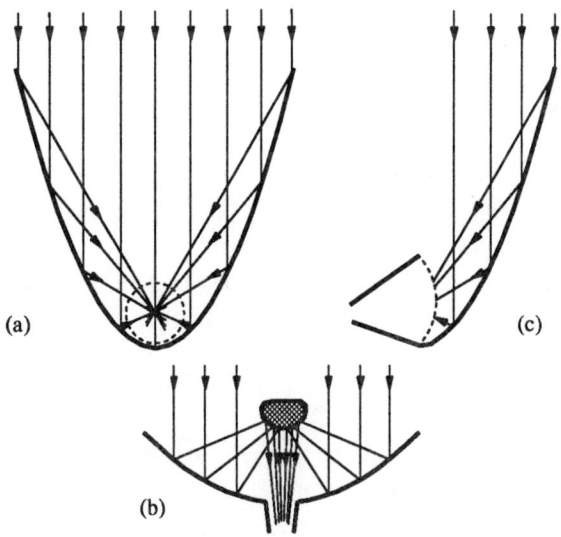

reflector (or "parabolic" horn) when the wavelength is small compared with the horn dimensions. Sound waves incident parallel to the axis are focused to a point above the base of the paraboloid, and thence out into the free air again, so that such a horn with a small aperture in its base would not have significant gain at the high frequencies for which the optical model applies. In an optical telescope this problem would be resolved by having a secondary mirror near the focus to redirect the light to the exit port, and indeed there is a small fleshy flap in some vertebrate ears that may perhaps serve this purpose to some extent, as illustrated in Fig. 10.13(b). Such an arrangement is possible, however, only for quite shallow ears, since the reflector must be near the focus of the main paraboloid, which is at a distance from its base equal to half of the radius of a sphere that matches its curvature. The focusing action of a parabolic reflector is, however, not perfect as one would expect geometrically, for the properties of waves mean that the energy cannot be focused into a sphere smaller than about one wavelength in diameter. Both the reflecting flap and the entry to the auditory canal must therefore be at least a wavelength in diameter for such a device to work efficiently, so that its use is confined to high frequencies.

As an alternative, and more common, approach to further concentrating the sound energy in the diffraction-limited focus sphere, we can use a subsidiary horn as shown in Fig. 10.13(c). The mouth of the horn must be about one wavelength in diameter in order to collect most of the focused sound energy, and can then guide it down to a much smaller throat. The model strictly implies a roughly rectangular horn section, but clearly a circular horn would work nearly as well. In animal pinnae the transition from a reflecting to a horn-like structure is understandably less well defined than in this simple model.

The extra acoustic gain achievable with a horn-and-reflector structure such as this, above that of the horn itself, is $G_R = 10\log_{10}(S_R/S_1)$ where S_R is the projected area of the reflector in the direction of the paraboloid axis and S_1 is the mouth area of the horn. This result applies, however, only at frequencies high enough that the horn mouth is larger in diameter than the wavelength and the reflector much larger. At lower frequencies where this condition is no longer met, the gain reduces to essentially that available from the horn itself, together with the small enhancement from the pinna flap discussed in Section 10.7.

At high frequencies where focusing is efficient, the acoustic axis of the hybrid horn is parallel to the axis of the original paraboloid. At lower frequencies the shape of the reflector becomes much less important, and the acoustic axis moves away from this direction, the detailed behavior depending upon the relative sizes of horn and reflector.

Hybrid horns of this type, in which the outer part functions as a reflector rather than a horn, are perhaps the basis of the pinnae of animals such as bats that use ultrasonic frequencies. The focusing action of the outer part of the pinna can be

PIPES AND HORNS

particularly effective if the frequency is high enough for ray-like reflection to occur, but qualitative discussion using such a reflection model at low frequencies is misleading.

References

Propagation in pipes: [1] Ch 6; [8] Ch 8; [9] Ch 5; [10] Ch 8; [11]
Horns: [1] Ch 6; [6] Ch 9; [9] Ch 5; [10] Ch 8; [25]
Short horns: [9] Ch 5; [10] Ch 8; [25]
Obliquely truncated horns: [25]
Shallow asymmetric horns: [23]; [24]

Discussion Examples

1. Calculate the resonance frequencies of a pipe of length l and cross section S, closed at one end by a slack diaphragm of thickness d.
2. Repeat question 1 for the case in which there are identical diaphragms on each end of the pipe.
3. Write a computer program to evaluate the impedance coefficients Z_{ij}, and to plot them in the form $\log(|Z_{ij}|/Z_0)$ for a cylindrical pipe and for a conical horn (neglecting wall losses). Note the infinities and zeros.
4. Write a computer program to evaluate the pressure in the throat of a horn (a conical horn is simplest) exposed to an axially incident plane wave, when the throat is rigidly stopped [equation (10.23)], and to plot this as a function of frequency. Examine the effect of changing the length of the horn, keeping mouth and throat diameters constant.
5. Write a computer program to calculate the pressure in the throat of a horn (a conical horn is simplest) exposed to an axially incident plane wave, when the throat is blocked by an arbitrary real impedance Z_L [equation (10.25)], and to plot this as a function of frequency. Find the effect of changing Z_L from a value much less than $\rho c/S_2$ to a value much greater than this quantity.

Solutions

1. The system will oscillate at its resonance frequency even after an external exciting force is removed. Write down the two network equations, neglecting radiation for simplicity, and require consistency. This gives $Z_{22}(Z_{11}+Z_T) = Z_{12}^2$. Substitute $Z_T = j\omega\rho_s d/S$ and expressions for the Z_{ij}. Simplifying gives $\tan kl = -\omega\rho_s d/S$. As $d \to 0$ resonances are $kl = n\pi$ as for an open pipe. For large d, resonances approach $kl = (n - 1/2)\pi$ as for a pipe stopped at one end. All intermediate values are possible.
2. Proceed as in example 1. Resonance condition is $Z_{11}+Z_T = \pm Z_{12}$. As $d \to 0$, resonances are $kl = n\pi$ as for open pipe. For large d resonances approach $(n-1)\pi$ with the diaphragms moving appreciably only for the lowest resonance $n = 1$.

11 HIGH-FREQUENCY AUDITORY MODELS

SYNOPSIS. When the frequency is high enough that the sound wavelength is comparable to the dimensions of an auditory system, it is necessary to use the more sophisticated analysis set out in Chapter 10. Of course it must give the same results as the simple analysis of Chapter 8 at sufficiently low frequencies.

Figure 11.1 shows a simple cavity-backed ear loaded by a short exponential horn, thus making a reasonable model for a mammalian ear, though the extra effects of an obliquely truncated or curved pinna should be added. The calculated response is shown in Fig. 11.2, together with the response of the same ear without the horn. It is clear that this addition makes a substantial improvement in the system performance. Also shown is the effect of adding a short auditory canal or meatus between the horn and the tympanum, as in Fig. 11.3. While the effect is detectable, it does not make much difference to total system response.

Animals such as birds, and some reptiles, have an auditory canal directly connecting the two ears, as shown in Fig. 11.4. The analog network for this system is also given in the figure. The calculated system response, for the parameter values in Table 11.2, is shown in Fig. 11.5. This system has very useful directional discrimination over about one octave near the center of its band of maximum sensitivity. Experiments with the model show that the presence of porous walls in the auditory canal does not have a great effect on system performance, and can be simply compensated for by small adjustments to other parameters.

There are, of course, auditory systems more complex than these, as illustrated in Fig. 11.6, which is a typical insect system as found in a cricket. The tympana are thin patches of cuticle on the forelegs, and these are connected by the internal tubes of the respiratory system, which terminate in more or less horn-shaped sections leading to the spiracles. While it is simple to write down the network equations describing the response of such a complex system, their solution is best carried out numerically rather than analytically, using one of the commonly available computer programs for solving simultaneous linear equations.

11.1 High Frequencies

Many auditory systems operate up to frequency ranges at which their dimensions are no longer small compared with the sound wavelength, and for these the simple analysis of Chapter 9 is no longer adequate. For such cases the more accurate formulation of acoustic behavior and electric analog components given in Chapter 10 provides a ready solution without a great deal more labor.

In this chapter we derive the performance of several model systems, both as a guide to the procedures involved and to illustrate the properties of real systems. At sufficiently low frequencies the results must, of course, agree with those we would have obtained using low-frequency analogs, and indeed this is guaranteed by the way in which we derived the formalism.

11.2 Horn-Loaded Simple Ear

The ears of many higher animals are, as we have remarked before, essentially uncoupled from an acoustic point of view, and so can be analyzed individually.

HIGH-FREQUENCY AUDITORY MODELS

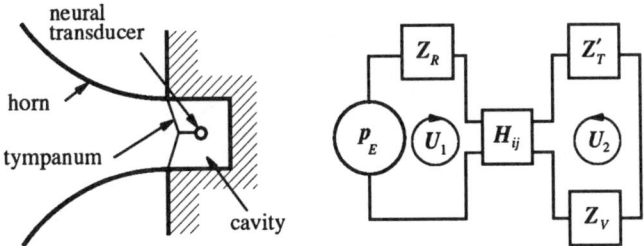

Figure 11.1 A horn-loaded ear and its associated network analog. H_{ij} are the impedance coefficients Z_{ij} for the horn.

Whatever comparison is made between inputs at the two ears is made at a neural level, and so lies outside the scope of this book. As our first example, therefore, we return to the simple ear of Section 9.5 and supplement it by the addition of a simple horn, representing the external ear or pinna. We realize, of course, that a real pinna is not a simple horn, but has an asymmetric extension as discussed in Sections 10.7 and 10.10. The main effect of this is a frequency-dependent addition to the horn gain, together with a modification of its directional properties. To an adequate approximation these refinements can be simply added on to the calculated behavior of the simple case.

The system to be analyzed appears as in Fig. 11.1, with a horn leading to a tympanum backed by a cavity. It is straightforward to draw the analog network, which is also shown in the figure. For clarity, since both horns and pipes have impedance coefficients Z_{ij}, we shall denote these instead by H_{ij} and P_{ij} respectively. (Readers familiar with electrical network theory should note that the impedance coefficients H_{ij} are *not* the hybrid parameters usually denoted by those symbols, but simply the Z_{ij} in disguise.) The impedance of the tympanum is written as Z_T', to include the loading effect of the neural transducer. The effective magnitude p_E of the pressure generator is given by the expression (10.30), with (10.19) and (10.31) to include wavefront curvature and directionality effects, and this could be modified by a further factor to account for the more complex geometry of a real pinna. The network equations are simply written down to be

$$p_E = (H_{11} + Z_R)U_1 + H_{12}U_2$$
$$0 = H_{12}U_1 + (H_{22} + Z_T' + Z_V)U_2 \qquad (11.1)$$

where we have taken care with the symmetry of the currents U_1 and U_2 relative to the horn, the mouth of which is port 1 and the throat port 2. These equations are easily solved to give the acoustic current U_2 through the tympanum as

$$U_2 = \left[\frac{H_{12}}{H_{12}^2 - (H_{11} + Z_R)(H_{22} + Z_T' + Z_V)} \right] p_E. \qquad (11.2)$$

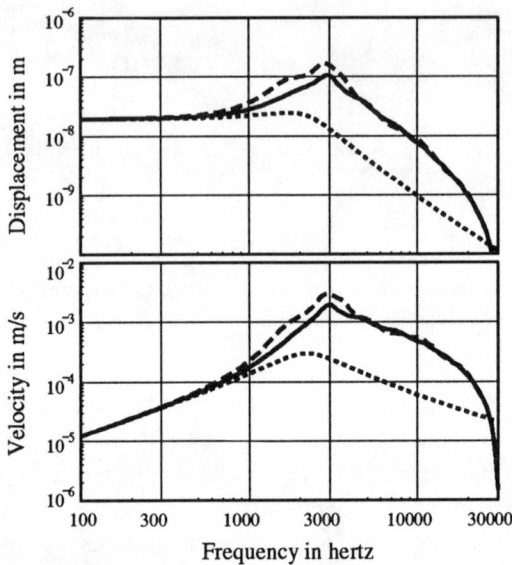

Figure 11.2 Response of the cavity-backed ear of Fig. 11.1 and Table 11.1 (a) alone (dotted curve), (b) coupled to a short exponential horn (full curve), and (c) coupled by a cylindrical meatus to the same exponential horn (broken curve). In each case the resonance frequency of the tympanum is 2 kHz and the free-field driving pressure is 0.1 Pa or 74 dB re 20 µPa. In a real ear, a further correction for the effect of the Eustachian tube, as shown in Fig. 9.5, is required.

While it is not difficult to evaluate this complex quantity once the parameters of the system are known, the task is made simpler by noting that only the absolute value U_2 of U_2 is required. The acoustic current can be simply converted to a displacement velocity or to a displacement at the neural transducer link, as discussed in Section 9.5.

Figure 11.2 shows the response for a cavity-backed ear, such as that discussed in Section 9.5, coupled to a short exponential horn, as discussed in Section 10.5. Details are given in Table 11.1. In order to bring the tympanum resonance into the pass-band of the horn we have had to raise the bare tympanum resonance frequency to 2 kHz, and curves giving the response of the ear in the absence of the horn are included for comparison. It is clear that the presence of the horn increases the gain in the mid-frequency range by a factor between 3 and 10 (10 to 20 dB), flattens the response in this mid-frequency region, and gives a sharper cutoff at high frequencies.

If we mentally invert the lower plot of velocity response to give a picture of sensory threshold, we find that it is quite similar to the threshold curve for human hearing shown in Fig. 6.3. This resemblance could be further improved by including the low-frequency effect of a Eustachian tube leak, as illustrated in

Table 11.1 Assumed parameters for a horn-loaded ear

Tympanum diameter	$2b = 5$ mm
Tympanum thickness	$d = 10$ μm
Loaded tympanum resonance	$\omega_T/2\pi = 2000$ Hz
Effective transducer mass	$m_N = 1$ mg
Q of tympanum	$Q = 1$
Cavity volume	$V = 1$ cm^3
Horn mouth diameter	$2a = 50$ mm
Horn length	$l = 50$ mm
Meatus length (if present)	$l_M = 20$ mm

Fig. 9.5, and by scaling all dimensions up (and hence all frequencies down) by a factor of about 1.5. We have not yet included any detail of the ossicle lever connecting the tympanum to the neural transducer (the cochlea in this case). The discussion of Section 8.3 shows that this lever itself has a high-frequency cutoff behavior and, since this is associated with details of the linkages involved, it could well change with age. The cochlea itself, of course, could also well have a high-frequency cutoff, as discussed in more detail in Chapter 12.

We can easily elaborate the model of Fig. 11.1 by including a cylindrical tube coupling the horn throat to the tympanum and representing the auditory canal or meatus. The network then has the more elaborate form shown in Fig. 11.3. There are three network currents to be considered, drawn as shown so that the symmetry of both horn and pipe currents is maintained, and the network equations can be easily written down as

$$p_E = (H_{11} + Z_R)U_1 + H_{12}U_2$$
$$0 = H_{12}U_1 + (H_{22} + P_{11})U_2 + P_{12}U_3$$
$$0 = P_{21}U_2 + (P_{22} + Z_T' + Z_V)U_3 . \tag{11.3}$$

The formal solution for the acoustic current U_3 through the tympanum is

$$U_3 = \left[\frac{P_{12}H_{12}}{(P_{22} + Z_T' + Z_V)[(H_{11} + Z_R)(H_{22} + P_{11}) - H_{12}^2] - H_{11}P_{12}^2} \right] p_E . \tag{11.4}$$

Once again we can take p_E to be real and simply calculate the absolute value of the current U_3.

Curves showing the calculated effect of a cylindrical meatus of length 20 mm

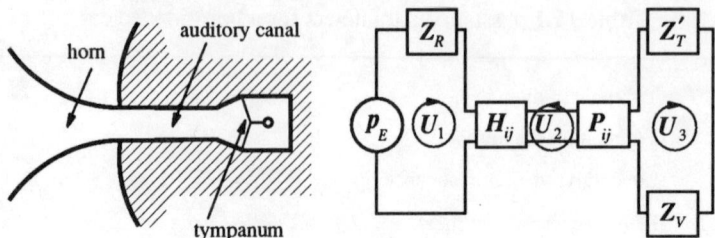

Figure 11.3 The horn-loaded ear of Fig. 11.1 with the addition of an auditory canal or meatus. In the analog network, H_{ij} and P_{ij} are the impedance coefficients for the horn and the pipe respectively.

between the tympanum and the horn are included in Fig. 11.2. Clearly the meatus makes rather little difference, though its effect is detectable. No significant pipe resonances occur in this case because the pipe is moderately well matched to the tympanum at one end and to the horn at the other, and the resistive parts of these matching impedances make the Q value of any pipe resonances very low.

Once again, we should note that in a real ear the internal cavity is connected to the atmosphere by a long narrow Eustachian tube. This has the effect of reducing the low-frequency response, as discussed in Section 9.5 and illustrated in Fig. 9.5. The method of analysis used there can easily be combined with our discussion above.

11.3 Pipe-Coupled Ears

Many animals have direct acoustic coupling between their two ears, and we have already examined, in Section 9.7, the case in which this coupling is through a simply cavity. At frequencies sufficiently high that the distance between the two ears is an appreciable fraction of a wavelength, it is no longer an adequate approximation to treat the cavity simply as an enclosed volume; we must take more care to specify its shape. Fortunately in many cases of interest, such as birds and some reptiles, the cavity is essentially a straight pipe joining the two ears, and this system is quite easy to analyze.

The prototype is shown in Fig. 11.4, together with its analog network. The pressures p_1 and p_2 at the two ears, supposed individually to be small compared with the sound wavelength, are given in terms of the incidence angle of the sound and the distance D between the two ears by (9.1). If we write the coefficients Z_{ij} for the pipe as P_{ij} and take Z_T' to include the loading of the neural transducer, then the network equations are

$$p_1 = (Z_R + Z_T' + P_{11})U_1 + P_{12}U_2$$
$$p_2 = P_{21}U_1 + (Z_R + Z_T' + P_{22})U_2. \qquad (11.5)$$

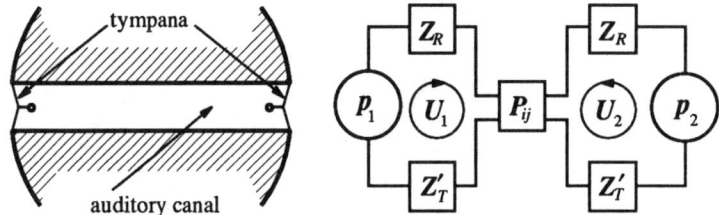

Figure 11.4 An auditory system consisting of a pair of ears coupled by a simple tubular auditory canal. In the analog network the impedance coefficients for the pipe are denoted by P_{ij}.

Remembering that $P_{12} = P_{21}$ and that, for a pipe, $P_{22} = P_{11}$, we find the acoustic current U_1 through ear 1 to be

$$U_1 = \frac{(Z_R + Z_T' + P_{11})p_1 - P_{12}p_2}{(Z_R + Z_T' + P_{11})^2 - P_{12}^2} \tag{11.6}$$

and this is readily converted to a mechanical displacement or velocity at the neural transducer link, as discussed in Section 9.5.

The significance of this expression is not immediately apparent, and it is necessary to solve the model for specific cases in which the parameters can be varied. This is not difficult, since the range of anatomically reasonable variation is not large. It helps, however, to recognize two distinct cases. In the first place we can consider auditory systems of animals such as birds or lizards in which the length of the auditory canal is perhaps 20 mm and its diameter a few millimeters. Leaving aside for the moment the possibility of a canal with porous walls, we can expect attenuation in the canal not to be important in this case. In the second place we can consider the auditory systems of insects, in which the length of the passages connecting the two ears is again 10 to 20 mm, but the pipe diameter is more like 0.1 mm. In this case, wall attenuation in the pipe can be significant. The pipe coefficients P_{ij} can, in either case, be readily expanded to include wall losses, as discussed in Section 10.3 and Appendix B.

Figure 11.5 shows the results of a calculation using the anatomical data of Table 11.2. As in the low-frequency case, we have neglected amplitude effects of diffraction around the head or body of the animal, which gives acoustic shadowing as discussed in Section 6.9 at very high frequencies, and included simply the phase shift. This additional complication can be included if we have information on the shape and size of the animal. Our calculation shows that, for reasonable parameter values, the system shows up to 30 dB of directional discrimination near its frequency of maximum sensitivity. The frequency range over which 10 dB of discrimination is maintained is about an octave (a factor 2), which is very useful. The

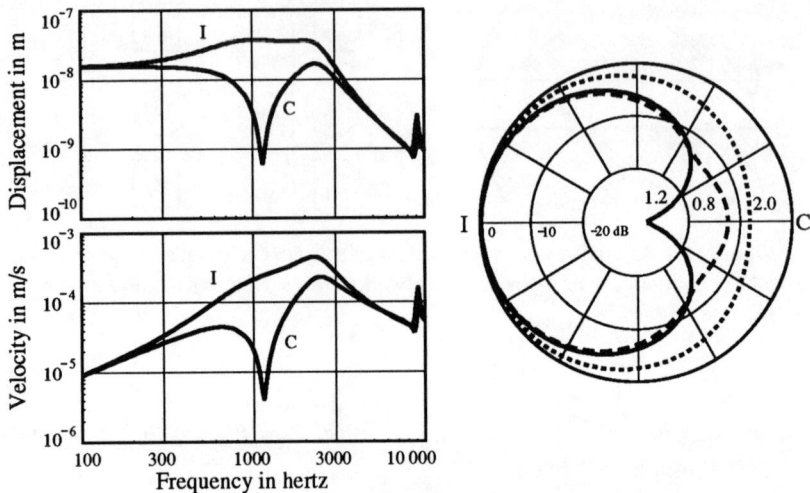

Figure 11.5 Response of the pipe-coupled auditory system of Fig. 11.3 and Table 11.2. The first part of the figure shows the displacement and velocity response at the point of connection to the neural transducer for ipsilateral (I) and contralateral (C) sound presentation with a free-field driving pressure of 0.1 Pa, or 74 dB re 20 µPa. The second part of the figure shows the relative directional sensitivity of the system at 0.8 kHz, 1.2 kHz, and 2.0 kHz.

overall behavior is very similar to that of a cavity-coupled ear pair, as calculated in Section 9.7 and displayed in Fig. 9.7. Indeed the only effect of the more accurate pipe calculation is seen in the presence of several small resonances at higher frequencies. These have little effect on the behavior of the system.

To examine the effects of porous walls on the performance of a pipe-coupled system we need only to replace the sound velocity and attenuation constants within the pipe by their modified values, as given by (10.10). If we assume that $c' \approx 0.5c$, to settle on a reasonable approximation, then we find that this change makes very little difference to the behavior of the system, and that such change as there is can be compensated for by relatively small changes in other parameters. The attenuating effect of the porous walls, which can be quite large, turns out essentially just to add to the damping effect of the low Q of the tympana, and these two quantities can be traded against one another with little effect on the response of the system. If the Q of the tympana is made rather larger, say 2, and this is compensated by increasing the pipe damping by a factor of about 15 over that of a smooth-walled tube, then the magnitude of the small tube resonances seen in Fig. 11.5 decreases, and there is a difference between the behavior of ipsilateral and contralateral tympana in the immediate vicinity of the resonance. The level of this effect is too small to be of much significance. We conclude that the porous nature of the bone

Table 11.2 Assumed parameters for pipe-coupled ears

Tympanum diameter	$2a = 5$ mm
Tympanum thickness	$d = 10\,\mu\text{m}$
Loaded tympanum resonance	$\omega_T/2\pi = 1000$ Hz
Effective transducer mass	$m_N = 1$ mg
Q of tympanum	$Q = 0.8$
Separation of ears	$D = 20$ mm
Auditory canal diameter	$2a = 5$ mm

is incidental to the functioning of the auditory system and essential only for the more important principle of weight reduction in flying animals.

Similarly, we find that a system with very small tympana and very narrow tubes, as might occur in an insect, functions in essentially the same way as does the larger-scale system characteristic of a bird. The main difference is that more damping is provided by the narrow tubes and less by the tympana themselves. If the tubes are too narrow then directional effects are reduced but, provided their diameter is greater than a few tenths of a millimeter, in a small insect, they can provide enough coupling to give a directional response very much like that of Fig. 11.5.

It is tempting to try to explain the acoustic behavior of a pipe-coupled system in terms of standing waves in the auditory canal, and indeed the explanation is moderately successful, but makes certain hidden assumptions. The argument goes somewhat as follows. Suppose we look for a situation in which the contralateral tympanum does not move at all in an incident free field. This means that the internal pressure must exactly balance the external pressure at the contralateral tympanum, both in magnitude and phase. This suggests that the auditory canal is operating as a stopped-pipe quarter-wave resonator, with the ipsilateral tympanum being effectively an open end. The phase relations would appear to be correct (at a glance) and, to make the pressure amplitude at the contralateral tympanum equal to the external free-field pressure, we require only an appropriately large value of the wall damping in the pipe. This argument is, in fact, valid if we can assume the acoustic impedance of the tympana to be small compared with the characteristic impedance of the canal. A numerical check, however, shows that this can be achieved only if the tympanic membrane is substantially less than 10 um in thickness and the moving parts of the linkage to the neural transducer are substantially less than about 0.1 mg in total mass. The first of these conditions could reasonably be met in a real system, but the second fails to be met by at least a factor

10. The ipsilateral end of the canal is therefore by no means in an acoustically "open" state, and the approximate correctness of the prediction relies upon the hidden assumption of appropriate magnitudes for both the mass and damping of the tympana.

11.4 Complex Systems

Just as the frog-like system that we discussed in Section 9.8 was an elaboration of a pair of cavity-coupled ears, so we can consider pipe-coupled systems with more than two pressure ports and more complex topology. An example is the insect system shown in Fig. 11.6, which is typical of that found in various types of grasshoppers and crickets. This auditory system is closely connected anatomically with the system of internal tubes of the respiratory system, which lead air from spiracle openings in the thorax to the muscles of the legs. The tympana are very thin areas of cuticle on the forelegs near the ends of the tube branches. The relative diameters and lengths of various parts of the system vary greatly from one species to another, but the topology of their interconnection remains the same. Indeed this topology is also the same as that of the human auditory system if we consider just the ears, the Eustachian tubes, and the nostrils!

The analog circuit for this system is also shown in Fig. 11.6. The tubes leading down the legs are of fairly uniform cross-section and can be treated as cylindrical,

Figure 11.6 The more complex 4-port auditory system of an insect such as a cricket. The shapes of the various elements can vary widely between species, but the same analog network applies, with only the numerical values of the impedances being altered.

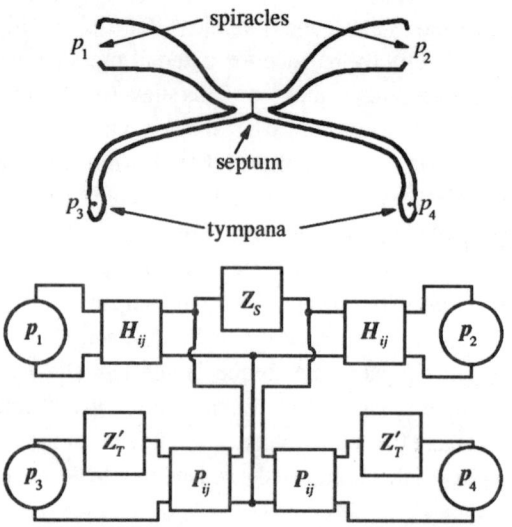

HIGH-FREQUENCY AUDITORY MODELS

with impedance coefficients P_{ij}, while the tubes leading to the spiracles form large horns in some species, and so are given the coefficients H_{ij}. The tympana have simple resonant properties, and the same is true of the septum symmetrically separating the system, though this latter may vary from a very light membrane to a heavy wall of tissue essentially dividing the system into two parts. The amplitudes of the acoustic pressures at the four ports of the system are nearly the same, and the relative phases can be readily evaluated by considering the path differences for an incident plane wave, as we did in Section 9.2 for the two-port case.

We shall not attempt to analyze this system in any detail. There are seven independent loops in the network, which we draw as symmetrically as possible for convenience—the central loop passing through the septum is necessarily antisymmetric—so that we have seven complex equations to solve. This formal solution can be achieved by careful algebra, or with the aid of a symbolic algebra computer program, but it is generally easier simply to use a standard program for the numerical solution of simultaneous complex linear equations once numerical values have been assigned to the various coefficients.

It is appropriate to emphasize that this model represents a whole collection of real systems that can have quite distinct properties. In some insects the tubes are fairly uniform and the central septum is very thin, so that the system behaves as a full four-port network as set out in the figure. In other insects, on the contrary, the septum may be thick, dividing the system into two virtually unconnected parts, and the tubes leading to the spiracles may form wide horns. While the complete network of Fig. 11.6 can still be used in this case, it is much simpler to analyze two independent networks, each corresponding to a single horn-loaded two-port ear.

Even this example does not exhaust the possible complexity that can be built into models. We can consider the addition of sacs of appreciable volume behind the tympana, or flaps partially closing the spiracles, and these may have significant acoustic effects. The techniques we have presented provide the machinery for investigating any of these cases.

References

System models: [21]

Discussion Examples

1. Derive an expression for the velocity response of a resonant tympanum closing the mouth of a tube, the far end of which is rigidly stopped. Insert typical numerical values and plot a frequency response curve.
2. Derive an expression for the velocity response of the auditory system of a bird with one tympanum missing.
3. Write a computer program to evaluate and display graphically the tympanum response for one of the systems discussed in this chapter, as given by equations (11.2), (11.4), or (11.6). [Take the magnitude of the response.] Investigate the behavior of the

response as the parameters of the system are varied away from those given in the tables.

Solutions

1. $U_2 = 0$ so $p = (Z_T' + Z_{11})U_1$ where Z_T' includes radiation and Z_{11} is a pipe coefficient. Hence U_1. In the plot, note the significant displacement of the resonances from the stopped-pipe frequencies.

2. Use the network of Fig. 11.4 with the right-hand $Z_T' = 0$. The final result is

$$U_1 = \frac{(Z_R + P_{11})p_1 - P_{12}p_2}{(Z_T' + Z_R + P_{11})(Z_R + P_{11}) - P_{12}^2}$$

since $P_{22} = P_{11}$.

12 THE INNER EAR

SYNOPSIS. The basic mechanical problem of the transduction organ is the achievement of frequency selectivity, so that some sort of spectral analysis of the acoustic signal can be made. At low frequencies, the firing of transducer cells is statistically synchronized with the exciting signal, so that the frequency analysis can be made at a higher neural level. At high frequencies there is no synchronization and neural impulses are correlated simply with signal intensity. If the individual transducer cells have mechanical resonant properties of their own, then they can form a series of secondary resonators in cascade with the resonance of the tympanum, as shown in a mechanical analog in Fig. 12.1. This results in an increase in frequency selectivity only if the mass of the transducer cells is small in relation to that of the primary resonator, otherwise a double-humped response curve is produced. The fraction of incident energy transferred to the transducer cell is then small.

In most animals, the transducer cells are coupled to a mechanical device that tends to separate, in space, vibrations of different frequencies imparted to it. An example of such a separation device is a simple tapered membrane, as discussed in Section 5.4. Transducer cells are then embedded in or attached to the membrane and are preferentially excited at different frequencies. Some insect systems bear a resemblance to this model.

In vertebrates, the analyzing membrane (basilar membrane) forms a length-wise partition in a fluid-filled tube, the cochlea, as shown in Fig. 12.2. Vibrations from the tympanum are communicated through a membrane-covered window to the fluid of the cochlea by the lever-like action of the ossicles of the middle ear. A second window in the base of the cochlea serves for pressure relief and imparts a transverse character to the fluid motion. This transverse fluid motion near the base of the cochlea generates waves in the basilar membrane that travel towards the cochlear apex. Because the stiffness or tension of the membrane decreases by a large factor from the base to the apex end, a wave of given frequency increases in amplitude and decreases in wavelength as it progresses along the membrane, as shown in Fig. 12.3. Near the point where the membrane is in resonance with the wave frequency, the wave energy is largely absorbed, and beyond this point there is only an exponentially damped non-propagating disturbance. The energy in waves of different frequencies is therefore absorbed at different places along the basilar membrane, and transducer cells distributed along its length are able to make an approximate frequency analysis of the incident signal.

The acoustic impedance of the cochlea as seen at its input window is essentially resistive and has a numerical value of order 1 mechanical ohm, depending on the dimensions of the cochlea. This is typically about 100 times the impedance of the tympanum near its resonance. The middle ear, which consists of one or more ossicle levers in cascade, is able to effect the necessary impedance transformation to achieve good matching if the total lever-arm ratio is about 10:1. Because of the mass and slight flexibility of the ossicles, the middle ear has a low-pass filter characteristic and attenuates high frequencies. A network analog for a complete vertebrate ear can be constructed as shown in Fig. 12.4.

12.1 Neural Transducers

This book is primarily concerned with the acoustics of the peripheral auditory system (and the vocal system) rather than with the inner ear, but it is appropriate to give a brief sketch of the acoustics of this neural transduction organ so that we can see how it fits into the total system. So far we have considered it simply as a mechanical load linked to the tympanum by a lever, and indeed this is an adequate description as far as the peripheral auditory system is concerned. However we would like to have some understanding of the factors determining the mechanical impedance of the transducer organ so that reasonable numerical values can be used in analyzing the whole system.

More than this, of course, the inner ear is the core of the whole system, for it connects the external world of acoustic pressures and flows to the internal world of neural impulses and sensory perception. Again, the variety of transducer organs that have evolved is considerable. Much research attention has focused upon the vertebrate cochlea, and most current auditory research is aimed at understanding the detailed mechanics and neurophysiology of that organ. We shall draw a firm line at the cellular level, but it is possible to make some useful generalizations about the behavior of the mechanical parts of the transducer organ.

With the exception of those belonging to fish, all the auditory structures we have discussed so far operate entirely in air. The neural transducers, in contrast, are always submerged in fluid, either simply in that of the surrounding tissues or, more usually, in a special volume of fluid that is intimately connected with the operation of the transducer organ. Part of the task, therefore, involves understanding the motion of dense fluids in confined spaces, a subject that is a good deal more complex than anything we have yet considered. We shall therefore be content with a brief and semiquantitative discussion, recognizing that many of the issues are still unresolved.

12.2 The Transduction Problem

In our discussion of auditory systems we have taken the analysis to the stage of calculating the motion of the point on the tympanum to which is attached the neural transducer when the ear receives a sound wave of given frequency and intensity incident from a specified direction. We have seen that many auditory systems can provide a broad yet spectrally shaped frequency response that is quite similar to the general sensory capacity of the animal in question, and it is therefore tempting to regard the neural transducer as simply a broad-band linear device that converts this motion into nerve impulses. The real situation appears to be more complex than this, however, in even the simplest animals.

The basic reason behind the added complication is to enhance the frequency discrimination of the ear. Neural transducers produce nerve impulses on a statistical basis, with the probability of firing being related to the mechanical deformation of the transducer cell. At low frequencies the impulses tend to be concentrated at a particular phase of the sensory signal, so that the summed response of several nerve fibres conveys information on the frequency, and to some extent the waveform, of the exciting sound. At higher frequencies this phase locking may be lost, so that the summed response indicates only the intensity level. Most animals enhance auditory discrimination by having a number of transducer cells that are excited in such a way that each responds preferentially to some part of the frequency spectrum. This gives immediate information on the frequency distribution of the incident signal, and may also allow the detection of a signal at one frequency in the presence of an intense signal at some other frequency. The physical mechanism for frequency separation, which we shall discuss in this chapter, ranges from having just a few cells of different sizes embedded in a membrane, in the case of some simple insects, to tens of thousands of hair cells distributed along a frequency-selective membrane structure in the case of the vertebrate cochlea.

In addition to the frequency-selective transduction problem, there is also the problem of efficient matching between the transduction mechanism and the vibrating tympanum. In virtually all cases this is accomplished by some sort of lever mechanism, either in the form of a simple oblique attachment to the tympanum or, in the case of vertebrates, through a chain of small ossicles that constitute several levers in series. We saw in equation (8.11) of Section 8.3 that a lever can act as an impedance transformer with a low-pass frequency characteristic determined by the mass and compliance of its arms, and indeed this is part of its function in auditory systems. We reserve discussion of this role until we have some estimate of the magnitude of the impedance transformation required.

12.3 A Frequency-Dispersive Membrane

It is tempting to think of a frequency-dispersive transduction system in terms of individual transducer cells, each with high-Q resonance behavior, driven by the rather low-Q peripheral auditory system. Each cell then forms a so-called "second filter" for the response of the system, and we might hope that the frequency selectivities might be compounded. Unfortunately the situation is not as simple as this. Figure 12.1 shows one possible mechanical version of such cascaded tuned circuits. The primary resonator, equivalent to the tympanum, is represented by a mass L_1 supported on a spring of compliance C_1. We assume the motion of the mass to be accompanied by viscous losses equivalent to a resistance R so that the Q value is finite. The secondary resonators, representing the tuned transducer cells, consist of smaller masses $L_2, L_3, \ldots$ attached to the primary resonator by their own springs $C_2, C_3, \ldots$. The transducer cells are then assumed to respond to the velocity of the small masses through the surrounding fluid, detected for example by protruding hair cells. Other arrangements are possible, but this one brings out the main features.

The electric analog circuit for this arrangement is also shown in Fig. 12.1. The primary resonator, which is driven by a force applied to the mass L_1, appears as a series resonant circuit, as we discussed in relation to Fig. 8.2. The secondary resonators, on the other hand, are driven through their attached springs and appear as parallel resonant circuits, as also shown in Fig. 8.2. The equivalent network for the case of two secondary resonators is thus as shown in Fig. 12.1.

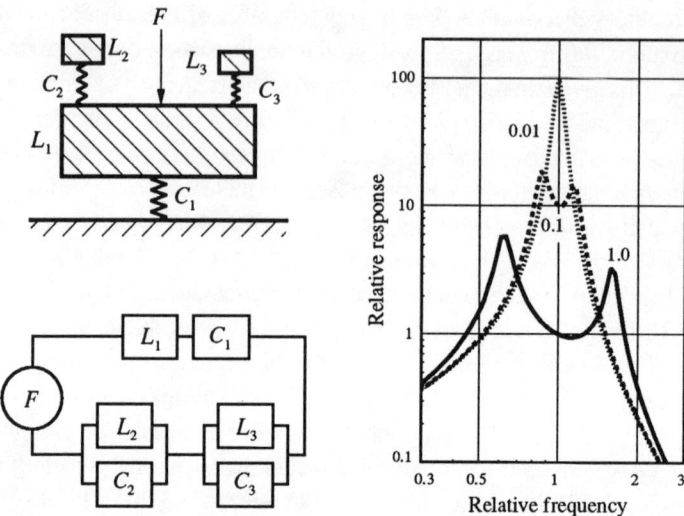

Figure 12.1 An arrangement of cascaded resonators, represented by small spring-mounted masses attached to the large mass of the primary resonator, which is itself supported on a spring. Below is shown the electric analog network for this arrangement. The graph illustrates the velocity response of a single secondary resonator, tuned to the same frequency as the primary, when both have $Q = 5$ and their masses have the ratio given as a parameter.

We shall leave the explicit solution of this network as a problem for the reader and simply illustrate the results. Suppose that we have just one secondary resonator and that its resonance frequency and Q are both the same as those of the primary resonator. The only variable is then the ratio of the masses of the two resonators. The response of the secondary resonator, corresponding to the transducer cell, for the particular case $Q = 5$ is shown in Fig. 12.1 for three assumed mass ratios, 0.01:1, 0.1:1 and 1:1. When the mass of the secondary resonator is very small, the resonances do in fact appear in cascade, and the effective Q, which shows up as the numerical value of the peak height on this normalized diagram, is just the product of the individual resonator Qs. In this case, however, very little energy is actually transferred to the secondary resonator and most is dissipated in the losses of the primary one. As the mass of the secondary resonator is increased, the resonant peak first broadens and then splits, so that when the mass ratio is 1:1 there are two distinct peaks of nearly equal height, and the response amplitude is less than that which would have been reached by the primary resonator alone. A common feature of all the curves however is that, well above the resonance, the velocity response declines at 18 dB/octave rather than at the 6 dB/octave characteristic of a simple resonator. The decline below the resonance is 6 dB/octave in both cases. A mechanical multiplication of the frequency selectivity is thus possible, but relies

THE INNER EAR

upon the secondary resonator having a reasonably high Q value and a very small mass. The individual cells of the neural transducer are indeed small in mass compared with the tympanum, but it is not at all evident that they could achieve a significantly high Q value, because they are generally immersed in fluid.

A more sophisticated approach, which is what is found in all real systems, makes use of some sort of frequency-dispersive mechanical element to separate the points of attachment of the different transducer cells and to match them, through its transfer impedance, to the primary circuit. A simple example is the tapered membrane discussed in Section 5.4 and shown in Fig. 5.5. The membrane can be excited at a position close to its apex where its input impedance is high (since the stiffness is large and the motion always small), and appropriate transducer cells are attached to or embedded in the membrane at the positions shown by arrows, where the membrane motion is maximal for a given frequency. The actual membrane discussed in Section 5.4 would not make a very suitable dispersive element, since it is itself resonant and responds only close to its own mode frequencies. This could be an advantage if only a very few cells were involved, but a propagating-wave system in which waves originating near the apex O were absorbed by damping material on the edge A would be acoustically more flexible. We might expect cells designed to respond to high-frequency signals to be smaller than those responding to low frequencies, but it is not clear that they could be tuned mechanically to a high-Q state because of the damping effect of the surrounding cellular fluid.

12.4 Cochlear Mechanics

In higher animals such as mammals, the frequency-dispersive transducer system, known as the cochlea, is developed to a high degree of sophistication. We have space for only a brief discussion, but this is enough to show the general principles.

As shown schematically in Fig. 12.2, the cochlea consists basically of a fluid-filled tapered tube, divided longitudinally by a membrane, the basilar membrane, upon which are mounted thousands of transducer hair cells in close proximity to the neighboring tectorial membrane. Motion of the basilar membrane excites these hair cells by shear deformation. Vibration of the tympanum is communicated by an ossicle link to the fluid inside the cochlea through the membrane-covered entry window (oval window) in the base of the cochlea, while a smaller relief window (round window) on the other side of the basilar membrane allows pressure relief. There are other membranes in the cochlea that need not concern us. The basilar membrane itself is narrow near the base of the cochlea and broadens towards its apex, where there is an aperture, the helicotrema, allowing communication between the two halves of the duct. There is thus a superficial resemblance between the basilar membrane and the simple tapered membrane of Section 5.4, but the behavior in the real case is strongly influenced by the motion of the surrounding fluid. The cochlea is coiled into a spiral shape, but this does

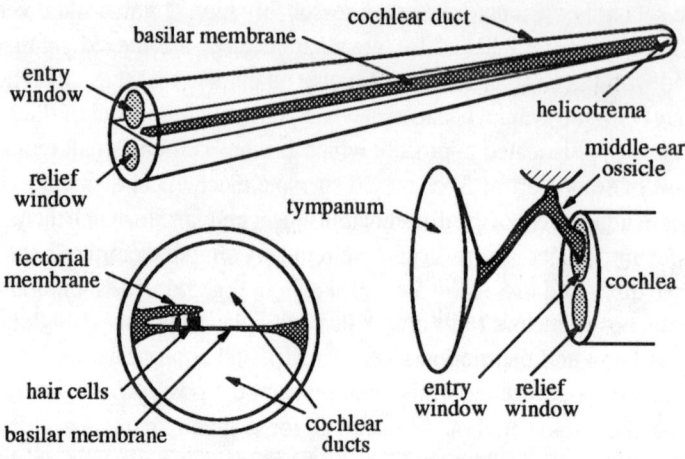

Figure 12.2 (a) Schematic diagram of the structure of a vertebrate cochlea, straightened to remove the spiral; (b) a section through the cochlea; and (c) a schematic diagram of a complete auditory system.

not have any significant effect upon its functioning. The length of the cochlear duct depends on the size of the animal, ranging from about 5 mm in the chicken to 60 mm in the elephant.

Experiments by Georg von Békésy and later workers have shown that tension or stiffness of the basilar membrane is not uniform, but decreases almost exponentially as one progresses away from the base of the cochlea, where the membrane is narrow, to the cochlear apex, where it is broad. The acoustic compliance per unit length typically varies by a factor of nearly 10^5 over the membrane length, while the membrane width varies by only about a factor 5. Von Békésy's measurements and experiments preceded any analytical discussion of cochlear mechanics, but for our present purposes an elementary theoretical treatment is the best way to proceed. With this in view, we neglect the taper of the cochlea and simply suppose the elastic stiffness of the basilar membrane to vary smoothly from the base to the apex of the cochlea. Typical values of the model constants involved are given in Table 12.1. Although the cochlear response exhibits nonlinearities at the neurophysiological and perceptual levels, it is unlikely that the amplitude of the vibrations within it is sufficient to involve any mechanical nonlinearity. A linear treatment is therefore appropriate in the first instance.

In our model we are concerned with the behavior of two long volumes of liquid separated by a flexible elastic partition of graded compliance. Even at the upper limit of hearing, the wavelength of sound in water is large compared with the length of the cochlear spiral, so that we can regard the fluid motion as essentially incompressible. It is also reasonable, as a first approximation, to neglect variations

THE INNER EAR

Table 12.1 Assumed parameters for a cochlear model

Channel length	$l = 30$ mm
Channel diameter	$d = 5$ mm
Membrane stiffness at base	$\beta_0 = 10^{14}$ Pa m^{-2}
Membrane stiffness at apex	$\beta(l) = 10^9$ Pa m^{-2}
Stiffness coefficient	$\alpha = 400$ m^{-1}
Membrane acoustic inertance	$m = 500$ kg m^{-1}
Input acoustic impedance	$Z_{IN} = 10^9$ acoust ohms
Input mechanical impedance	$Z_{IN}^{(M)} = 1$ mech ohm

in flow velocity across either channel, and to take the duct cross-section to be uniform, so that the problem becomes one-dimensional and analogous to an electrical transmission line with properties changing along its length.

As a first approximation again, we can concentrate attention on the behavior of the liquid, since its mass is very much greater than that of the membrane and its attached structures. Because the liquid is incompressible, its volume flow U along the upper channel, represented as usual by a complex number, is exactly balanced by the return flow along the lower channel, but there may be an oscillating volume exchange (though not an actual fluid exchange) caused by displacement of the basilar membrane. Suppose that x measures distance along the cochlea and that $Z(x)$ is the acoustic impedance per unit length for flow along the two channels (regarded as being in series) and $Y(x)$ the acoustic admittance per unit length for volume flow through the membrane. Both these quantities will, of course, depend on the excitation frequency ω. Then if $p(x)$ is the acoustic pressure at point x, we can write

$$\frac{\partial p}{\partial x} = -Z(x)U(x) \qquad \frac{\partial U}{\partial x} = -Y(x)p(x). \qquad (12.1)$$

The first of these equations describes the pressure gradient required to move the fluid along the channels, and the second gives the loss of channel flow caused by displacement of the membrane. If we differentiate the first of equations (12.1) with respect to x and use the second equation, then, since we have assumed the cross-sectional area and hence the series impedance Z to be independent of x, we find

$$\frac{\partial^2 p}{\partial x^2} - ZYp = 0. \qquad (12.2)$$

If we had not assumed a duct of constant cross-section, then there would have been an additional term involving $\partial Z/\partial x$, but neglect of this does not change the essential behavior.

We see that (12.2) looks very much like a simple wave equation, but its solution is complicated by the fact that the product ZY depends upon the coordinate x. We shall not go into the detailed solution of the equation, since fortunately a quite simple discussion reveals the essentials of the behavior. Thus, if we consider just a small length of the duct and a rather high frequency, ZY will not change significantly over this length or over a wavelength of the disturbance, and it is reasonable to regard the solution as a propagating wave of the form $a(x)e^{j(\omega t - kx)}$, where $k = \omega/c'$ is the wave vector associated with a wave velocity c' and we expect the amplitude $a(x)$ to vary slowly with x. Substituting this into (12.2) we find

$$k(x) = [-ZY(x)]^{1/2} \qquad c'(x) = \frac{\omega}{[-ZY(x)]^{1/2}}. \qquad (12.3)$$

We now need to substitute simple expressions for the channel impedance Z and the membrane admittance $Y(x)$. Suppose the total cochlear duct cross section is S and the density of the aqueous liquid in the duct ρ_W, then if we neglect wall friction the series impedance is a simple inertance

$$Z = \frac{4j\omega\rho_W}{S} \qquad (12.4)$$

per unit length. The factor 4 comes from the fact that we really have two half-channels in series, as far as the flow is concerned. If we assume that the membrane has a constant mass per unit length, then this contributes an acoustic inertance m that is numerically equal to the mass of unit length (1 meter) of membrane, divided by the square of this membrane area. Similarly, the acoustic stiffness $\beta(x)$ is numerically equal to the pressure required to cause a volume distortion of 1 m³ in unit length of the membrane. We also allow for a damping resistance denoted by γ. Then the admittance per unit length of membrane can be written as

$$Y(x) = [j\omega m + \gamma - j\beta(x)/\omega]^{-1} \qquad (12.5)$$

so that, from (12.3),

$$k(x)^2 = -\frac{4j\omega\rho_W}{S[j\omega m + \gamma - j\beta(x)/\omega]}. \qquad (12.6)$$

The behavior of this wave on the basilar membrane divides conveniently into three regimes. Close to the base of the cochlea near source of the excitation at the

THE INNER EAR   223

oval window, the membrane is very stiff and the term $j\beta/\omega$ dominates the denominator. We can then simplify (12.6) to the form

$$k(x) = \left[\frac{4\rho_w}{S\beta(x)}\right]^{1/2} \omega . \qquad (12.7)$$

The propagation in this region is in the form of a simple undispersed wave, since k is real and proportional to ω, but the wave velocity $c' = \omega/k(x)$ decreases as $\beta(x)^{1/2}$. If we represent the measured stiffness of the membrane by the exponential relation

$$\beta(x) = \beta_0 e^{-\alpha x} \qquad (12.8)$$

then the wave speed becomes

$$c' = \left(\frac{S\beta_0}{4\rho_w}\right)^{1/2} e^{-\alpha x/2} . \qquad (12.9)$$

The decrease in wave speed with increasing x in this region has two effects. In the first place, the wavelength decreases but, rather more importantly, the amplitude increases. The simple reason for this is that the energy passing any point along the channel must be constant and therefore the energy density, measured by the amplitude, must be higher in regions of low wave velocity. The shape of the wave on the membrane must therefore look like that in the region OA of Fig. 12.3. A more rigorous solution of equation (12.2) verifies these conclusions, though the details and the exact form of the mathematical result are a little different.

The second propagation regime to be considered is that where ω is very nearly equal to the local resonance frequency of the membrane. Near this point the imaginary terms in Y nearly cancel, so that $Y \approx \gamma$ and, from (12.6)

$$k^2 \approx -\frac{4j\omega\rho_w}{\gamma S} \qquad k \approx \left(\frac{2\omega\rho_w}{\gamma S}\right)^{1/2} (1-j) . \qquad (12.10)$$

The second form of writing comes from the general rules for simplifying complex expressions set out in Appendix A. In this resonance region, therefore, the wave is highly damped and its energy is absorbed by the membrane and its associated structures. It is obviously important to know the location $x^*(\omega)$ of this resonance region on the basilar membrane, shown as A in Fig. 12.3. This is given by (12.8) and the resonance requirement to be

$$x^*(\omega) = \frac{1}{\alpha} \ln\left(\frac{\beta_0}{m\omega^2}\right) . \qquad (12.11)$$

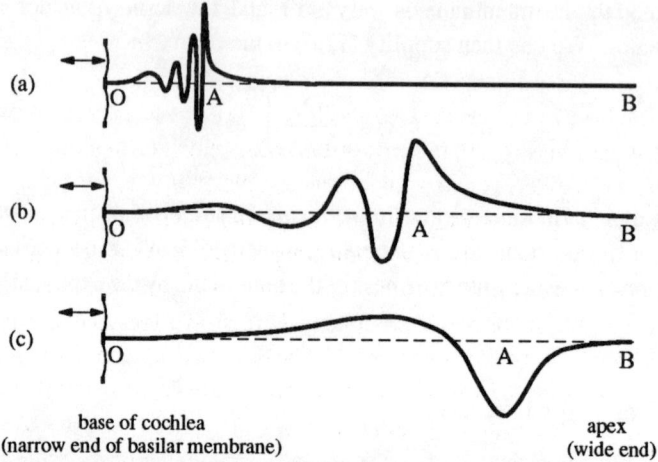

Figure 12.3 Wave propagation on the basilar membrane. The input is at the base end O of the cochlear spiral, the membrane comes to resonance near the point A, and the disturbance is then attenuated towards the cochlear apex B. The drawings show schematic responses for (a) a high frequency, (b) a medium frequency, and (c) a low frequency.

For the physical parameters given in Table 12.1, the local membrane resonances cover a frequency range from about 30 kHz near the oval window where the membrane is stiff down to about 100 Hz at the apex of the cochlea where the membrane is very slack. It is significant that the distribution of resonance positions depends upon the logarithm of the frequency, and that such logarithmic discrimination is characteristic of human auditory perception.

The third propagation regime is that for $x > x^*(\omega)$. In this region $Y \approx j\omega m$ and, from (12.6),

$$k^2 = -\frac{4\rho_W}{Sm} \qquad k = -j\left(\frac{4\rho_W}{Sm}\right)^{1/2}. \qquad (12.12)$$

The wave no longer propagates, but we have an in-phase displacement which is sharply attenuated with distance, as shown in the region AB of Fig. 12.3.

This completes the description of the behavior of the cochlea in this approximation. It accords quite well with the results of experiments on real cochleas. We see from Fig. 12.3 that basilar membrane motion is concentrated in a characteristic position for each exciting frequency, and that this characteristic region moves towards the base of the cochlea as the frequency is increased. Since the base of the cochlea corresponds to the narrow end of the basilar membrane, this behavior is similar to that found in Section 5.4 for a simple tapered uniform membrane, but basilar membrane behavior is altogether more complex because of the effect of the surrounding fluid and the variation of membrane properties along its length.

The success of this model in accounting semiquantitatively for most of the mechanical behavior of the cochlea should not be allowed to obscure the fact that some of its assumptions are rather far from being realistic. In particular, the mass of the basilar membrane is an important parameter in the theory, while a little consideration shows that its inertia will be greatly influenced by the neighboring liquid that moves with it in a direction transverse to the axis of the cochlea. A more realistic model must include these transverse fluid motions in its initial formulation. The details are too complicated to consider here, but the same general picture of cochlear mechanics emerges. The effective mass, or more properly the acoustic inertance m, given in Table 12.1, makes a reasonable but rather arbitrary allowance for this co-moving fluid mass.

The important feature of cochlear mechanics is that both basilar membrane motion and energy absorption are concentrated in the region close to the membrane resonance at the frequency concerned, so that for a wide-band signal there is a rough spectral sorting along the length of the basilar membrane. Different clusters of sensory cells, and ultimately of auditory nerve fibers, are thus excited primarily by different frequencies. This "place theory" of auditory spectral analysis was first put forward by the great German physicist and physiologist Hermann Helmholtz (1821–1894). In the case of human hearing it is certainly not the whole story from a psychophysical point of view, particularly at low frequencies when neural impulses are statistically synchronized with the exciting wave, but it does represent a most important attribute of the neural transduction organ. If we assume sensory hair cells of comparable sensitivity and number density to be distributed along the whole length of the basilar membrane, then we should expect the total sensitivity curve to be quite similar to that determined at the input to the cochlea by the peripheral auditory system, but in addition we should expect a spectral resolution that is roughly logarithmically compressed at the high-frequency end of the spectrum in accord with (12.11). This is, in essence, what is found for real auditory systems. There is also the possibility, evolved in animals such as bats, for having a special section of basilar membrane over which the properties change more gradually so as to give increased neural output and frequency separation in an acoustically important spectral region.

12.5 The Middle Ear

There is one further important thing that can be deduced from our simple model, and this is the mechanical load presented by the cochlea to the peripheral parts of the auditory system. The analysis of the model suggests, but does not prove, that waves traveling along the basal membrane are completely absorbed near the resonance point, so that there is no reflected wave. This seems to be true to a good approximation for most cochlear systems. If we assume it to be true, then the impedance presented at the oval window can be calculated from a simpler model in which the cochlear structure extends to infinity and the membrane has uniform properties similar to those in the region within about one duct radius of the basal end. From (12.1), if we divide the first equation by the second, we find

$$Z_{IN} = \left[\frac{Z}{Y(0)}\right]^{1/2} \approx \left[\frac{4\rho_W \beta(0)}{S}\right]^{1/2} \tag{12.13}$$

where the second form comes from (12.4) and (12.5) on the assumption that membrane motion is dominated by stiffness near the base of the cochlea. We see that this input impedance is a simple real resistive quantity, and independent of frequency.

If we insert the physical values given in Table 12.1 for a medium-sized animal, then (12.13) gives an acoustic input impedance of order 10^9 acoustic ohms. If we assume the oval window to have an area S_W equal to about half that of the total duct cross section, then this converts to a mechanical input impedance of magnitude

$$Z_{IN}^{(M)} = Z_{IN} S_W^2 \approx 1 \text{ mechanical ohm.} \tag{12.14}$$

This estimate is fortunately independent of the mass of fluid assumed to move with the membrane, and depends only on the compliance of the membrane and the dimensions of the system. It will, of course, vary with the overall size of the animal and its auditory system.

We can now look at the matching problem between this load and the tympanum. We discussed the acoustic impedance of the tympanum in Section 8.7 and gave explicit expressions in Table 8.1. If we assume, again for a medium-sized animal, a tympanum of area $S_T = 1$ cm^2 and thickness $d = 10$ μm, then Table 8.1 shows that the mechanical inertance of the tympanum has magnitude

$$L_T^{(M)} = L_T S_T^2 \approx 1.4 \rho_s d S_T \approx 0.01. \tag{12.15}$$

If we take $Q \approx 2$ for the unloaded system, then this gives $R_T^{(M)} \approx 0.005$ mechanical ohms. For optimal matching, as discussed in Section 9.6, we should expect that the load applied by the neural transducer should be resistive and about equal to $R_T^{(M)}$ in magnitude, thus reducing the loaded Q value to 1.

From our discussion above, the transducer load is indeed primarily resistive, but its magnitude is perhaps 100 times as large as is required for matching. The ossicle links of the middle ear go a large way towards providing this impedance match. From our discussion of levers in Section 8.3, the impedance transformation achieved by a simple lever is equal to the square of the ratio of its arm lengths, as given in (8.11), and it is possible to cascade two or more levers to magnify the effect. A total arm-length ratio of about 10 is therefore adequate to provide the necessary matching. If the oval window is significantly smaller in cross section

THE INNER EAR

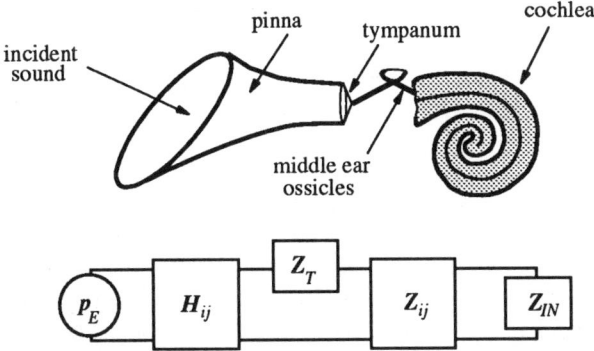

Figure 12.4 General network model of a complete vertebrate ear. An equivalent pressure generator p_E drives a tympanum of acoustic impedance Z_T through a horn with impedance coefficients H_{ij}. Motion of the tympanum is transmitted through the middle ear lever, with impedance coefficients Z_{ij} to the cochlea, which has an input impedance Z_{IN}.

than the cochlear half-duct area, then this reduces the mechanical input impedance of the cochlea, as expressed by (12.14), and further aids in the matching.

It is interesting to estimate the absolute basilar membrane displacement near the auditory threshold. At the normal human threshold of about 20 μPa r.m.s. at 1000 Hz, the vibration amplitude in air is only 10^{-11} m. The horn pressure gain in the pinna is 10–20 dB and the vibration amplitude of the tympanum is thus about 10^{-10} m, as can be derived from Fig. 11.2. This increase is reversed by the lever action of the middle ear, so that basilar membrane displacements are probably again about 10^{-11} m. This is about one-tenth of an atomic diameter, and therefore just sufficient to influence the opening of ion channels through the sensory cell membranes when motion displaces the sensory hairs.

It is now quite straightforward to formulate a model and an analog network for a complete vertebrate auditory system as shown in Fig. 12.4. An equivalent pressure source p_E, as discussed in Section 10.5 and given explicitly by (10.21), drives a horn with impedance coefficients H_{ij} given in terms of its dimensions by the formulae in Appendix B. This in turn is linked to a tympanum of impedance Z_T, given in Section 8.7 and Table 8.1. The levers of the middle ear are represented by one or more cascaded two-port networks with coefficients Z_{ij} given by (8.14), and this drives the resistive impedance Z_{IN} of the cochlea, given by (12.13). We do not usually have enough data to calculate all the impedances involved in this chain in detail, but the general behavior is fairly clear. We should note again, as discussed in Section 8.3, that the lever is not just a simple mechanical transformer, but acts in addition as a low-pass filter because of its mass and compliance. Each part of the system therefore contributes to the frequency response of the whole. The network calculation then gives us the mechanical motion of the input window to the cochlea, but we need a rather more detailed cochlear model than we have been able to describe here in order to convert this into a quantitative estimate of the motion imparted to the hair cells.

12.6 Active Cochlear Response

All our discussion above has assumed that the transducer cells of the cochlea simply provide a dissipative mechanical load to the basilar membrane. Such an assumption is certainly a reasonable first approximation, but there is increasing evidence that the transducer cells may have a more active role. This could come about if, for example, the mechanical stimulation of a cell gave rise to an electrical response that caused further mechanical deformation of the cell or of its neighbors. This response could be linked either to the small continuous potentials produced by cell deformation—the cochlear microphonic—or could have a nonlinear threshold-type response associated with the generation of neural impulses. In either case the phase of the mechanical response is important, and it could lead either to further mechanical dissipation, to negative damping, or to a change in the reactive component of mechanical impedance. Experimental observation of spontaneous or stimulated oto-emissions suggests that the neural response may contribute a negative damping under some circumstances, but the picture is not yet clear.

Another aspect of cochlear behavior that appears to have its origin in the transducer cells themselves is the nonlinearity observed in many neurophysiological and psychophysical experiments. This includes auditory masking of one tone by another of neighboring frequency, difference tones, and many other nonlinear effects. It is quite unrealistic to expect that the mechanical vibration of elements of the peripheral auditory system or of the inner ear should exhibit any such nonlinearity at the extremely small vibrational amplitudes involved, but cellular phenomena are quite another matter. Relying, as they do, upon mechanical opening or closing of ion channels in the cell membrane, these phenomena are inherently nonlinear. When we include the electrical interaction between neighboring transducer cells, the range of possible phenomena becomes very complicated. Any discussion would take us outside the range of this book.

References

Cochlear physiology: [38], [41], [44]
Cochlear models: [38] Ch 7

Discussion Examples

1. The ossicle lever in a middle ear consists of a bent bony rod, 1 mm in diameter, pivoted at its bend to form two arms of lengths 5 mm and 2 mm respectively. Estimate (a) the impedance transformation achieved, (b) the mass-like load contributed by the ossicle at the tympanum at low frequencies, and (c) the cut-off frequency of the middle ear.

Solutions

1. (a) The bend has no effect on the lever operation. By (8.11) the impedance transformation is $(2/5)^2 \approx 0.16$.

    (b) From the discussion in example 5 of Chapter 8, or the result in (8.12), the load contributed by the long arm of the lever is equivalent to a mass equal to one-third that of the lever arm, and thus about 2 mg if we assume the density of bone to be 1500 kg m^{-3}. The mass load of the short arm is equal to one-third of its mass (thus about 1 mg), multiplied by the impedance transformation achieved by the lever (0.16). This load is negligible compared with that contributed by the long arm.

    (c) As discussed in example 5 of Chapter 8, the cutoff frequency is approximately that of the first resonance of the long arm when it is clamped at the pivot. By (3.36), assuming a Young's modulus of 10^{11} Pa for bone, this resonance frequency is about 70 kHz. We get the same result from the simple LC circuit of Fig. 8.3, using (8.12).

13 MECHANICALLY EXCITED SOUND GENERATORS

SYNOPSIS. An active animal can typically produce a sustained mechanical power output of up to 10 watts per kilogram of body weight, or 10 mW/g, though humans can work at only about one tenth of this rate. Humans and other vertebrates can typically employ only a small fraction, 1 to 10%, of this power for sound production, but insects may be able to use as much as 50%. These figures serve as a guide to the available power input to the acoustic system but, as we see later, the efficiency with which it is converted to radiated sound is very small.

Many animals, particularly insects, produce sound by the direct mechanical excitation of a resonant radiating structure. There are several varieties of such excitation mechanisms. In the simplest case the excitation is impulsive and the sound is a click or more resonant "pop". The excitation mechanism usually involves a steady muscular strain against a notched member which then slips suddenly. The excited structure may typically be a simple shell or a vented air-filled cavity. This excitation is analogous to plucking the string of a guitar or striking a bell, though the damping is generally much more rapid. In aquatic animals, the most efficient structure of this type is a closed air-filled bladder, the radial vibrations of which radiate very efficiently.

Shell, plate or membrane elements can also be excited by a frictional mechanism analogous to the bowing of a violin, as illustrated in Fig. 13.1. Even though the mode frequencies of the vibrating element may not have any simple relationship, the nonlinearity responsible for controlling the amplitude of the vibration effectively locks all overtones into simple harmonic integer relationship with the frequency of a dominant fundamental. The efficiency of this mechanism is low, partly because of frictional losses but principally because of the dipole nature of the radiator, which is usually part of a wing. An estimate of mechanical to acoustic conversion efficiency is as low as 0.002%.

In a more efficient version of this mechanism, frictional excitation is replaced by relative motion between a sharp pick on one element and a ribbed file on the other, as shown in Fig. 13.2. If the frequency generated by passage of the ribs under the pick is approximately equal to the resonance frequency of the driven element, then this determines the vibration frequency. More generally, if the drive frequency is considerably below the fundamental frequency of the resonator, then nonlinear effects come in to ensure phase coherence between successive impulsive excitations by the teeth of the file. Because little friction is involved, this mechanism is more effective than that above, and a conversion efficiency of order 0.01% is probably typical.

In both these cases, however, if one side of the source is shielded by the animal's body so as to reduce the amplitude radiated by say 30%, then the source behaves like a monopole of strength 0.3 superposed on a dipole of strength 0.7. The monopole is typically 50 times more efficient than the dipole and consumes 10% of the input energy, so that the overall efficiency is raised by a factor of about 5.

The third common form of mechanical excitation depends upon the sequential buckling of a ridged shell, or tymbal, closing the opening to an air-filled cavity. The combination constitutes a sort of Helmholtz resonator and, being a monopole source, it is particularly efficient. The principal losses are internal to the buckling shell, and conversion efficiencies as high as 5% may well be achieved.

13.1 Sound Production

Just as important as hearing for the purpose of communication between animals of the same species is the need for a means of producing controllable sounds. The variety of possible sounds is very great, with frequency and time variation being the two major parameters. We shall return to look at this in more detail in Chapter 15; here we are concerned with the mechanics of sound production.

Before going on to consider mechanisms in detail, it is helpful to have a general idea of the power available to an animal for muscular activity or for sound production. Humans are rather sedentary animals and have a background metabolism rate of about 100 W or about 1 watt per kilogram of body weight, which is perhaps typical of warm-blooded animals. As a measure of sustainable muscular exertion we take the rate of doing work against gravity, which is easier to quantify than steady horizontal motion. A 100 kg human can walk rapidly up an incline of about 1 in 10 at a speed of about 1 m/s. This represents a steady power of just 100 W or about 1 watt per kilogram of body weight. The peak power production by a human, corresponding to running up stairs, is perhaps 5 times this level. A more active animal, such as a dog or a bird, can run or fly up such an inclined path nearly 10 times as quickly as a human. This represents a sustainable power of about 10 W/kg or, expressed more usefully for small animals, of 10 milliwatts per gram. We get the same answer for horizontal bird flight at 10 m/s if we assume a ratio of aerodynamic lift to drag of about 10:1.

In mammals, and indeed in vertebrates generally, only a small fraction of this power can be used to drive the respiratory system, and thus the voice. A figure of 1% is typical, with a maximum vocal effort using up to 10% of the available steady muscular power. There is, however, a good deal of difference between species, and some animals, such as humans in quiet conversation, typically use much less than their possible maximal vocal effort. In insects, however, sound production assumes a much greater importance, and the energy expended can be comparable with that devoted to flying—say 30 to 50% of the maximum sustainable power, and thus perhaps as much as 5 mW/g. These figures will provide a useful reference point in our discussion in this and the following chapter, but we should caution at the outset that vocal systems are not very efficient at converting muscular energy to sound, the acoustic output power typically ranging from about 0.01% to 1% of the muscular input power.

There are two principal means by which animals produce sound. The first is by means of mechanical excitation of some structure that couples well to the surrounding air, generally a membrane or plate, while the second involves the pneumatic power of air released from the lungs. Insects have only the first option available to them, while higher animals can use either or both. In this chapter we consider only mechanically produced sounds, leaving pneumatic sounds to Chapter 14.

Among the mechanical sound sources we can make a further simple distinction. In one case the sound generator is controlled directly by neuromuscular action, which produces an oscillating force at the desired frequency and applies this to the vibrating structure. While this is conceptually straightforward, it is difficult to implement except at quite low frequencies, and is not generally used for the primary purpose of sound production. A good example is the rapidly beating wing of an insect, the purpose of which is not primarily to produce sound but rather aerodynamic lift and thrust, but which nevertheless does radiate appreciably. Details of the sound radiated by a set of flapping wings are complicated, but for our present purposes we can obtain an estimate of the radiated power by the methods set out in Chapter 7 for simply vibrating plane bodies. The second and more important class of generators is that in which the muscular power is supplied essentially as a steady force or motion, and the mechanism of the sound generator converts this by mechanical means to a high-frequency vibration. Steady in this context means, of course, steady during stroke of the exciting apparatus: the force is generally supplied by a low-frequency reciprocating motion. The song of the insect will therefore generally consist of a repetitive series of nearly steady tone bursts. It is with this second class of generator that we shall be primarily concerned.

One general comment is appropriate before we commence the more detailed discussion, and this is that the action of any generator that converts steady motion into oscillating motion almost always, in practice, depends for its operation upon some sort of negative-resistance behavior. Usually this will be associated with some sort of nonlinearity in the generating process, and certainly such a nonlinearity turns out to be essential for determining the amplitude of the driven vibration. The vibrating part of the system will generally be quite linear in its behavior.

13.2 Impulsive Excitation

The first system we consider appears to be an exception to our statement about negative resistance and nonlinearity, but does not in fact involve a steady force. The animal simply applies an impulsive force to the vibrating element to set it into motion, and its vibration then dies away in a linear fashion as the imparted energy is dissipated by radiation and by internal losses. The radiated signal generally contains components at frequencies corresponding to those of all the normal modes of the vibrator, and these are individually exponentially damped.

The simplest case to discuss is one in which the vibrator is actually not mechanical, but consists of the air in a cavity forming a Helmholtz resonator. The mechanical excitation can be applied by causing a sharp deflection of the wall of the cavity, thus changing its volume and hence its internal pressure or, more effectively, by closing its mouth with an elastic stopper that is suddenly withdrawn. This latter procedure is just the one that we use to make a "pop" sound with a finger withdrawn from the mouth or from the neck of a bottle. The initial amplitude of the pressure excursion is determined by the impulsive volume change δV and has the value $\delta p = \gamma p_0 \delta V/V$ where p_0 is the static air pressure and V the cavity volume. The subsequent behavior is determined by the analog circuit in the lower part of Fig. 8.5 with the current generator inactive and the values of L, C, and Z_R given by Table 8.1. The initial amplitude of the acoustic current is $U_0 = \delta p/L\omega_H$, where ω_H is the Helmholtz frequency of the resonator, given by (8.20), and the radiated intensity is

$$I(t) = R_R U_0^2 e^{-(\omega_H/Q)t} \qquad (13.1)$$

where Q is the quality factor of the oscillator, taking all damping sources into account. The radiated sound is a simple highly-damped resonant "pop".

We cannot expect to find an analogous source consisting of a water-filled vented cavity in aquatic animals because of the impossibility of producing an adequately rigid container. Animals with swim bladders, or aquatic mammals, can however produce closely related sounds by applying an impulsive force to the wall of an air-filled cavity. We discussed the oscillatory behavior of such an air-bubble in Section 9.10. It can act as a very efficient, though highly damped, impulsive acoustic source because of the large magnitude of the radiation resistance under water. The resonance frequency of the air chamber is given by (9.23) and is typically several hundred hertz, depending upon the volume.

The other major possibility for impulsive excitation is that in which an impulse is applied to a resonant mechanical structure. The impulse could be applied in the form of a hammer blow, but this is generally not very efficient and would require a rather unusual anatomical structure. It is much simpler for the animal to deflect the vibrating element against its internal stiffness using a muscular element with a notch. At some suitable displacement the vibrator slips out of the notch and thereafter vibrates and radiates freely. This form of excitation is analogous to plucking the strings of a guitar or, an even better analogy, to the mechanical plucking action of a harpsichord.

The initial amplitudes of the various normal modes of the structure are determined by the shape and magnitude of the initial displacement. In a very flexible structure such as a string or a membrane, plucked somewhere away from the rigid edge supports, the deflected shape is very sharply angled at the point of plucking, and many higher modes have significant initial amplitudes. The actual amplitudes can be evaluated by a spatial Fourier analysis (in which we take the waveform along a line with coordinate x rather than over a period of time with coordinate t), as discussed for the time domain in Chapter 15. The vibrating elements commonly subject to such impulsive excitation are, however, much more likely to be plates or shells, and these are sufficiently stiff that the shape of the initial deflection is fairly smooth, and not greatly different from that of the lowest or perhaps the next-to-lowest vibration mode. Indeed a combination of these two modes is often likely if the structure is plate-like and plucked at one edge. It is then quite easy to estimate the initial mode amplitudes, either by assigning all the displacement to a single mode or by dividing it nearly equally between two modes. In either case, the radiated intensity can then be estimated following the methods discussed in Chapter 7. The sound radiated will generally have the nature of a sharp "click", because the structure is small and the resonance frequency high, but detailed analysis will show it to be either one, or a superposition of a small number of, exponentially damped sine waves.

13.3 Frictional Excitation

As the first and simplest case in which a steady force is converted to a steady oscillation, we consider the excitation of a resonant structure, such as a plate, by a steady frictional force. This force might be applied to the edge of the plate by rubbing it with a leg that is rough only on a very microscopic scale. In more familiar terms, the excitation mechanism is analogous to the bowing of a violin string, or perhaps more directly to the bowing of a metal plate to excite Chladni figures, in which dust collects along the nodal curves of the vibration. Even more commonly, it is the cause of squeaking in poorly oiled door hinges!

Consider the situation shown in Fig. 13.1 in which a mass, forming part of a simple vibrator, rides on a moving belt. It is common experience that the frictional force between two objects is greatest just before one begins to slide on the other, and that it decreases further as their relative velocity increases. This is illustrated by the curve of Fig. 13.1, the stationary block being represented by the point A.

Figure 13.1 (a) Conceptual diagram of a simple vibrator driven by a frictional energy source. (b) The nonlinear relation between frictional force F and velocity v for the block in (a) when the belt speed is v_B.

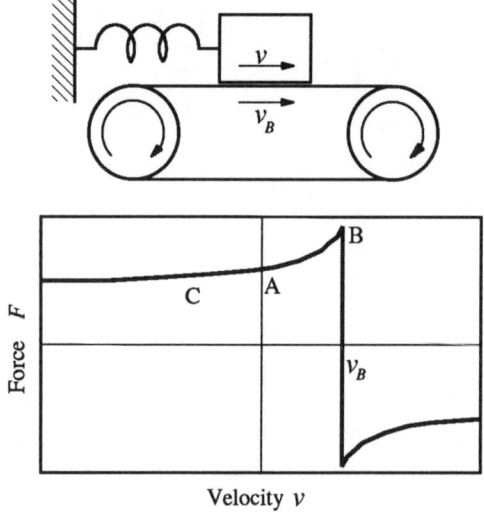

If v_B is the speed of the belt and v that of the block, then the relation between the frictional force F and the velocity v can be represented by the power series expansion

$$F = F_0 + a_1 v + a_2 v^2 + a_3 v^3 + \ldots \tag{13.2}$$

where a_1, a_2, a_3 are constants and $a_1 > 0$ for $v < v_B$. If we neglect all terms higher than linear in v, then the equation of motion of the mass becomes

$$m\frac{d^2x}{dt^2} + \gamma\frac{dx}{dt} + m\omega_0^2 x = F_0 + a_1\frac{dx}{dt} \tag{13.3}$$

where we have written x for the displacement coordinate of the mass. The first term on the right-hand side simply gives a steady displacement, which is not important. The second term, however, which arises from the nonconstancy of the friction characteristic, has just the same form as the damping resistance $\gamma dx/dt$ but, since it is on the opposite side of the equation, it is effectively a negative resistance. If $a_1 < R$ any motion of the mass will be damped, but if $a_1 > \gamma$ then an oscillation of frequency ω_0 will grow without limit.

In fact other nonlinearities intervene to prevent this catastrophic increase in amplitude. It is physically clear that, once the amplitude becomes sufficiently large that the speed of the mass overtakes that of the belt, the characteristic point will reach B on the curve of Fig. 13.1 and there will be a sudden force in the opposite direction to stop further growth. This is what happens in a violin string, the operating point moving between a situation in which it is caught by the bow at the point B and an alternate situation in which it slides back at constant velocity at a point such as C, lower down the curve. For a vibrating element heavier than a string, the situation is generally less extreme than this, and the amplitude is limited before the point B is reached. To see how this occurs, consider the cubic term in (13.2) and suppose that $a_3 < 0$, which it must be to give the general S shape of the curve. For simplicity we neglect the quadratic term, since it can be shown to have only a small effect on the behavior. The final equation then looks like

$$m\frac{d^2x}{dt^2} + \alpha\frac{dx}{dt}\left[1 - \beta\left(\frac{dx}{dt}\right)^2\right] + \omega_0^2 = 0 \tag{13.4}$$

where α and β are new constants related to a_1, a_3 and γ. If $a_1 > \gamma$ then $\alpha < 0$ and the oscillation will begin to build up from nearly zero amplitude. Once the amplitude becomes large enough that $\beta(dx/dt)^2 = 1$, however, further growth is prevented and the oscillation becomes stable. Indeed if impulsive excitation starts the oscillation off with an amplitude greater than this stable value, the amplitude will decay until it is reached. Equation (13.4) is a minor variant of an important and widely studied relation called the Van der Pol equation.

The situation idealized in Fig. 13.1 is much simpler than that in an actual sound generator, the main difference being that the driven element is not a simple mass-and-spring vibrator but an extended multi-mode vibrator such as a plate. It is important to understand what complications this introduces. As a first approximation we might represent the motion of the complex vibrator by a series of equations like (13.3), one for each mode. In this linear approximation (we omitted all terms of higher than first order) there is little interaction between the different modes, and we expect each mode to be driven in much the same way, leading to an inharmonic vibration with all mode frequencies represented. Indeed this is what happens while the amplitude of oscillation is very small. Once the amplitude of the most favoured mode grows sufficiently, however, we must include at least the third-order terms, as was done in (13.4) to limit the amplitude. The inclusion of a term of order n has other very significant effects, for the vibrations of the form $\cos \omega t$ then produce cross terms $(\cos \omega t)^n$ that give rise to terms of the form $\cos m\omega t$ with $m \leq n$. These terms drive the other modes strongly and, in almost every situation, the whole motion finally locks into one harmonic vibration with all modes driven at harmonics $m\omega$ of the dominant mode.

The mathematics of this mode-locking process is rather complicated—a brief discussion is given in the next section—but we shall find that it is extremely common in nonlinearly driven systems. Musical instruments with sustained tones rely upon it to produce concordant sounds—even the modes of musical instruments are not precisely harmonic in frequency relation—and the process is responsible for the exact harmonics found in the steady sounds produced by most animals.

If we knew enough about the form of the friction curve, then we could use the method outlined (though including more terms) to calculate the amplitude of the vibration. This information is generally not available, but we can deduce a few general results from the equations above.

The driven frequency is approximately equal to one of the natural frequencies of the vibrator, generally the fundamental, so that the main parameter of interest is the amplitude. An absolute upper limit to the amplitude is determined by the condition that the velocity at the driving point should not exceed the velocity of the frictional driver. This suggests that a large drive velocity should be used to produce a loud sound. Often, however, the limitation will be imposed by the available acceleration force, and thus by the absolute magnitude of the frictional force, as expressed by α in (13.4). This in turn suggests that there should be a large normal force between the drive and the vibrator. These two factors taken together imply a large expenditure of driving energy, which is as we should expect. In an extended vibrator the situation is a little more complex, for the location of the driving point must also be taken into consideration. The animal could equally excite a mode near a position of maximum amplitude and use a large velocity and relatively small force, or close to a nodal line using a relatively small velocity and a correspondingly larger force. The difference between the results is not great.

The efficiency of this type of drive in converting steady mechanical energy into acoustic energy can be roughly estimated. In the best possible situation, the vibrator moves with the drive at point B on the friction curve for something like half of each cycle and then slips back at a point such as C for the other half. The energy given to the string is the product of force times distance, so that the net energy given to the string each cycle is roughly the product of the peak-to-peak amplitude multiplied by the difference in frictional force between points B and C. Since sliding friction is typically about 0.7 times static friction, this means that only about 30% of the energy supplied is converted to vibrational energy. The energy given to the vibrator is then dissipated, partly in acoustic radiation and partly in internal mechanical losses.

The second part of an efficiency calculation concerns this distribution of power between radiation and internal losses. Internal losses are determined by the Q of the vibrating element, which is often a dished shell such as a wing cover. Suppose that its area is S, its thickness d, and its resonance frequency ω_0. Then the acoustic resistance contributing to energy dissipation is $R_I = \rho_s d \omega_0 / QS$, where ρ_s is the material density. Now the vibrating shell acts essentially as a dipole acoustic source, like the freely vibrating disc discussed in Section 7.5. If $2a$ is the smallest transverse dimension of the shell, then by (7.13) the radiation resistance is approximately $R_R \approx (ka)^4 \rho c / 8S$ in the frequency region where $ka < 2$, saturating at $2\rho c/S$ for higher frequencies. Here $k = \omega_0/c$. The conversion efficiency from vibrational to acoustic energy is thus

$$\mathcal{E} = \frac{R_R}{R_I + R_R} \approx \frac{\pi^3 \rho Q a^4 f_0^3}{\rho_s c^3 d} \qquad (13.5)$$

where $f_0 = \omega_0/2\pi$ is the resonance frequency in hertz. This expression applies only if $ka < 2$; if $ka > 2$ then the efficiency saturates at the value for $ka = 2$. To achieve a large conversion efficiency the animal thus needs a resonant vibrator of large area, small thickness, and high Q, but above all of high resonance frequency. These requirements are competitive, in the sense that large area and small thickness imply low rather than high resonance frequency, though stiffness can be imparted if the element is an appropriately curved shell.

To consider a numerical example, suppose that the minimum dimension of the shell is 4 mm, its thickness 0.1 mm and its Q value 30. The maximum efficiency is reached when $ka = 2$, which corresponds to a frequency of about 55 kHz. This is generally beyond physical reach, and the highest likely resonance frequency for a wing shell of this size is about 10 kHz. Substituting these values into (13.5) gives a conversion efficiency of only $\mathcal{E} \approx 6 \times 10^{-5}$. When this is combined with an excitation efficiency of only 30%, the total conversion efficiency from muscular energy to acoustic output is only of order 2×10^{-5} or 0.002%. If the insect expends, say, 1 mW of muscular power in exciting the vibration, then the acoustic output is only 2×10^{-8} W. Although this acoustic power is extremely small, it is equivalent to a sound pressure level of 33 dB at a distance of 1 meter and would be clearly audible to a human listener and, more importantly, to another insect at this distance. Of course some insects might use considerably more input power than this.

Clearly the coupling between the radiator and the air is poor in this case, partly because of the small size of the radiator, but mostly because of its predominantly dipole character. If the acoustic situation can be modified to impart some asymmetry and thus some monopole component, then the output can be considerably increased. Suppose, for example, that absorption by the animal's body, or some other asymmetry, reduces the radiation amplitude from one side of the dipole source by 30%. Then the source can be considered as a superposition of a dipole of strength 0.7 times the original and a monopole of strength 0.3 times the original. We shall discuss the efficiency of monopole sources in Section 13.6, and the formula derived there shows that they are more efficient than dipole sources of the same strength by a factor $2c^2/\pi^2 a^2 f^2$. For the numerical example used in our previous paragraph, this factor is about 50. Applying this to the amplitude fraction 0.3, or power fraction 0.1, associated with the monopole gives a radiated power for it that is 5 times that of the dipole. The total power output is therefore increased by about 7 dB in this case, and the overall efficiency is about 0.01%. It has been reported that some crickets further overcome this problem by digging horn-shaped holes and positioning themselves in the throat in order to improve the coupling. An increase of acoustic output by 10 to 20 dB might reasonably be achieved in this manner.

13.4 Harmonics, Mode Locking, and Chaos

The harmonic nature of the spectra of most steady sounds produced by animals—the frequencies all being exact integer multiples of the fundamental frequency—often comes as something of a surprise, particularly when it is known that the vibrating structure has mode frequencies that are very far from harmonic in frequency relationship. This mode locking is a consequence of nonlinearity in the negative resistance driving the basically linear but inharmonic passive resonating structure. This is an important subject, since it applies to all nonlinear systems, and to both mechanical and pneumatic sound generators. The following brief discussion shows how this comes about.

Suppose we have a structure with several modes, denoted by x_n, having frequencies ω_n respectively. There is no simple relation between the ω_n but we shall suppose them to be arranged in ascending order. Suppose that the system is brought into oscillation by a negative-resistance drive that has the nonlinear characteristic $F(v_1, v_2, ...) \equiv F(v_m)$, where $v_m = dx_m/dt$. Then, simplifying where possible, the equation of motion for mode n has the form

$$\frac{d^2 x_n}{dt^2} + \omega_n^2 x = F(x_m) - \gamma v_n \qquad (13.6)$$

where $-\gamma v$ is the normal damping term transferred from the left-hand side. There is an equation similar to this for each mode, and we note that the forcing term $F(x_m)$ depends on contributions from all the modes, often through terms like $(x_1+x_2+x_3+...)^N$.

Now each mode has a nearly sinusoidal vibration of necessity, though it may be at a frequency somewhat different from its natural resonance, so we can write

$$x_n = a_n(t)\sin[\omega_n t + \phi_n(t)] \tag{13.7}$$

where the phase $\phi(t)$ depends upon time, so that the actual frequency of oscillation is $\omega_n + d\phi_n/dt$. We substitute the relation (13.7) back into (13.6) and require that the time variation is apportioned between $a_n(t)$ and $\phi_n(t)$ so that the velocity is just $v_n = a_n \omega_n \cos(\omega_n t + \phi_n)$. We can use the equation resulting from this condition, along with the original equation (13.6), to obtain the results

$$\frac{da_n}{dt} = \left[\frac{F(x_m)-\gamma v_n}{\omega_n}\right]\cos(\omega_n t + \phi_n)$$

$$\frac{d\phi_n}{dt} = -\left[\frac{F(x_m)-\gamma v_n}{a_n \omega_n}\right]\sin(\omega_n t + \phi_n). \tag{13.8}$$

So far this is all exact, but we now make the approximation of neglecting all except slowly varying terms on the right-hand sides of both these equations. This can be done by integrating over one cycle of the frequency $\omega_n + d\phi/dt$ and discarding all terms except those with nearly zero frequency. This gives

$$\left[\frac{da_n}{dt}\right]_0 \approx \left[\frac{F(x_m)}{\omega_n}\cos(\omega_n + \phi_n)\right]_0 - \frac{\gamma}{2}a_n$$

$$\left[\frac{d\phi_n}{dt}\right]_0 \approx -\left[\frac{F(x_m)}{a_n \omega_n}\sin(\omega_n + \phi_n)\right]_0 \tag{13.9}$$

where the brackets $[...]_0$ imply that all except slowly-varying terms are omitted. The factor 1/2 comes from the average value of $\cos^2$ over one cycle. Clearly if $F = 0$, the oscillation simply decays without any frequency change, but if the linear term in F is larger than $\gamma \omega_n a_n$ and positive, the amplitude a_n will increase.

Suppose, however, that some other mode x_m is dominant in amplitude, and that $\omega_m < \omega_n$. Then the nonlinearity of $F(x_m)$ will generate components of frequency $s\omega_m$, s being an integer, and of large amplitude, one of which will almost certainly be sufficiently close to ω_n in frequency that its contribution to (13.9) must be regarded as slowly varying. We can then have a possible steady-state solution if $\omega_n + d\phi/dt = s\omega_m$ and $da_n/dt = 0$, this being reached by suitable adjustment of the arbitrary additive constant in each ϕ_n to exactly balance the right-hand side of the first equation to zero. Exactly the same sort of thing can happen with all the other modes, since the phases ϕ_n are independent.

A completely phase-locked harmonic solution of this type is nearly always possible provided the nonlinearity is strong enough, in the sense that $F(x_m)$ has terms of large magnitude and high index, and that at least one of the excitations x_m is of large magnitude. It must be said, however, that while such a harmonic solution is usually possible, and results in an exactly repetitive waveform, other solutions are also possible, depending upon the exact nature of the nonlinearity, the mode frequencies and amplitudes, and the initial conditions. No general method is known for predicting exactly what will happen for a particular form of the nonlinear function $F(x)$ of (13.6), but the behavior of particular systems can be examined by integrating the equations of motion numerically, using a computer.

One such common solution has two modes of nearly equal amplitude and disparate frequencies, in the sense that there is no approximate relation $r\omega_m \approx s\omega_n$ between them involving two small integers r and s. It may then be possible to excite both modes, and the nonlinearity will generate all sum and difference frequencies $r\omega_m \pm s\omega_n$ for all pairs or small integers r and s. Such a sound is generally steady, but with pronounced beating effects. An even more extreme possibility is that the system will degenerate into chaotic vibrations, in which the frequency spectrum is continuous, and such regularity as there is occurs only at a more abstract level. The cries of some cockatoos suggest a behavior of this type.

Quite generally, we should note that the amount of energy in the higher harmonics in a phase-locked vibration regime is determined largely by the extent of the nonlinearity, and therefore by the amplitude of the dominant mode. In general terms, the amplitude of the nth harmonic increases as the nth power of the amplitude of its fundamental, so that "loud" sounds have high harmonic development and "soft" sounds are dominated by their fundamental. This simple rule breaks down when the nonlinearity becomes extreme, for example when the vibrating element is actually captured by the static friction of the drive for a large part of each cycle. The waveform then tends towards a constant shape, implying constant relative harmonic content.

The mathematics developed above, and particularly equations (13.9), can be used to calculate the build-up of vibration amplitude as the generator mechanism is activated, and also the decay when the activation ceases. Obviously the behavior will vary from case to case, but typically it takes at least a few cycles of the dominant frequency for the oscillation to build to full amplitude and achieve phase locking. In some situations this initial transient may extend over tens of cycles of this fundamental. The decay behavior, in contrast, is rather simply determined by the damping γ.

13.5 Pick and File Excitation

In some cases the frictional drive mechanism of Section 13.3 is an oversimplification, because one of the rubbing surfaces is ribbed, rather than being smooth. A

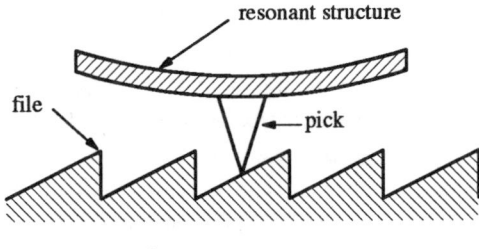

Figure 13.2 Conceptual diagram for a vibrator driven by a pick-and-file energy source. The exact angle of vibration of the pick relative to the file is important to the action of the generator.

better model is then one in which a sharp pick, or plectrum, on one surface moves across a ribbed file on the other, as shown in Fig. 13.2. There are now two possibilities: either the spacing of the ribs and the speed of motion are such that the frequency of the resultant driving force is about equal to that of the fundamental resonance of the structure, or else it is lower than this frequency by a factor 2 or more. In the first case it is reasonably possible for the animal to control the rubbing speed to get maximal acoustic output by bringing the forcing frequency into resonance with the structural vibration. In the second case a rather different mechanism operates, as we now discuss.

File excitation basically involves a stick-slip process in which the pick is caught on one tooth of the file, moved along, and then suddenly released to catch on the next tooth. If the tooth spacing is large, so that the slip frequency is very much smaller than the natural resonance frequency ω_0 of the driven structure, then each slip event will simply generate in the structure a pulse of resonant vibration that will substantially die out before the next slip event. Such a characteristic time record will be clearly evident in the sound waveform.

If the file frequency is only a few times less than the resonance frequency, then slip from one tooth of the file may take place before the vibration from the previous slip event has died away. We can see that, if the time relation between the two slip events is unrelated to the vibration frequency, the second slip excitation is just as likely to be out of phase with the first, and thus to be partially cancelled by it, as it is to be in phase and thus reinforced. If the angle between the direction of steady motion of the file and the direction of vibration of the pick is correctly chosen, however, the pick itself will have a component of vibration in a direction parallel to the file, and this will enhance the probability of slip at a particular phase of the vibration cycle. The slip impulses will then all tend to occur near the same phase of the vibration cycle and thus to reinforce each other, producing a nearly steady and coherent vibration. Such a dependence on file angle is commonly observed when filing metal in a vise.

The pick-and-file mechanism is inherently more efficient than the frictional mechanism discussed in Section 13.3, for the reason that in the slip part of the driving cycle the pick jumps freely rather than sliding under friction. There is some loss of efficiency, however, for two other reasons: the phase correlation between jumps is not likely to be perfect, and the impulsive nature of the drive excites higher modes for which the phase problem is more severe. Overall, however, we might expect an increase in ultimate efficiency by a factor of about 3 relative to the friction drive. The total efficiency is still very low because of the dipole nature of the radiation, as discussed in Section 13.3, and probably does not exceed 0.01%. Again, however, this efficiency might be improved by a factor 5 or so if radiation from one side of the vibrator can be partially shielded.

13.6 Buckling Tymbals

The final source we should mention derives its excitation from the muscular buckling of a ribbed tymbal that is set over a resonant cavity. We discussed this buckling behavior in some detail in Section 5.11, and the oscillatory behavior of a shell backed by a cavity is, of course, essentially the same as that of a membrane stretched over a cavity, as discussed in Section 9.5. The membrane and cavity constitute a resonator with a single mode, the frequency of which is determined by the cavity volume and the elastic properties of the shell. When the shell reaches its first bucking point, under the influence of muscle tension, it collapses slightly and excites the cavity-and-shell combination into oscillation. The vibrations of the shell clearly influence the time of occurrence of the next buckling event, as with the pick-and-file mechanism, so that successive bucklings are correlated in phase and combine to give a coherent tone burst, repeated with each buckling cycle of the tymbal.

This type of excitation mechanism, employed by insects such as cicadas, is particularly efficient. All the steady mechanical energy is converted to vibrational energy of the tymbal and enclosed air, and the source, being monopole in nature, is a much more efficient radiator than the dipole sources discussed above. There are, it is true, significant losses associated with the shell buckling, but these are more than balanced by the other efficiencies of the system. Following the same analysis as set out in Section 13.3 but noting that, for a monopole radiator, the radiation resistance is $R_R \approx (ka)^2/4$, we find an acoustic conversion efficiency of

$$\mathcal{E} \approx \frac{\pi \rho Q a^2 f_0}{2 \rho_s c d} \tag{13.10}$$

for a tymbal of radius a, thickness d, and resonance frequency f_0.

Choosing values appropriate for a large insect such as the Australian cicada, we set $a = 5$ mm, $d = 0.2$ mm, $Q = 20$, and $f_0 = 3$ kHz, which yields a conversion efficiency $\mathcal{E}$ of about 4%. Note the huge improvement over the value calculated for the dipole radiation from a vibrating wing cover. Indeed we shall find that few other sources can surpass this efficiency. Some species of cicadas have a very large acoustic output; typical sound pressure level may be as high as 70 dB at a distance of 3 meters, which corresponds to an acoustic source power of about 1 mW. This in turn implies that the animal expends a power of about 25 mW in singing. This is not unreasonable, though very large for an insect some 50 mm in length and weighing only about 5 g.

It is interesting to remark that low-frequency adaptations of this mechanism have also evolved, specifically in the cicada *Cystosoma saundersii* (Westwood), the male of which species has an enormously enlarged air-filled abdomen, typically 7 cm^3 in volume, and a song frequency of about 800 Hz. The tone burst repetition frequency is about 40 Hz, corresponding to the rate of tymbal muscle contraction. The wall of the abdomen, which is very light in structure, appears to act as the primary radiator in this case, with the tymbals providing almost an internal acoustic current drive, as discussed in Section 8.6.

13.7 Conclusions

Mechanically driven sound generators can provide a useful means for communication within a particular insect species. Clicks and pops can be produced in many ways, and are rather less interesting than more sustained songs. When a readily available resonant structure, such as a wing cover, is used to produce the sound, and is excited by rubbing or scraping, the efficiency is very small, but still adequate to provided location and identification to members of the same species over distances of a few meters. If a specialized organ is employed, such as a ribbed tymbal over a resonant cavity, which constitutes a monopole source, then the efficiency can be quite high and the resulting sound pressure level very great, from even a small insect.

There are two features of sustained song produced by the mechanical devices discussed here. The first is that it necessarily has a pulsed structure, corresponding to repetitive movements of the animal's legs or to successive cycles of tymbal muscle contraction. The rhythm of these pulses, which can be easily varied by the insect, presumably is a part of species identification, and may convey other information as well. The second characteristic is that, in order to be efficient, the sound-producing system necessarily uses a resonant structure with high Q. For simple communication, the resonant frequency, which provides the carrier for any time-varying rhythmic pulses, is characteristic of the species, and again serves as a partial identification. It would be possible for this frequency to be varied, in those animals possessing specialized sound generators, but this does not seem to have evolved in most insects.

References

Impulsive excitation: [10] Ch 18
Frictional excitation: [10] Ch 10
Harmonics, mode-locking, chaos: [10] Ch 5; [14]; [15]

Discussion Problems

1. An insect generates sound by scraping a leg-file over a pick on a wing case. The file has 500 teeth, each 0.1 mm high, and a leg stroke takes 100 ms. If the normal force between the pick and the file is 0.1 gram weight, estimate the mechanical power expended and the likely acoustic power output.
2. The tymbal of a cicada buckles through a distance of 1 mm when it produces a call. The pulse repetition frequency is 100 Hz. If the sound pressure level produced is 73 dB at a distance of 1 meter, estimate the muscular force used to buckle the tymbals.

Solutions

1. Force = $10^{-4} g = 10^{-3}$ N. Work to lift pick over each tooth = force × distance ≈ 10^{-7} J. The stroke produces 5000 pulses per second, so power expended is of order 0.5 mW. From discussion, efficiency for this case is of order 2×10^{-5}, so acoustic output is of order 10^{-8} W = 10 nW. Note that "of the order of" implies a total uncertainty of about a factor 10, and so an uncertainty of about a factor 3 in either direction. In this case, the uncertainty is probably greater than this.
2. SPL of 73 dB at 1 m implies acoustic power of about 0.2 mW. Typical efficiency for such a system, from discussion, is about 4%, so mechanical power expended is about 5 mW. Mechanical energy expended per pulse is thus about 5×10^{-5} J. Distance is 10^{-3} m, so force is of order $5 \times 10^{-2} N$ or 5 gram weight.

14 PNEUMATICALLY EXCITED SOUND GENERATORS

SYNOPSIS. Animals with active respiratory systems usually use the expired air as a power source for a vocal system. There are two different types of sound-generating mechanisms, the first associated with "voiced" sounds and relying upon a vibrating mechanical element, and the second associated with "whistled" sounds and having an entirely aerodynamic mechanism.

The essential component of such a voiced system is a flow valve, the opening of which depends upon the pressures acting upon its two sides, as illustrated in Fig. 14.1. Three types of valve are shown here, the first and third having realization in the larynx and syrinx of animals, while the second type is found only in woodwind musical instruments. Analysis of the flow through the valve, and its motion under the influence of oscillating pressures and flows, shows that the acoustic conductance can have a large negative value either just below or just above the resonance frequency of the valve itself, as shown in Fig. 14.3. When the acoustic conductance is negative, the whole system can oscillate near the valve resonance, drawing energy from the expired air stream. The resonance frequency of the valve is normally under muscular control, and determines the fundamental frequency of the sound produced. The air flow through the valve is quite nonlinear, and this nonlinearity generates a complete set of exact harmonics of the fundamental in the radiated sound. The only exception to this rule occurs when the valve motion becomes chaotic, as in the screech of certain cockatoos.

The trachea and mouth cavity have an important influence upon the spectral envelope of the radiated sound. Their behavior can be summarized in a reflection function, as shown in Fig. 14.5, any resonances in the mouth cavity showing up as oscillations in the reflected pulse. The bronchi connecting the lungs to the vocal valve are also an important part of the system, and their acoustic impedance influences the spectrum of the sound.

As a specific example, we calculate the acoustic output from an idealized avian vocal system, with the simplified anatomy shown in Fig. 14.4 and the parameter values given in Table 14.1. These parameters are appropriate for a bird such as a raven. The calculated waveforms for tracheal and bronchial pressures, syrinx membrane motion, and acoustic flow are shown in Fig. 14.6 and give a complete description of the system. From these we can calculate the radiated acoustic power, which varies according to lung pressure between 1 µW and 10 mW, with a conversion efficiency between about 0.2% and 2%. This agrees well with measurements on birds. The radiated frequency spectrum calculated from the model is shown in Fig. 14.7, where it is compared with a Sonagram of the call of a real raven. The agreement is good. In particular, the harmonic upper partials of the syrinx fundamental and the formant bands associated with tracheal resonances are clearly evident. Similar models can be constructed for other vocal systems. Frogs and bats are particularly interesting and, of course, great attention is given in the literature to the human vocal system.

Aerodynamic systems rely upon the instability of a flowing airstream to either aerodynamic or acoustic disturbances. A jet emerging from an aperture develops turbulent instability, as illustrated in Fig. 14.8(a) and radiates a wide-band hissing noise with very low efficiency. An obstacle placed in a steady flow generates a street of vortices of alternating sign, with a shedding frequency dependent upon flow speed, and radiates a sound with a more nearly tonal character. These mechanisms may play some part in the hissing noises and consonants in speech and other vocalizations.

More efficient generators rely upon the interaction of a jet with an edge, as shown in

Fig. 14.9(a), which generates sinuous instability on the jet. This instability is amplified as it travels along the jet and radiates an edge-tone. If the jet is acoustically coupled to a Helmholtz resonator, as in Fig 14.9(b), then the sound generation may be very efficient. Similarly, a jet interacting with an aperture, as in Fig. 14.10(a) undergoes varicose (radial) instability and generates an aperture tone. Once again this may be coupled to a Helmholtz resonator as in Fig. 14.10(b) with consequent increase in efficiency. In both these situations the frequency of sound generated by the jet itself is determined largely by jet velocity and the spacing to the obstacle. When the jet is coupled to a cavity, the sounding frequency is closely locked to that of the resonator, and can be varied by changing its volume. It seems likely that these mechanisms account for the ways in which animals produce whistling sounds, though little direct anatomical evidence is available.

14.1 Air-Driven Systems

Animals that have developed an actively aspirated respiratory system have available a ready source of power in the form of the expired air, and most make use of this in a vocal system. There are two basic types of sound production possible with the use of pneumatic power. In the first, which is the primary vocal system for most animals, the air flow induces motion in some sort of mechanical valve, which then acts back on the flow to maintain a steady oscillation. This is the sort of system used by humans in producing vowel sounds. The second mechanism is entirely aerodynamic and does not involve vibration of any part of the anatomy. It is responsible for the production of unvoiced consonants in human speech and for whistling sounds. The precise anatomy of the vocal system differs from one species to another—birds, for example, are thought of as having a very different vocal system from humans—but we shall see that these are all variants of one basic system and can be described in terms of a single unified model.

Some people have been tempted to describe the operation of a pneumatic vocal system in terms of an analogy with the operation of wind-driven musical instruments. Unfortunately such an analogy can be misleading at the qualitative level, since there are important differences between the two types of systems, but it is very productive at the analytical level, since the same model elements occur in each case. What differ are the numerical values of the parameters describing the detailed operation of the elements and their mutual couplings, and with this the identity of the primary controlling agent. We shall not make any very specific use of the analogy here, but the points of similarity will be mentioned as they arise.

We have already mentioned, in Section 12.2, the impulsive sounds that can be produced by a sudden change in the volume of a resonant cavity. In an active pneumatic system these sounds can be intensified by building up air pressure on one side of a controllable stop, such as the tongue pressing against the hard palate, and then suddenly opening the stop to release the air into a Helmholtz resonator, in this case the mouth, which responds with a sharp transient oscillation. The discussion in Section 12.2 can readily be extended to encompass this case, and we shall not bother to do this explicitly.

In the discussion to follow we first examine the operation of mechanically valved vocal systems, since these are the most important, and then give briefer consideration to purely aerodynamic systems.

14.2 Larynx and Syrinx

A pneumatically powered vocal system necessarily has at least two components—the muscularly compressed lungs or air sacs that provide a steady supply of air at a pressure a little above that of the surrounding atmosphere, and some sort of valve (the larynx or syrinx) within the respiratory tract that can be set into spontaneous oscillation. Most systems have another important element in the form of a pipe (the trachea), with various attached cavities (the mouth and nose), through which the air flow and most of the sound energy is passed to the surroundings. Although the vibrating valve is the key element of the system, all three parts must be considered if we are to understand in detail how the system functions. Indeed the valve cannot be set into oscillation without the cooperation of at least one of these other elements.

In discussing the action of the valve, it is helpful to define two different pressures: p_B, the pressure in the bronchial passages leading from the lungs to the valve, and p_T, the pressure in the base of the trachea immediately above the valve. We usually think of these both as gauge pressures, that is as pressures relative to the surrounding air, though the equations are correct also for absolute pressures, since only pressure differences are involved. The valve itself may be called the larynx, in animals such as humans, or the syrinx, in animals such as birds, but we develop our equations in a generalized way to apply to all cases.

Figure 14.1 illustrates three possible configurations of pneumatic valves that are distinguished on the following basis. The valve in Fig. 14.1(a) tends to be blown open by a steady excess pressure p_B in the bronchus and to be blown closed by a steady excess pressure p_T in the trachea. The valve in Fig. 14.1(b) is blown closed by excess pressure in the bronchus and blown open by excess pressure in the trachea. The valve in Fig. 14.1(c) is blown open by excess pressure either in the bronchus or in the trachea. It would be possible to devise a valve that is blown closed by excess pressure in either bronchus or trachea, but this is not of practical significance. Valve configuration (a) corresponds to the human vocal folds, as shown in Fig. 14.1(d), the restoring spring action being supplied by muscular ligaments in the folds themselves. Configuration (c) corresponds to the avian syrinx, illustrated in Fig. 14.1(e), the restoring force being supplied by excess air pressure in sacs surrounding the syrinx. There does not seem to be a biological example of configuration (b), which is understandable since it is not "fail-safe" and can be jammed closed! It is interesting to note that configuration (a) also corresponds to the valve action of the player's lips in a trumpet or other brass instrument, while (b) corresponds to the valve action of the reed in a clarinet or oboe. There does not seem to be any musical analog of (c).

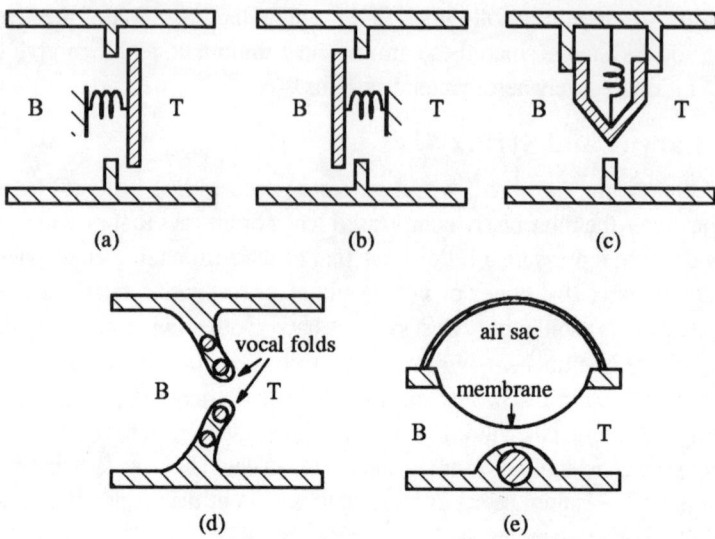

Figure 14.1 Three generic pressure-controlled valves with configurations (a) (+,−), (b) (−,+), and (c) (+,+) respectively. Below are shown schematic cross sections of (d) the human larynx and (e) the avian syrinx. Shaded circles represent elastic tensioning tissue, while the letters B and T denote respectively the bronchial and tracheal sides of the valve.

We can include all the valve types in Fig. 14.1 in a single mathematical treatment by inserting parameters (σ_1, σ_2) that are assigned values of ± 1 depending upon the opening configuration of the valve. If a positive sign indicates that the pressure forces the valve open, then the three configurations in Fig. 14.1 are (+,−), (−,+) and (+,+) respectively. We shall be concerned here only with the biological configurations (+,−) and (+,+).

In all the valves illustrated in Fig. 14.1 there is also another aerodynamic driving force to be considered, and this is the Bernoulli force. When a fluid flows rapidly through a constricted aperture, its flow velocity must increase. If energy is conserved in the fluid, this increased kinetic energy must come from work done on the fluid by a pressure difference between the region upstream, in the bronchus in this case where the pressure is p_B, and in the aperture itself where the pressure is p. Consider a small mass δm of air in the bronchus, occupying a volume δV. When this is pumped by the pressure differential $p_B - p$ to a position in the constricted valve aperture, the amount of mechanical work done, ignoring the slight expansion of the gas, is $(p_B - p)\delta V$. If the area of the aperture is small compared with that of the bronchus, then the flow velocity in the bronchus is negligible relative to the flow velocity u in the aperture, and all the expenditure of energy is accounted for by the increase in kinetic energy from zero to $\frac{1}{2}\delta m u^2$. Balancing these two quantities, and recognizing that $\delta m/\delta V$ is just ρ, the density of air, we find the result, originally due to Bernoulli,

$$p = p_B - \rho u^2/2. \tag{14.1}$$

This means that the pressure in the aperture of a valve is always lower than the static pressure in the bronchus upstream from the valve. If we know the volume flow U through the valve, then the velocity u is simply related to this through the area of the valve aperture at the position considered. Bernoulli pressure thus always tends to draw the valve closed in a situation of steady flow.

We can find the total force acting on the valve flaps by integrating the pressure given by (14.1) over their area, recognizing that pressure acts normal to the surface and that we must take the component of force in the direction of motion of the flap. In most cases, flow velocity through the valve is rapid enough that the air emerges into the trachea as a jet that takes some distance to expand to fill the pipe, as shown in Fig. 14.2(a). This has the effect of making the static pressure on the downstream side of the aperture of the valve essentially equal to the pressure p_T at the base of the trachea, and exposing the downstream side of the valve flaps to the tracheal pressure p_T rather than to the pressure that would be calculated from (14.1) if we assumed the flow to expand again immediately to fill the cross-section. Substituting p_T for p in (14.1) then allows us to calculate the flow velocity through the valve for a given pressure difference across it. Multiplying by the area of the constriction gives the volume flow U. In vocal systems this pressure difference is typically a few hundred pascals, corresponding to a few centimeters height difference on a water gauge (1 cm WG $\approx$ 100 Pa) so that, from (14.1), flow velocities are typically

Figure 14.2 (a) In flow through a valve, the airstream usually separates from the walls and forms a jet. (b) Coordinates used to describe the valve. θ is the angle between the direction of motion y of the valve flaps and the axis of the pipe, while S_B and S_T are effective flap areas on the bronchial and tracheal sides of the valve.

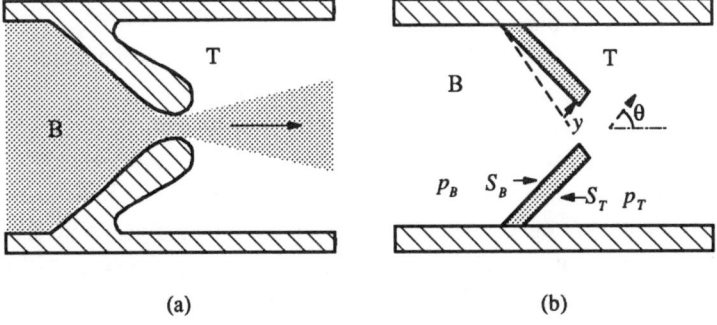

several tens of meters per second. The volume flow depends upon the size of the animal.

For an accurate model of the vocal system we must include an appropriately detailed description of the dynamics of the motion of the valve. This motion can be quite complex, for the valve flaps are pieces of membrane or flexible plates, which have many modes. For the purposes of our discussion here, it will be adequate to concentrate on the fundamental mode and treat the flaps as simple vibrators each with mass m, resonance frequency ω_V, and quality factor Q. These parameters then define the elastic stiffness and internal damping as elaborated in Chapter 2. We take the displacement of the flaps of the valve from the closed position to be y in a direction making an angle θ with the axis of the surrounding pipe, as shown in Fig. 14.2(b). The motion of each flap is then described by the equation

$$m\frac{d^2y}{dt^2} + \frac{m\omega_V}{Q}\frac{dy}{dt} + \beta^2(y - y_0) = F(t) \tag{14.2}$$

where y_0 is the equilibrium value of the opening coordinate y, $\beta = m\omega_V^2$ is the elastic stiffness of the flap motion, and $F(t)$ is the effective force acting on the flap. F is actually the force distribution weighted by the shape of the vibrational mode and y is then the amplitude of this mode, as given formally by equation (3.24). We shall not bother with this subtlety, since we do not generally know numerical values sufficiently well for a factor of 2 or so to make a significant difference. If there is only one moving flap, then the clear opening of the valve is $y \sin \theta$, as shown in Fig. 14.2(b). If there are two identical flaps, then the second obeys a similar equation with y replaced by $-y$, and the clear opening of the valve is $2y \sin \theta$. We shall assume such a symmetric valve in our discussion, though it is easy to revert to a one-sided valve, as found in the syrinx of most birds, if desired. The valve is, of course, a three-dimensional object, and we can approximate this by assuming it to have a width W in the direction perpendicular to the plane of the drawing, though we recognize a refinement to this simple assumption presently.

This linear vibrator model would be a reasonable approximation if we were able to assume that the amplitude of the motion of the valve flaps was small compared with the width y_0 of the static opening. This is, however, almost never a valid approximation. Not only is the vibration amplitude quite large, but the valve flaps usually close nearly completely once in each cycle. It is both the nature of this closure and the stresses in the converse fully-open condition that limit the amplitude of the vibration.

The elastic Young's modulus of biological materials generally increases dramatically as they are stretched, because the elastic force derives essentially from straightening out biological molecules. This effect introduces at least a cubic nonlinearity so that the elastic term $\beta(y - y_0)$ in (14.2) should be replaced by

$\beta(y-y_0)[1+\gamma_1(y-y_0)^2]$, where γ_1 measures the extent of the nonlinearity. We might expect γ_1 to be of order 10^6 m^{-2}, so that nonlinearity becomes appreciable for displacements of more than about 1 mm in an animal of moderate size. This nonlinearity in the elastic force has another important effect, since the elastic distortion of the valve flaps is greatest near their vibrating edges. With increasing amplitude these edges therefore tend to stiffen, with the result that areas of the flap further from the edge are increasingly involved in the motion. We should therefore replace m in (14.2) by something like $m[1+\gamma_2(y-y_0)^2]$, where γ_2 is a parameter measuring the extent of this nonlinearity, and should be of the same order of magnitude as γ_1. The large-amplitude resonance frequency then becomes

$$\omega_V(y) \approx \omega_V \left[\frac{1+\gamma_1(y-y_0)^2}{1+\gamma_2(y-y_0)^2}\right]^{1/2} \tag{14.3}$$

which may either increase or decrease with increasing amplitude y, depending upon the ratio of γ_1 to γ_2.

If the flaps of the valve close hard against one another for part of the cycle, then there is also likely to be a great increase in the viscous forces during the time of contact, giving a high and nonlinear damping for part of each cycle. This effect can be included in a parametric way through additional nonlinear terms in (14.2), though it is simpler to include it numerically during the actual calculation of a particular case by decreasing the effective Q for the closed part of each cycle.

Our purpose in this discussion is to elicit general principles, rather than to achieve a very close simulation of reality. It is necessary to construct a much more detailed physical model if realistic comparisons to an actual system are contemplated. In the case of models of the human glottis, which has understandably attracted a great deal of attention, mathematical models with as many as 16 mass elements and 16 nonlinear springs for each of the vocal folds have been constructed to achieve such an improved approximation to reality, but such detail is clearly outside our present discussion.

We can get a first approximation to the force acting on the valve flaps by defining S_B to be the flap area on the bronchial side of the valve up to the point of separation of the flow as a jet. The pressure on this region is given by (14.1), which includes the Bernoulli force. Similarly we define S_T to be the area exposed to the pressure p_T at the base of the trachea following separation of the flow as a jet. The total force arising from these two pressures and tending to open the valve by increasing y is then approximately

$$F \approx \sigma_1 p_B S_B - \int_{S_B} \frac{1}{2}\rho u^2 dS + \sigma_2 p_T S_T \tag{14.4}$$

where the parameters σ_1 and σ_2 are set equal to $+1$ or -1 depending on the configuration (σ_1, σ_2) of the valve. If the geometry of the valve is complicated we must perform this calculation more carefully, taking into account the direction of action of the pressure force relative to the direction of motion of the valve and, as we remarked before, allowing for the mode shape of the valve vibration.

The relative contributions of the Bernoulli force, represented by the integral in (14.4), and the simple pressure forces, represented by the other terms, depend upon geometry. If the channel through the valve is long and narrow—the musical equivalent is the double reed of an oboe—then the Bernoulli force may be quite important. The direct pressure forces are, however, always important and usually provide the dominant forcing term for valve motion.

In the steady state, the flow U through the valve can be derived directly from the Bernoulli equation (14.1) and the valve aperture $S_V = 2Wy \sin\theta$, but for an oscillating acoustic flow we need to allow for the acoustic impedance of the narrow passage between the valve flaps. This adds a pressure drop $L_V dU/dt$, where L_V is the inertance of the valve passage, given in terms of its dimensions, according to Table 8.1, as $L_V = \rho l_V/S_V$, where l_V is the length of the valve channel. The relation (14.1) between pressure and flow then becomes

$$p_B(t) - p_T(t) = \frac{\rho U^2}{2S_V^2} + \frac{\rho l_V}{S_V}\frac{dU}{dt}. \tag{14.5}$$

In this equation we have written dU/dt explicitly, rather than expressing everything in the frequency domain as $e^{j\omega t}$, because the flow $U(t)$ has the nonlinearity associated with the valve deflection y, together with other complications of its own.

The equations (14.2) through (14.5) adequately describe the behavior of a pressure-controlled valve in isolation if we add appropriate terms, as discussed, to take account of elastic nonlinearity. The flow nonlinearity is already included in (14.5), which shows that the acoustic volume flow is essentially proportional to the square root of the pressure difference multiplied by the area of the valve aperture, if we neglect the dU/dt term. The nonlinearity of vocal systems is, in fact, entirely confined to the behavior of the valve. All other components are very nearly linear in behavior.

It is not really possible to treat the acoustic behavior of the valve in isolation from the other parts of the vocal system, but a brief discussion shows how its motion gives rise to the negative conductance necessary to feed energy into the system. If we ignore the small Bernoulli term in (14.4), then the force tending to move the valve flaps is $F = \sigma_1 p_B S_B + \sigma_2 p_T S_T$, where (σ_1, σ_2) defines the configuration of the valve. The valve opening y is therefore $y = y_0 + B'F$, where y_0 is the static opening and, at low frequencies, B' is simply the mechanical compliance of the valve flaps. At higher frequencies we have to be concerned with phase shifts, as we discuss in a moment. For simplicity, let us suppose that $S_B = S_T = S$ and write $B \equiv B'S$. Then, if we ignore the small dU/dt term and take the square root of both sides, (14.5) can be rewritten as

$$U \approx A(p_B - p_T)^{1/2} y \approx A(p_B - p_T)^{1/2} [y_0 + B(\sigma_1 p_B + \sigma_2 p_T)]. \quad (14.6)$$

where $A \equiv (2/\rho)^{1/2} S_V$ is a new constant. Now the behavior of B as we increase the frequency was discussed in Section 2.5, where we treated the sinusoidally-driven simple oscillator, and is illustrated in Fig. 2.3. At low frequencies B is essentially real and positive (phase = 0), while at high frequencies it is real and negative (phase = $-180°$). Just at the resonance, B is purely imaginary (phase = $-90°$), so that its real part vanishes, but the real part actually goes through a positive maximum just below resonance and a negative minimum just above the resonance. These two extreme values are involved in the self-excitation of pneumatic valve oscillators of the type we consider here.

The acoustic conductance G_V of the valve is just the real part of the acoustic admittance. It has two different values, depending upon whether we excite the valve by an acoustic signal in the bronchus (B) or in the trachea (T) and, taking account of the direction of flow of U, we can write

$$G_{V(B)} = \frac{\partial U}{\partial p_B} \qquad G_{V(T)} = -\frac{\partial U}{\partial p_T}. \quad (14.7)$$

In both cases we can let $p_T \to 0$ after the differentiation, since the tracheal pressure is relatively small. The results are quite simple to calculate, and we find

$$G_{V(B)} \approx \frac{A[y_0 + 3\sigma_1 B_{\text{Re}} p_B]}{2 p_B^{1/2}} \qquad G_{V(T)} \approx \frac{A[y_0 + (\sigma_1 - 2\sigma_2) B_{\text{Re}} p_B]}{2 p_B^{1/2}} \quad (14.8)$$

where B_{Re} is the real part of B. We see immediately that, for valves of either $(+,+)$ or $(+,-)$ configuration, $G_{V(B)} < 0$ provided that $B_{\text{Re}} < 0$ and $p_B > y_0/3B_{\text{Re}}$. This means that, to be an acoustic generator driven by pressure fluctuations in the bronchus, the valve must operate somewhat above its resonance frequency, and the blowing pressure must exceed a particular value which is set by the equilibrium valve opening y_0. This condition is most fully satisfied just above valve resonance where the value of B_{Re} is large. Similarly, $G_{V(T)} < 0$ in the case of a $(+,+)$ valve if $B > 0$ and $p_B > y_0/B_{\text{Re}}$, implying operation just above the resonance for a tracheally excited valve, while a $(+,-)$ valve requires that $B_{\text{Re}} > 0$ and $p_B > y_0/3B_{\text{Re}}$, implying operation just below the resonance. Note that acoustic generation by each of these types of valve is facilitated if they are nearly closed in equilibrium, so that $y_0 \to 0$.

We should emphasize again that this is a linearized calculation, and applies only while the vibration amplitude is very small. A careful calculation must, of course, also treat the motion of the valve in more detail, with proper concern for the exact phase shifts and amplitudes involved. Figure 14.3 shows the results of such a more careful calculation, though again still in the linearized approximation. It is clear that, in the case of a valve controlled by variations in the bronchial pressure, there is only a small frequency range just above valve resonance in which

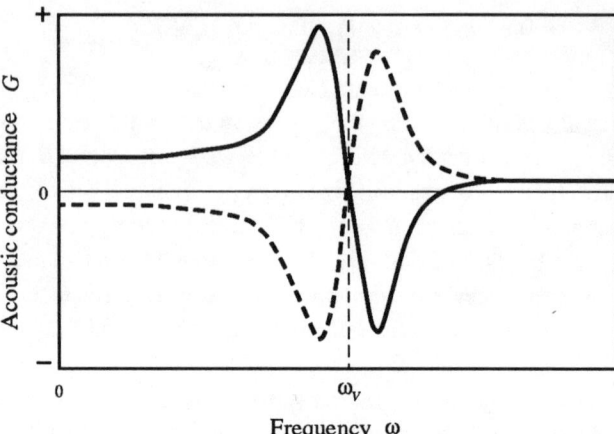

Figure 14.3 Acoustic conductance G of pressure-controlled valves above the threshold bronchial pressure. ω_V is the resonance frequency for the mechanical vibration of the valve flaps. The full curve refers to control from the bronchial side for valves of configuration $(+,+)$ and from either side by valves of configuration $(+,-)$, while the broken curve refers to control from the tracheal side by valves of configuration $(+,+)$. G must be negative for the valve to oscillate.

the acoustic conductance is negative, so that the vocal system must operate in this range, and its actual sounding frequency can be controlled by changing the muscular tension and thus the resonance frequency of the valve.

14.3 Vocal Systems

There are two parts of the vocal system to be considered in addition to the larynx or syrinx. The lung upstream of the valve provides the air supply at controllable pressure p_L, and has associated with it a linear acoustic impedance, which we denote by Z_L. In a simple system Z_L might just be the acoustic compliance associated with a pressurized air reservoir, but in general it will be more complicated than this. The part of the vocal system downstream from the valve consists of a quite long and regular pipe, the trachea, terminated by a mouth cavity open to the air. As a first approximation the trachea and mouth can be considered as a simple horn, but generally the mouth and associated nasal cavities will add extra resonances and other complications.

Let us look first at the air supply part of the model system shown schematically in Fig. 14.4. The system is powered by muscular tension around the lungs, which provides a steady pressure p_L at the lower inlet to the bronchial system. What we need is a relation between the pressure p_B just before the valve and the flow $U(t)$. The internal impedance of the air supply from the lungs depends upon the anatomy of the animal. If the lungs are essentially air sacs, then they may behave as a simple

PNEUMATICALLY EXCITED SOUND GENERATORS

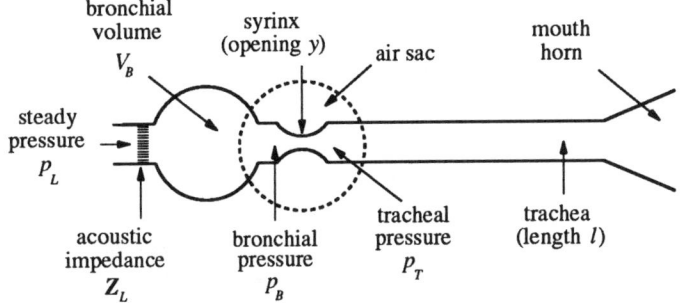

Figure 14.4 Simplified model for the vocal system of a bird, as discussed in the text. The air sac surrounding the syrinx can be pressurized during song so that the syringeal membranes protrude into the airway and constrict the air flow. (From [26].)

large compliance with a pipe-like impedance representing the bronchial tubes. In mammalian lungs, the branching tubular structure gives a much higher and generally resistive impedance leading to the wider bronchial tubes, which have significant length and volume. Each such system can be worked out in detail, but as an example we choose a lung with acoustic impedance Z_L feeding bronchial tubes that are short enough to be represented by a simple cavity of volume V_B and acoustic compliance $C_B = V_B/\rho c^2$. The flow from the lungs into this bronchial volume is $(p_L - p_B)/Z_L$ and there is a flow $U(t)$ out through the valve. The rate at which pressure builds up in the bronchial volume is therefore given by

$$\frac{dp_B}{dt} = \frac{\rho c^2}{V_B}\left[\frac{p_L - p_B}{Z_L} - U(t)\right]. \tag{14.9}$$

This equation describes all that we need to know about the upstream part of the system in this approximation. In practice Z_L is usually primarily resistive, and so written R_L, and is usually within a factor 5 either way of the magnitude of the characteristic impedance of the trachea. For the avian system model to be discussed in detail we have assumed that R_L is small compared with the characteristic impedance of the trachea, while the opposite assumption is generally more appropriate for mammalian systems. As we shall see later, this distinction has an effect on the formant structure of voiced sounds in the two cases, the voice with low lung impedance having each formant band split into two.

We need to treat the downstream part of the system in a little more detail, since its length is comparable with the sound wavelength of at least the higher harmonics of the valve vibration. There are several ways in which this can be done. The simplest from a computational point of view is based on work on musical instruments by McIntyre, Woodhouse, and Schumacher, referenced in the Bibliography. Suppose the trachea and mouth are regarded as a horn of some complex shape, and consider what happens when a short pulse of air $\delta U(t)$ is injected by the valve at time t. If Z_0 is the acoustic impedance at the input to the horn ($Z_0 = \rho c/S$ for a cylindrical horn, but has a more complex value for a conical horn), then this causes an immediate increase in the tracheal pressure p_T equal to $Z_0 \delta U(t)$. The horn, however, has a memory for previous pulses of air, which will have traveled to its mouth and been partially reflected. This reflection behavior can be described by a reflection coefficient $r(t')$, which tells us how the pressure at the base of the trachea at time t is affected by the memory of a unit pulse injected at an earlier time $t - t'$. We need to integrate over all previous times, and the formal result is

$$p_T(t) = Z_0 U(t) + \int_0^t r(t') [Z_0 U(t - t') + p(t - t')] dt'. \qquad (14.10)$$

The first term on the right-hand side is the immediate response of the horn pressure to the injected flow $U(t)$, the first term in the integral is the gradually decaying memory of the pressure produced by previous flow events, reflected from the mouth of the horn, and the second term in the integral is the memory of pressure reflections from the input end of the horn. The introduction of an integral into the equations defining the behavior of the system appears rather complex, but in fact it is very simply handled when we solve the equations numerically on a computer.

The reflection coefficient $r(t')$, is actually just the inverse Fourier transform from the frequency domain to the time domain of the reflection coefficient $r(\omega)$ at the horn throat as measured for steady sinusoidal waves. This is itself identical in form to the wave amplitude reflection coefficient between two media, as discussed in Section 6.4. The formal expression is

$$r(\omega) = \frac{Z_{IN} - Z_0}{Z_{IN} + Z_0} \qquad (14.11)$$

where Z_{IN} is the input impedance of the horn. If we know $Z_{IN}(\omega)$ for the horn in question, by measurement or calculation, then we can calculate $r(t)$ by taking the inverse Fourier transform of (14.11), as discussed in Chapter 15. More directly, $r(t)$ can be measured in real cases by applying a flow impulse from a generator that is acoustically matched to the input impedance of the horn, so as to eliminate pressure echos from the input end, and examining the reflected echo.

For a simple cylindrical pipe of length l and with an ideally open mouth, a flow pulse injected at time $t = 0$ gives rise to a simple inverted pressure pulse after a delay time $\tau = 2l/c$. This can be expressed mathematically in terms of the Dirac delta function $\delta(t - \tau)$, which is zero except when $t = \tau$, and is then infinite in such a way that its integral is equal to unity, giving an extremely sharp pulse. Thus, in this case,

$$r(t) = -\delta(t - \tau) \qquad (14.12)$$

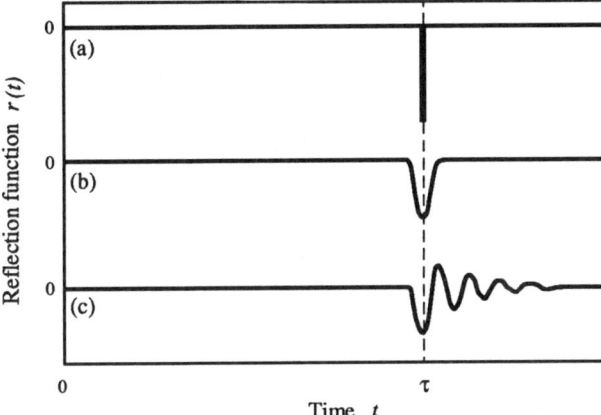

Figure 14.5 The reflection function $r(t)$ for (a) an ideally open cylindrical pipe, (b) a real cylindrical pipe with finite end correction, and (c) a cylindrical pipe terminated by an open Helmholtz resonator. In each case $\tau = 2l/c$ is the pulse travel time to the open end and back.

as illustrated in Fig. 14.5(a). In a real situation we need to modify this result to take account of radiation losses out the end of the pipe and wall losses during propagation along the pipe. The first effect attenuates high frequencies predominantly, with the result that the returning pulse is broadened and reduced in peak height, though still having nearly unit integral. The width of the reflected pulse turns out to be equal to the propagation time through the end-correction distance at the open end of the pipe, and so is greater for wide pipes, which lose more of the high frequencies, than for narrow pipes. Wall losses reduce the amplitude of the reflected pulse by a factor $e^{-2\alpha l}$ where α is the attenuation coefficient for wave propagation in the pipe, as discussed in Section 10.3. The final form of the reflection function for this simple case is thus as shown in Fig. 14.5(b).

In a more realistic case, the mouth constitutes a sort of Helmholtz resonator at the end of the trachea, which the pulse excites into a damped vibration. Some of this radiation propagates back down the trachea, so that the reflection function has the form illustrated in Fig. 14.5(c). In some animals, particularly humans, the mouth cavity is large and has very flexible geometry, so that the nature of the reflection can be greatly modified. In birds, the possibilities are rather more limited. Whatever the form of $r(t)$, this affects the behavior of the whole system through (14.10).

It is worthwhile to mention one other approach to calculation of the behavior of a vocal system, though we shall not pursue it in any detail. We recall that the nonlinear behavior of the flow through the valve is relatively straightforward to analyze in the time domain in terms of the coupled differential equations (14.2) through (14.5), while the linear behavior of the tracheal resonator is simply converted between the frequency domain and the time domain by taking the inverse Fourier transform of the reflection coefficient $r(\omega)$ of (14.11) to give the time-domain reflection coefficient $r(t)$. The analysis outlined above is therefore carried out in the time domain, and describes both the initial transient and the steady state of the sound. It is possible to carry out the whole analysis in the frequency domain, using the techniques outlined in Section 12.4, but this is quite complicated if we wish to include many overtones. If we concentrate attention just on the steady-state sound output, however, much of the complication falls away. The analysis is then known as harmonic balance.

Essentially what we do is to assume a vibration frequency and amplitude for the valve and calculate the flow in the absence of any variation in the pressures p_B and p_T. We then take the Fourier transform of this flow, which gives us flow components at every harmonic of the fundamental frequency. We multiply these flows by the impedances of the bronchus and trachea, both specified in the frequency domain, and this then gives a first approximation to the time-varying pressures p_B and p_T. Transforming these pressures back to the time domain by an inverse Fourier transformation and substituting in the flow equations then gives a new and better approximation for the flow. We simply repeat this process, with adjustment of the fundamental frequency of valve oscillation, until successive cycles repeat, and we then have a consistent solution for the steady state—the method is clearly inapplicable to the initial transient. This approach, essentially in the frequency domain, should give just the same results as the time-domain approach. Which one should be used depends on individual preference and experience, and on the ready availability of appropriate computer routines.

14.4 An Avian Vocal System

The vocal system is clearly quite complicated, and it is really necessary to consider a model for a specific system in order to appreciate its operation. Any system could have been chosen for this illustration, but we have selected the avian vocal system, partly for its intrinsic interest and partly because detailed discussions of the human vocal system are available elsewhere in the literature.

The essential anatomy of the avian vocal system is shown in Fig. 14.4. We have already discussed the various components, and it is simply a matter of choosing numerical values for the relevant parameters and then integrating the equations (14.2) through (14.5) and (14.9) through (14.11) describing the behavior of the system. These equations constitute a set from which the entire performance of the vocal system, including starting transients and nonlinear behavior, can be calculated. It is necessary to do this numerically on a computer by integrating the differential equations step by step, and this integration gives the time variation of all the pressures, flows, and valve motions involved.

The parameters chosen for the model system are given in Table 14.1, and are appropriate for a bird such as a raven. The calculated results are summarized in Fig. 14.6. There are several things to notice about the curves in this figure. In the first place, all the waveforms repeat with a period of about 5 ms, corresponding to a frequency of about 200 Hz, which is close to the natural resonance frequency of the valve, in this case a syrinx. The regular repetitive waveforms imply, from the discussion of Fourier transforms in Chapter 15, that the frequency spectrum consists simply of a series of exact harmonic overtones with a regular spacing of 200 Hz. These harmonic overtones are generated naturally as a consequence of the nonlinearity of flow through the valve. The waveforms also show prominent oscillations with frequencies around 1200 and 2200 Hz, the first in the tracheal pressure when the valve is closed and the second in the tracheal flow when the valve is open, though the amplitude of this latter vibration depends greatly on the assumed impedance of the lungs. These correspond to formants generated by the resonances of the trachea, and the waveform oscillations are generated from groups of harmonics of the syrinx oscillation close to these resonance frequencies. There are actually also other higher formant bands, as can be established by calculating the acoustic output and performing a Fourier transformation. We return to this in a moment.

Once we have calculated the pressure and flow waveforms internal to the system in this manner, we can also calculate the radiated acoustic power. We can do this by transforming from one of the internal variables, say the flow into the trachea near the syrinx, to the acoustic flow at the mouth of the horn. This flow acting on the radiation resistance then gives the radiated spectrum and, of course, the total radiated power. Suppose we confine our attention to a cylindrical trachea and let x measure distance along the trachea so that the mouth is at $x = 0$, and the syrinx is $x = -l$. Then, as discussed in Section 10.1, the pressure and flow in the pipe associated with any particular frequency component ω can be written

$$p = Ae^{-jkx} + Be^{jkx} \qquad U = Z_0^{-1}(Ae^{-jkx} - Be^{jkx}). \qquad (14.13)$$

where $Z_0 = \rho c/S$ is the characteristic impedance of the tracheal pipe. If Z_R is the radiation resistance at the mouth, then we know that $p/U = Z_R$ at $x = 0$, which gives $B/A = (Z_R - Z_0)/(Z_R + Z_0)$. If we know the flow $U_\omega(-l)$ at frequency ω at the syrinx position $x = -l$, then a little algebra gives the flow at the mouth $x = 0$ as

$$U_\omega(0) = \frac{Z_0 Z_R U_\omega(-l)}{(Z_R + Z_0)(Z_R \sin kl + Z_0 \cos kl)}. \qquad (14.14)$$

The radiated power at this frequency is $\frac{1}{2} R_R U_\omega(0)^2$, where R_R is the resistive part of Z_R, and we must sum over all spectral components to get the total radiated power. This is a rather tedious calculation, but in practice we can ignore all but the strongest components when calculating the radiated power.

In our avian system model we have included a simple correction for the bird's

Table 14.1 Parameters for an avian vocal system

Length of trachea	$l = 70$ mm
Diameter of trachea	7 mm
Diameter of mouth horn	20 mm
Bronchial volume	$V_B = 1$ cm^3
Diameter of syrinx membrane	7 mm
Thickness of syrinx membrane	100 µm
Membrane resonance frequencies	$f_V(1) = 150$ Hz
	$f_V(2) = 250$ Hz
Membrane Q value (open)	2
Membrane Q value (closed)	0.2
Nonlinear coefficients	$\gamma_1 = \gamma_2 = 10^6$ m^{-2}
Initial membrane separation	0
Lung resistance	$R_L = 10^6$ acoust ohms
Lung pressure	$p_L = 300$ Pa

Figure 14.6 Calculated waveforms for tracheal pressure p_T, volume flow U, syrinx membrane motion y, and bronchial pressure p_B for the model avian vocal system of Fig. 13.4 and Table 13.1. (From [26].)

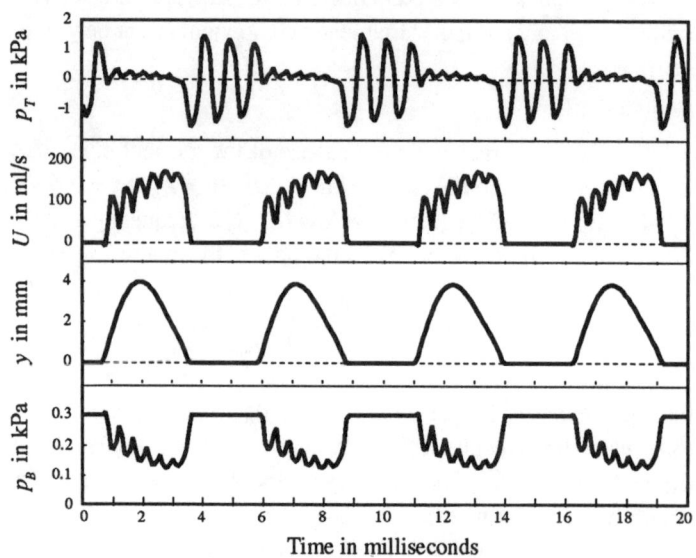

PNEUMATICALLY EXCITED SOUND GENERATORS

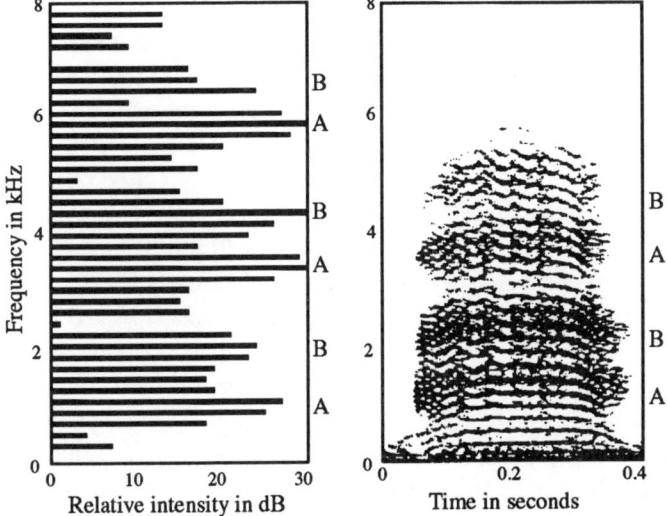

Figure 14.7 Frequency spectrum of the acoustic radiation calculated for the avian vocal system of Fig. 13.4 with the parameters of Table 13.1. The formant bands derived from the tracheal resonances with the syrinx closed are marked A, and those from resonances with the syrinx open are marked B. On the right is shown a Sonagram of the cry of the raven *Corvus mellori*, for which the model parameters were estimated. Note the good match of spectral features. The calculation over-emphasizes high frequencies. (From [26].)

mouth by treating it as a short horn. Provided we are dealing with frequencies within the pass-band of the horn, it is an adequate approximation to treat it as a simple acoustic transformer, which effectively increases the radiation resistance at the end of the trachea by a factor S_M/S_T, where S_M is the area of the open end of the horn and S_T is the cross-sectional area of the trachea. It is possible to perform a more careful calculation using the analysis of Chapter 10, but this is not really justified.

A display of the frequency spectrum of the total radiated power, as calculated from this type of model, is informative. Such a spectrum for the particular case under discussion is shown in Fig. 14.7(a). The regular harmonic series based on a fundamental near 200 Hz is clearly visible, as are the formant bands in which the amplitude of the harmonic partials is relatively high. The formant bands marked A derive from the resonances of the trachea with the syrinx closed, and those marked B from the resonances with the syrinx open. If the impedance of the lungs were to be assumed to be large compared with the characteristic impedance of the trachea, instead of small, then the frequencies of the B bands would be reduced so that they effectively overlapped the A bands. This type of spectrum, with its formant bands, is characteristic of all voiced vocal sounds. It is most commonly seen as a Sonagraph display, in which amplitude is coded on a gray scale, frequency is plotted vertically, and the time-variation is displayed along the horizontal axis. A Sonagraph recording of the call of the raven *Corvus mellori* for which the model was constructed is shown in Fig. 14.7(b). Clearly the agreement between the measured and calculated spectra is quite good.

The parameters of the model, such as lung pressure, can be varied to find their effect on acoustic output. For the particular model investigated, air sac pressure was varied from 50 to 2000 Pa above atmospheric (0.5 to 20 cm WG). The system in fact stopped operating for a lung pressure much below 50 Pa. Flow volume was found to increase from 10 to 300 ml/s over this range, and radiated acoustic power from 1 µW to 10 mW. The pneumatic power expended by the bird is just the product of gauge pressure in the lungs and volume flow, and so ranges from 0.5 to 600 mW in the model. Dividing output power by input power shows that conversion efficiency ranges from about 0.2% at the low-power end to about 1.7% at the high-power end. The major power loss occurs in flow through the syrinx, with significant but smaller losses in the lung resistance and in wall effects in the trachea. These calculated output powers and efficiencies agree well with those measured for birds of medium size by Brackenbury, though there is clearly a large range between different species. The efficiencies are comparable with those achieved in air-powered musical instruments, and not very different from those of the best of the monopole mechanical generators discussed in Chapter 13.

It is possible to experiment with the model by varying all of its parameters in reasonable ways and calculating the acoustic behavior. For example, we can vary the resonance frequency of the syrinx to change the fundamental pitch of the song, as do birds with more varied calls than the raven. The model operates satisfactorily over a frequency range of more than 8 to 1 (3 octaves), though the higher frequencies require rather high lung pressures. We can also examine the effect of varying the initial opening of the syrinx, and experiment with different membrane thicknesses and nonlinearities. The tracheal length is essentially fixed in a given animal but, if we elaborate the model to include mouth resonances, which are certainly important in determining formant frequencies in some birds, then this too can be encompassed within the model.

The model described by our equations is clearly the simplest possible, since we took the valve mechanism to possess only one degree of freedom and so to have only one resonance frequency. In reality the syringeal membranes of a bird or the vocal folds of a mammal are extended objects with many possible vibration modes and natural frequencies, generally without any simple relation between them. It is reasonable to ask how this multiplicity of possible vibration modes might affect the operation of the system. Fortunately the answer is relatively simple, and has already been discussed in Section 12.4. Because of nonlinearities in the mechanical vibration of the valve, particularly if it closes completely for part of each cycle, the motion tends to lock into a repetitive cycle close to the frequency of one of the modes, and other modes then lock to harmonics of that basic frequency. This is just the behavior demonstrated in Fig. 14.6, which was in fact calculated for a membrane system with two inharmonically related modes. The existence of several modes of valve vibration does, however, allow the possibility that the oscillation might lock onto a different mode, depending on parameter values and initial conditions. Indeed the animal might be able to change at will from one mode to another, giving a sort of yodeling effect. As discussed in Section 12.4, there is also the possibility of chaotic vibration, which seems to occur in some of the raucous cries of birds such as cockatoos.

Many extensions and improvements can be made to the general model outlined above. The most obvious extension, which applies to models of all vocal systems, is to make the valve model properly three-dimensional by including structural detail in the direction normal to the plane of the page in Fig. 14.1, rather than simply multiplying by a dimension. As well as allowing for more complex vibrational modes, this extension also allows the valve to close gently, rather than abruptly cutting off the flow when $y = 0$. This in turn decreases the content of very high frequencies in the sound, which is a defect in the simple calculation.

Another refinement that might be contemplated is the recognition that the trachea expands slightly in diameter as we go from the syrinx towards the mouth, so that its form is that of a conical frustum. The mathematics of conical systems is more complicated than that of cylindrical systems since the characteristic impedance is complex. In fact the refinement makes very little difference to the behavior except that the formant resonances are shifted from their cylindrical values $(n - \frac{1}{2})c/2l$ to about $(n - \frac{1}{2} + \delta)c/2l$ where $0 < \delta < \frac{1}{2}$. For a cylinder $\delta = 0$, while for a cone complete to the apex $\delta = 0.5$. For a typical tracheal tube $\delta < 0.1$, so that the formant frequencies are shifted only a small amount away from the cylindrical values.

An interesting and important feature of the avian vocal system is that it generally has two symmetrical valves in the syrinx, one in each of the bronchi, rather than a single valve at the base of the trachea as in most mammalian systems. Since the two valves are under separate muscular control, this allows the bird to sing two notes of different pitch simultaneously, though not all birds do this. The acoustic coupling between the two valves arises from the common tracheal pressure acting upon them, and this is enough to lock them into synchronism if they are closely matched in resonance frequency. The coupling is, however, not usually strong enough to enforce such locking if the frequencies are appreciably different. It would be relatively easy to incorporate this complication into the model discussed above.

Finally we should mention an unresolved problem in bird song. This is the occurrence in many birds of snatches of song, or even the whole repertoire, in a voice with a nearly pure sinusoidal character. The fundamentally nonlinear character of the syrinx mechanism makes it most unlikely that a relatively loud sound with so little harmonic development could be produced by the normal vibrating syrinx membrane valve, even if the membrane were to be held well away from the fully closed configuration. Experimentation with the model certainly gives no indication of such a possibility. It may be that such sounds are genuine aerodynamic whistles of the type to be discussed later in this chapter, but the matter is still unresolved.

14.5 Other Vocal Systems

Not surprisingly, the human vocal system has been the subject of more attention than that of any other animal, and quite complex models have been developed for the motion of the vocal folds of the larynx, which constitute a valve of type $(+,-)$. The human voice shows many similarities to the avian vocal system as calculated above. The vocal folds normally vibrate at a fundamental frequency in the range 100 to 300 Hz, depending on sex and age, and in singing this frequency can be varied by muscular control over a range of about 4 to 1 or 2 octaves. The spectrum of steady vowel sounds shows multiple harmonics of the vocal-fold frequency just as for the bird, and several formant bands corresponding to resonances of the vocal tract. The acoustic impedance of the lungs is larger than the characteristic impedance of the trachea, so that the open-larynx resonances are very close in frequency to the closed larynx resonances. The formant resonances are therefore essentially those of the horn formed by the trachea and mouth, closed by a rather high impedance at its larynx end. In the simplest model, taking the trachea to be a cylinder of length about 15 cm, these resonances are at about 500, 1500 and 2500 Hz. The large mouth cavity, however, has flexible walls and a large flexible tongue, and the lips can also be opened or closed, with the result that the geometry of the upper half of the tracheal horn can be very greatly modified at will. This allows independent modification of the frequencies of these three principal formants over frequency ranges of nearly one octave (2:1) each, thereby producing the different vowel sounds upon which human language is built.

With the availability of specialist texts on the subject, we shall not discuss human vocalization in detail here. The human voice is flexible in pitch range, practised performers having a fundamental frequency range of more than two octaves (4:1), which is comparable with that of birds. In normal speech, however, basic pitch variations are usually no more than ±20%. The bronchial pressure is typically about 300 Pa and the power output about 0.1 mW in normal speech, though higher output levels are easily possible, particularly in singing. The efficiency is typically about 0.1%, increasing at higher power levels.

Another interesting animal is the frog, which generally has two valves acting in series on the air flow. The lower valve is a normal set of vocal folds with a resonance frequency between about 500 and 2000 Hz, depending on species, while the upper set is heavier and has a resonance frequency typically around 100 Hz. The vocal utterances are thus roughly of the form of bursts of carrier wave at the higher frequency, amplitude modulated at the lower frequency, a pattern quite similar to that produced by the mechanical generators of insects. The frog is also unusual and primitive in terms of its pulse-fed lung-sac respiratory system, and because it usually sings with its mouth and nostrils closed. The air pressure in the mouth causes the throat to distend to form a large thin-walled sac that acts as an efficient sound radiator.

The bat is also of great interest because of the high frequency of its cries, and its use of echo ranging for orientation during flight and for the capture of prey. Neither the vocal system nor the auditory system appears to be particularly unusual in basic structure, apart from delicate construction and consequent high resonance frequency, but this need not apply to the precise tuning parameters, such as damping, or to the neural transducer. Indeed some "constant frequency" bats have a very sharp resonance for the auditory system as a whole near the frequency of their cry. The signal processing and pattern recognition techniques that have evolved in these and other sonar-locating animals are particularly impressive. We shall return to some of these matters in Chapter 15.

14.6 Aerodynamic Systems

Sound can also be produced by purely aerodynamic means without the intervention of any vibrating solid surface. Aerodynamic sounds play a part in many vocal systems in the form of whistling or hissing noises, the former being characterized by a well-defined fundamental frequency with or without harmonic overtones, and the latter by a broad-band and generally high-pitched continuous frequency spectrum. The mechanisms by which such sounds are produced are often complicated and not yet well understood, but a brief survey will show what is possible.

The whole basis of aerodynamic sound generation arises from the fact that the steady flow of a fluid becomes unstable if it is forced to change direction or to undergo shear deformation near a solid surface, or even against a body of the same fluid moving with a different speed. A familiar visible example is the flapping of a flag, while an audible demonstration of sound generation by this mechanism is the howling of wind around the sharp corners of a building.

A particularly simple sound generator is the turbulent jet, which appears to be responsible for at least some part of the hissing noises that form components of many vocal utterances—aspirates and sibilants in human speech. If air flows through an aperture at moderate to high speed, as shown in Fig. 14.8(a), then the jet that it forms is unstable and becomes turbulent. The turbulence is mostly internal to the jet, but the fluctuating velocities on its boundaries generate sound in a rather inefficient manner. Because the turbulence is more or less random on all scales below the width of the jet, the radiated sound has a wide-band frequency spectrum. The power radiated by a turbulent jet of speed u and cross-section S is approximately

$$P \approx 10^{-4} \frac{\rho S u^8}{c^5} \tag{14.15}$$

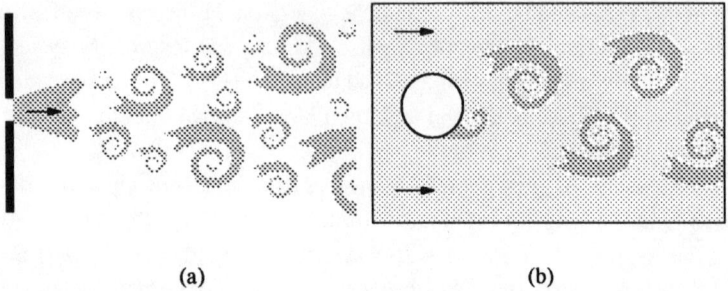

Figure 14.8 (a) Homogeneous turbulence in a fast air jet emerging from a flue aperture. (b) Turbulence in the wake behind a stationary cylinder in a uniform flow takes the form of a double row of counter-rotating vortices. Both diagrams are simply schematic.

corresponding to an acoustic efficiency of about $10^{-4}(u/c)^5$. Clearly this is very small for biological systems, in which $u \sim 0.1c$. Indeed, inserting this figure and an appropriate value for S gives $P \sim 10^{-9}$ W, which corresponds to a sound pressure level of only about 30 dB at 1 meter, even allowing for substantial directionality.

Rather more efficient as a generating mechanism are the aeolian tones (after Aeolus, Greek god of the winds) produced when an airstream flows around a small obstacle. The situation is simplest if the obstacle is long and narrow, like a wire, and it is then found that a double row of counter-rotating vortices is produced in the flow downstream from the obstacle, as shown in Fig. 14.8(b). This pattern is often called a Kármán vortex street. The separation between the vortices depends on the flow speed u and the diameter d of the obstacle, and the quasi-periodic nature of the vortex shedding leads to sound emission with a predominant peak near the shedding frequency f, given by

$$f \approx \frac{0.2u}{d}. \tag{14.16}$$

For an obstacle of millimeter dimensions in a normal vocal airstream with velocity about 30 m s^{-1}, this leads to a dominant frequency in the low kilohertz range. For a short cylinder of length l, the acoustic power produced is about

$$P \sim 10^{-2} \frac{\rho l^2 u^6}{c^3} \tag{14.17}$$

which might amount to about 0.01 mW in a typical case, giving a sound pressure level of about 60 to 70 dB at 1 meter, depending on directionality. Note that the frequency of the tonal component can be controlled by altering the breath pressure and thus the air flow velocity.

The partially tonal sound produced by the aeolian mechanism acting on a

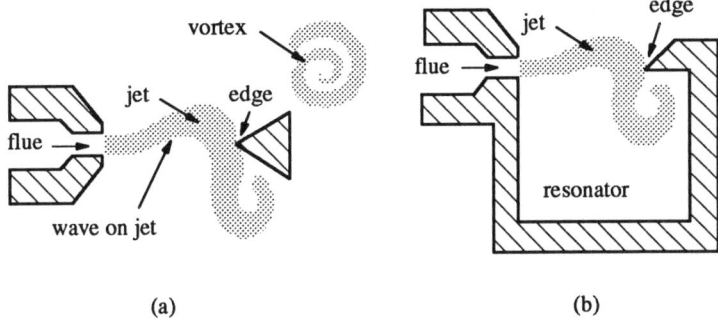

Figure 14.9 (a) Generation of edge tones by the sinuous instability of a jet emerging from a flue. (b) An unstable jet coupled to an acoustic resonator and edge forms a simple whistle.

distribution of small obstacles, superimposed on the broad turbulence noise and perhaps supplemented by other components generated by the tonal mechanisms to be discussed in a moment, probably accounts for most hissed sounds. In many animals, the sound will be generated by a jet and obstacles lying within the mouth cavity, which acts as an internally driven Helmholtz resonator, as discussed in Section 8.6, to modify the spectrum of the sound.

We should give brief mention to a related generator called the vortex whistle, which generally relies upon a jet of air injected at an angle into a short open tube to produce vortices. These emerge from the open end to produce sound. Vortex whistles can provide intense sound sources when very high blowing pressures are available, but do not appear to be important in the biological context.

These three generation mechanisms all depend upon incompressible aerodynamic effects for their operation. In the generators next to be considered, acoustic feedback is important to maintaining the flow oscillation, and the sound produced is tonal in character. Suppose an air jet emerges from a flue aperture and impinges on a rather sharp edge quite nearby, as shown in Fig. 14.9(a). It is the closeness of the flue aperture to the obstacle that differentiates this situation from that just discussed in relation to aeolian tones. If the major part of the jet flow flips from one side of the obstacle to the other, then this produces an acoustic dipole disturbance that has an acoustic flow normal to the direction of the jet at the flue. The jet is unstable against this type of disturbance, and undergoes a sinuous displacement that propagates along it with a speed v which is a little less than half the jet speed u, growing exponentially as it travels. The sinuous displacement of the jet arising from such a propagating wave is shown in the figure. When the displacement of the jet arrives at the obstacle, its direction may be just right to flick the jet across to flow on the other side. If it is not, then nothing happens until the reverse end of the pulse arrives, when the flip occurs. The system then goes into spontaneous oscillation, with the jet flipping from one side of the obstacle to the other at a rate determined by the jet-wave propagation time from the flue aperture. We need to give detailed consideration to the phase relations involved in initiating the instability at the aperture, but the result turns out to be that a phase delay of 180° is required for propagation along the jet in order to close the feedback loop. If l is the distance between the aperture and the edge of the obstacle and u the jet speed, then this implies that the sounding frequency is given by

$$f = \frac{v}{2l} \approx \frac{u}{4l}. \tag{14.18}$$

A higher mode with $3 \times 180°$ of phase shift along the jet and a frequency of $3f$ could, in principle, also be produced, but it turns out that the jet amplification mechanism becomes inefficient as the frequency is raised, so that the fundamental is usually the tone produced. The sound, however, can have higher harmonics present because the jet switches from one side of the obstacle to the other in a quasi square-wave fashion.

It is difficult to be precise about the acoustic output from an edge-tone generator of this type. If the obstacle is narrow in the propagation direction, then the output power should be similar to that from an aeolian generator as given by (14.17). If the obstacle is wider, however, then there is better separation between regions of high and low pressure, and the output should be greater. Note that the sounding frequency can be controlled by varying either the breath pressure, and hence the jet velocity, or the aperture to edge distance.

In an important variant of this type of sinuous-jet generator, we can imagine a resonator of some kind to be located adjacent to the aperture, with the edge against which the jet impinges forming one of its walls, as shown in Fig. 14.9(b). It is this sort of generator that drives musical instruments such as flutes and organ pipes and most simple whistles. The resonator can be either a simple closed cavity, forming a Helmholtz resonator with essentially a single mode, or else an open or closed pipe. Only the former case is relevant to the biological situation. As the jet flips into or out of the mouth of the resonator, it sets the air within it into vibration, and this perturbs the jet where it emerges from the flue aperture. It is the natural frequency of the resonator that now determines the frequency at which the generator must operate, and the blowing pressure must be adjusted so that the phase delay along the jet is just right to close the feedback loop. The associated phase shifts are slightly different from those for the edge-tone case, and oscillation can be maintained with only a small frequency shift for propagation phase delays between about 90° and 270°. The relation (14.18) applies approximately for the center of this range, but must now be interpreted as giving the necessary jet velocity (and hence blowing pressure) when the resonator frequency and jet-to-edge distance are both given. The lung pressure p_L must therefore satisfy

$$3\rho(lf)^2 < p_L < 30\rho(lf)^2 \tag{14.19}$$

if sound is to be produced. The sounding frequency can be changed by varying the volume of the resonator, or some other internal dimension if it is a more complex structure, and this relation shows that, while the lung pressure should be increased as the frequency is raised, there is a good deal of latitude available. Once again there is the possibility of another sounding regime in which there is an extra wavelength, or extra 360° of phase shift, along the jet. Sounds in this regime can indeed often be produced, the blowing pressure being lower than that in the normal regime by a factor close to 9, but their intensity is very low.

The sound output from a generator of this type can be quite large, because it is a monopole source in contrast to the dipole nature of the other sources discussed above. An expression for acoustic output can be derived, but is rather complicated. A good estimate of output can be obtained, however, by noting that the maximum acoustic efficiency is typically a few percent, so that

$$P \approx 10^{-2} p_L U \tag{14.20}$$

where p_L is the pressure in the lungs and U is the volume flow in the jet. This relation applies, however, only while the jet velocity remains within the range defined by (14.19). Clearly a high operating pressure within this range increases both p_L and U and hence the power output. While the sound source may have some small content of harmonics because of the nonlinearity of the jet drive, only the fundamental is reinforced by the Helmholtz resonator, so that the upper harmonics are relatively very weak.

Another type of generator driven by jet instability is the aperture whistle shown in Fig. 14.10(a). If the jet expands a little so that it no longer passes completely through the aperture, then this generates an acoustic pressure pulse that interacts with the jet where it emerges from the flue. The jet instability in this case is of varicose nature, with the jet cross-section varying, and this disturbance wave propagates along the jet with a speed that is rather more than half the jet speed. Again the disturbance grows exponentially as it propagates and, on reaching the aperture, switches the jet diameter so that it again passes through the hole. This system, which is similar to that found in whistling kettles, thus breaks into regenerative oscillations at a frequency determined by the propagation delay along the jet and given once again by a relation similar to (14.18). Experiments suggest that, in this case, the alternative regime in which the phase shift along the jet corresponds to one and a half wavelengths, rather than a single half-wavelength, can be efficiently excited, the pressure being lower by a factor 9 than that given by (14.19). Vortex rings are often generated surrounding the jet, representing a nonlinear extreme of fluctuation in its diameter, but the extent to which they are incidental or essential to the sounding mechanism has not yet been established.

Sound radiation from a source of this type is moderately efficient, since the

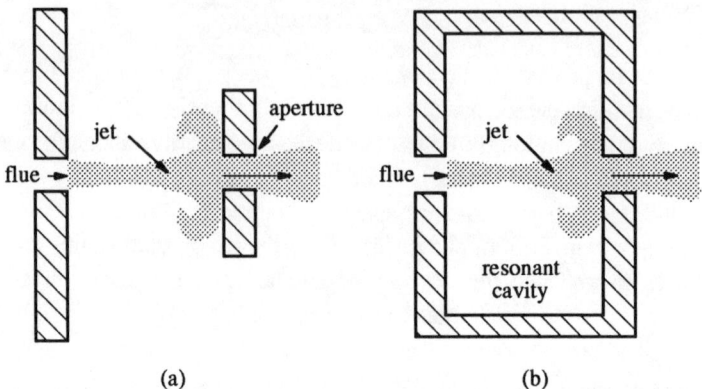

Figure 14.10 (a) Generation of aperture tones by the varicose (radial) instability of a jet emerging from a flue. (b) An unstable jet coupled to an acoustic resonator and aperture forms a simple whistle.

inner end of the dipole source is partially shielded. Again the source efficiency can be enhanced by coupling it to a Helmholtz resonator, or something more complex, usually by simply enclosing the space between the flue and the exit aperture as shown in Fig. 14.10(b). The operating frequency is then determined by the natural frequency of the resonator, and the blowing pressure must satisfy a relation similar to (14.19)—perhaps including an extra factor 1/9 for a weak low-pressure regime—if sound is to be produced. Such a generator is once again a monopole source, and has an acoustic efficiency of a few percent, which makes it very effective. Ordinary human whistling probably relies upon just this mechanism, with the pitch being controlled by varying the position of the tongue and hence the effective volume of the resonator cavity.

It seems likely that the pure-tone whistling sounds made by some birds are produced by one or other of these aerodynamic mechanisms. Just which one is responsible is a matter for speculation, since the anatomical features and physiological adjustments involved in whistled song have not yet been clearly identified.

References

Musical analogs for voiced systems: [10] Ch 13; [16]; [17]; [18]
Whistles: [10] Ch 16; [19]; [20]
Mechanism of bird calls: [26]; [27]; [36]

Discussion Problems

1. Calculate the speed of an air jet emerging from a small hole in a reservoir if the gauge pressure in the reservoir is (a) 3 cm WG, (b) 50 cm WG.
2. Each of the jets of question 1 is separately used to excite an edge tone from a thin wedge placed 10 mm from the hole. Estimate the edge-tone frequency in each case.
3. An empty bottle is excited to produce sound by blowing across its neck. If the volume

PNEUMATICALLY EXCITED SOUND GENERATORS

of the bottle is 300 ml and its neck is approximately a cylinder 10 cm in length and 2 cm in diameter, calculate (a) the frequency of the sound excited, (b) the approximate blowing pressure that must be used. (c) Approximately what blowing pressure must be used to excite the next mode?

4. An animal has a vocal-fold frequency of 100 Hz and a vocal tract length of 10 cm, including allowance for its open mouth. Calculate the approximate frequencies of the first three formant bands (a) if the lung passages below the larynx have an acoustic impedance low compared with that of the vocal tract, and (b) if their acoustic impedance is high.

5. In human speech, the sub-glottal pressure is about 3 cm WG and the air flow is about 100 $cm^3 s^{-1}$. If the SPL is 70 dB at 1 meter distance, estimate the efficiency of the system.

6. Construct a computer model for the operation of a syrinx-like sound generator with a reservoir upstream but no trachea downstream from the valve, and investigate its operation. For simplicity, use (14.2) to describe valve vibration, (14.4) with the Bernoulli integral omitted and $p_T = 0$ to describe the force, (14.5) with the inertial term dU/dt omitted to describe the flow, and (14.9) to describe reservoir pressure. Use typical numerical values and integrate the equations numerically to calculate the flow waveform and the behavior of other physical parameters as functions of blowing pressure. Note the necessity for nonlinearity (the valve opening distance must remain positive, and has no effect on flow if it exceeds the diameter of the trachea) to limit vibration amplitude. You may find a threshold pressure for operation of the system.

7. Extend the model of example 6 by including the trachea using (14.10).

Solutions

1. Pressure differential is (a) 300 Pa, (b) 5000 Pa. From (14.1), jet speed u is (a) 22 m/s, (b) 92 m/s.

2. From (14.18), frequency is (a) 550 Hz, (b) 2600 Hz.

3. (a) From (8.20) the Helmholtz frequency is approximately 170 Hz. (b) From (14.8) $u \approx 14$ m/s so from (14.1) $p \approx 110$ Pa or about 1 cm WG. (c) From Section 10.4 the next mode is at approximately the lowest open-open pipe frequency 1700 Hz. From (14.18) this requires $u \approx 140$ m/s and so $p \approx 10,000$ Pa or 100 cm WG. Try this with a bottle!

4. (a) From Section 14.4, the formants are approximately the open- and stopped-pipe mode frequencies 850, 1700, 2550, and 3400 Hz. (b) In this case the frequencies are approximately the stopped-pipe mode frequencies 850, 2550, 4250, and 5950 Hz. Vocal fold frequency is low enough to show up these bands, but only the lowest three are likely to be detectable.

5. Driving power = pressure × volume flow = 300 N × 10^{-4} $m^3 s^{-1}$ = 30 mW. Acoustic intensity = 10^{-5} W m^{-2}, so acoustic power ≈ 0.1 mW. Thus efficiency ≈ 0.3%.

15 SIGNALS, NOISE, AND INFORMATION

SYNOPSIS. By a result known as Fourier's theorem, any repetitive waveform with frequency f can be represented as the sum of a series of simple sinusoidal waves with frequencies nf where $n = 1, 2, 3, \ldots$. These components are called the harmonics of the fundamental f. This is illustrated in Fig. 15.1, from which it is clear that the first few harmonics define the general shape of the wave, while higher harmonics are associated with sharp changes in slope. The amplitudes of the harmonics define the frequency spectrum of the original wave. A few examples are shown in Fig. 15.2, which also illustrates the fact that the amplitude of the harmonics generally declines steadily with increasing frequency.

In the case of an isolated pulse we can also calculate a frequency spectrum but, instead of consisting of a set of harmonics, it involves a distribution of all frequencies within a band. If the pulse is a tone-burst, as in insect song, then the frequency band is centered on the frequency of the tone. Examples are shown in Fig. 15.3. If the pulse is short, then the associated frequency band is wide, and indeed the width of the band in hertz is numerically about equal to the reciprocal of the pulse duration in seconds. A wide frequency bandwidth is necessary if a system is to follow rapidly varying signals.

In an acoustic communication system, the signal must always compete with natural noise in the environment. The level of natural noise increases at low frequencies roughly as 1/frequency. This suggests that communication systems should use high frequencies, which are also easier for small animals to produce. However, high frequencies suffer attenuation in the air proportional to the square of the frequency, so that a compromise is necessary. The best frequency for auditory communication depends on available acoustic power and desired communication range. The combined effects of propagation attenuation and noise masking are illustrated in Fig. 15.4. High frequencies are best for communication between small animals over short distances, and lower frequencies for larger animals over greater distances. The propagation ranges achievable for different source powers are summarized in Fig. 15.5.

In all communications systems, some form of coding must be used to represent the information being transmitted. The total amount of information that can be transmitted over a noisy channel depends on the bandwidth of the channel and the ratio of signal to noise power. Insects use moderately narrow bandwidths and simple coding schemes to transmit limited messages of identity. Higher animals such as humans use sophisticated coding languages to communicate much more varied messages with a good deal of immunity from interfering noise. Analysis suggests a data rate of only about 100 bits per second for human speech, the limitation being imposed by the neural processing rate, rather than by physical channel capacity.

Some animals, such as bats, make use of echo-ranging sonar techniques to locate prey and to navigate in the dark. The echo strength varies as the sixth power of target diameter and as the fourth power of frequency for small targets, indicating that high frequencies are desirable for such applications. High frequencies also mean that the transmitted cry can be localized into a broad beam or, in the case of a bat emitting sound through its two nostrils, into a series of fan beams as shown in Fig. 15.6, aiding location of the prey. To increase the power in a transmitted pulse, some animals use a chirping technique, in which the frequencies composing the pulse are spread into a descending whistle, as shown in Fig. 15.7. The echo can then be reconstructed to a pulse by neural processing. Some animals also use changes in the frequency of the echo to estimate the relative speed of motion of the prey, and so facilitate its capture.

All these communication activities suggest the necessity for the neural system of the animal to carry out the equivalent of many complex computing operations to recover all the information in the signal. It is certain, however, that the brain operates in a manner very different from a conventional computer, which follows a defined set of computational instructions. Instead, the brain appears to operate by recognition of patterns and correlation of these with stored memories. It is also tolerant of errors and incomplete information. Recent developments in neural network computers, in which "learning" is achieved by the modification of parameters prescribing the strength of excitatory and inhibitory interconnections, may provide a suitable model for the operation of the brain in such tasks. An example of a three-layer neural network is shown in Fig. 15.8.

15.1 Communication

In considering the acoustics of vocal and auditory systems we must not overlook the fact that they have evolved because they provide survival advantages to their possessors. These advantages are of two types. The first is that hearing gives information about the environment and the location of prey and predators; the second is that communication with other members of the same species facilitates mating, definition of territory, and ultimately social behavior. All these applications involve the transmission and reception of information, and in this chapter we give some general attention to the nature of acoustic information and to the optimization of communication systems.

Technical understanding of these matters has developed greatly in the past fifty years because of the importance of electronic communications to modern human society, but the subject is wider than this and encompasses the neural processing power of the animal brain and ultimately the development of language. We must here, of necessity, keep close to the formal and technical end of this spectrum of interest, but the reader should not lose sight of the wider aspects.

15.2 Time and Frequency

The most natural way to view physical phenomena is in the time domain, in which the behavior of the system unfolds like the successive frames of a motion picture film. We can regard an acoustic signal in the same way by noting at each instant the sound pressure or acoustic flow and then assembling these values as a continuous time record. For many acoustic systems, when we do this we find that the time patterns repeat in a regular cyclic manner, as illustrated for example in the bird song waveforms of Fig. 13.6. It turns out that it is possible to analyze such cyclic patterns in terms of more elementary repetitive functions such as sinusoidal waves or square waves, and that this analysis helps our understanding of many phenomena. Analysis in terms of square waves, known with some generalizations as a Walsh transform, need not concern us further and is mentioned only to show that there are several possibilities from a mathematical point of view. Rather we concentrate on analysis in terms of sinusoidal waves, which is known as a Fourier transform after the French mathematician Jean Baptiste Joseph Fourier

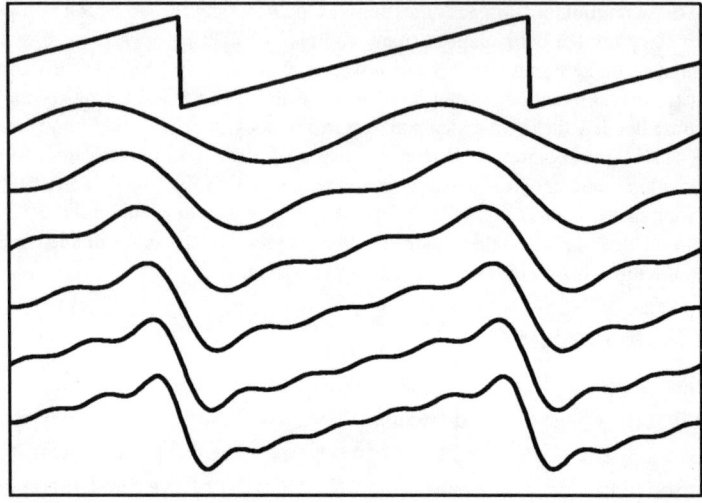

Figure 15.1 A saw-tooth wave and its successive approximation by Fourier series with 1, 2, 3, ..., 6 harmonics. Still higher harmonics are required to approximate closely the sharp corners of the waveform. The amplitudes of the harmonics decrease as $1/n$ where n is the harmonic number.

(1768–1830) who developed the method for the study of heat transfer, rather than wave motion. The reason that the method is so important and appropriate in acoustics is that sinusoidal waves arise automatically from the solution of the simple oscillator equation (2.1) that lies at the heart of all acoustic theory.

Figure 15.1 shows an example of the addition of sinusoidal waves to represent an arbitrary periodic function. Successive stages of the addition are shown, and it is clear that low-frequency or long-wavelength waves give the broad outline of the behavior while high-frequency waves are required to represent adequately any sharp corners on the waveform. If τ is the repetition period of the wave in time, then all the component sinusoids must also repeat with this periodicity, or with submultiples of it such as τ/n, where n is an integer. Since frequency f is simply the reciprocal of the period, and angular frequency $\omega = 2\pi f$, this means that the component waves have angular frequencies $\omega_n = n\omega_1$ where $\omega_1 = 2\pi/\tau$ is the angular frequency of the fundamental. These components with angular frequencies that are exact integer multiples of the fundamental frequency are called harmonics, the fundamental being the first harmonic and ω_n the nth harmonic. In more general situations such as bell sounds, in which integer frequency relations do not apply, the components with frequencies above that of the fundamental are called upper partials or overtones. In a harmonic sound the first overtone is the second harmonic.

We can write a general expression for the Fourier analysis of a periodic sound in the form

SIGNALS, NOISE, AND INFORMATION

$$p(t) = \sum_{n=0}^{\infty} A_n \cos(n\omega_1 t + \phi_n)$$

$$= \sum_{n=0}^{\infty} [B_n \cos n\omega_1 t + C_n \sin n\omega_1 t] \quad (15.1)$$

where A_n, B_n, C_n and ϕ_n are constants. We can see that both sine and cosine terms are required in general, since a simple shift of the time origin creates such a mixture. We can find the coefficients B_n and C_n, and from them A_n and ϕ_n, by multiplying the second line of (15.1) by either $\cos n\omega_1 t$ or $\sin n\omega_1 t$ and integrating over one period of the waveform from $t = 0$ to $t = \tau = 2\pi/\omega_1$. Remembering that

$$\int_0^{2\pi} \cos m\theta \cos n\theta \, d\theta = \int_0^{2\pi} \sin m\theta \sin n\theta \, d\theta = \pi \delta_{mn}$$

$$\int_0^{2\pi} \cos m\theta \sin n\theta \, d\theta = 0 \quad (15.2)$$

where $\delta_{mn} = 1$ if $m = n$ and $\delta_{mn} = 0$ if $m \neq n$, this leads to the results

$$B_n = \frac{\omega_1}{\pi} \int_0^{2\pi/\omega_1} p(t) \cos n\omega_1 t \, dt \qquad C_n = \frac{\omega_1}{\pi} \int_0^{2\pi/\omega_1} p(t) \sin n\omega_1 t \, dt$$

$$A_n = (B_n + C_n)^{1/2} \qquad \phi_n = -\tan^{-1}(C_n/B_n). \quad (15.3)$$

Quite often we are not greatly concerned with reconstructing the wave, but simply with finding the distribution of energy in frequency. This is given by the set of magnitudes $\frac{1}{2} A_n^2$. For all physically realistic waves the magnitude of A_n decreases with increasing n, at least for n greater than some fairly small integer. For the saw-tooth, square, and triangular waves shown in Fig. 15.2, for example, A_n decreases as $1/n$, as $1/n$ (with odd terms only), and as $1/n^2$ (with odd terms only) respectively, giving spectral slopes of -6 dB/octave, -6 dB/octave and -12 dB/octave.

We can treat non-repetitive waveforms in much the same way, simply by letting the period τ tend towards infinity. The fundamental frequency then approaches zero and its harmonics become closely packed, so that ultimately we can replace the sum in (15.1) by an integral over frequency. It is usual to employ the complex notation we introduced in Chapter 2 so that the expansion is in terms of $e^{j\omega t} = \cos \omega t + j \sin \omega t$ rather than in terms of the trigonometric functions individually. The analog of (15.1) then becomes

$$p(t) = \frac{1}{2\pi} \int_{-\infty}^{\infty} \tilde{p}(\omega) e^{j\omega t} d\omega \quad \text{or} \quad p(t) = \int_{-\infty}^{\infty} \tilde{p}(f) e^{2\pi j f t} df \quad (15.4)$$

where the complex quantity $\tilde{p}(\omega)$ or $\tilde{p}(f)$ replaces $A_n - jB_n$ and is called the Fourier transform of $p(t)$. Numerically $\tilde{p}(\omega) = \tilde{p}(f)$, and it is only their mathematical form

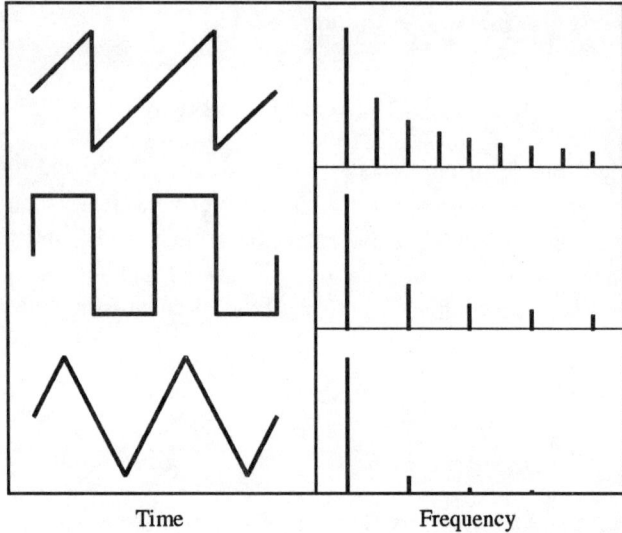

Figure 15.2 Three simple waves and their Fourier amplitude spectra. The phases of the harmonics must be correct in order to resynthesize the wave.

that is different. The value of $\tilde{p}$ can be found by performing an inverse Fourier transform, defined by

$$\tilde{p}(\omega) = \int_{-\infty}^{\infty} p(t)e^{-j\omega t}dt \quad \text{or} \quad \tilde{p}(f) = \int_{-\infty}^{\infty} p(t)e^{-2\pi jft}dt \quad (15.5)$$

which is analogous to the first two of equations (15.3). We have written each of (15.4) and (15.5) in two different ways. The first pair is expressed in terms of the angular frequency ω while the second pair uses the frequency f. There are clear advantages of symmetry in the f notation, and it is usually adopted.

While we must always preserve phase information in the transform if we later want to transform back, we are often concerned simply with the power spectrum of a signal, which is proportional to the distribution of the absolute value of $1/2\,\tilde{p}(f)^2$. To get an absolute value, let us suppose that the acoustic signal is propagating down an infinite pipe, and that we sample it for 1 second. If we imagine this sample to be repeated end-to-end, then the whole signal becomes periodic with period 1 second, and the Fourier transform breaks into harmonic components with spacing 1 Hz. This is just what is done in a typical digital spectrum analyzer, with a switchable sampling period to control frequency resolution. The integral in (15.4) becomes a sum, with $df = 1$, and the total power in the signal is $\Sigma[\tilde{p}(f)]^2/2Z_0$ where Z_0 is the acoustic impedance of the pipe and $\tilde{p}$ is the magnitude of $\tilde{p}$. If the sampling time is changed to τ then this value must be modified by a factor $1/\tau$, making the power $\Sigma[\tilde{p}(f)]^2/2Z_0\tau$, measured in watts. Because we are generally interested in

SIGNALS, NOISE, AND INFORMATION

the power spectrum, instruments such as spectrum analyzers are usually calibrated to measure $\bar{p}(f)$ in units of Pa s$^{1/2}$ or, equivalently, Pa Hz$^{-1/2}$.

Figure 15.3 shows the waveform and power spectrum for two signals of biological interest, corresponding to the short pulses emitted by insects and by frogs, as discussed in Chapters 13 and 14. We see that, in each case, a pulse with carrier frequency ω_0 and having a duration of about Δt has a power spectrum centered about ω_0, as would be expected, and with an angular frequency spread $\Delta \omega$ of about $2\pi/\Delta t$. This relation, which we write as

$$\Delta t \Delta \omega \approx 2\pi \quad \text{or} \quad \Delta t \Delta f \approx 1 \tag{15.6}$$

is analogous to Heisenberg's uncertainty relation in quantum mechanics. A system that has a good time resolution must necessarily have a wide bandwidth. Conversely, to make a precise measurement of frequency requires a long time. The more useful form of the relation (15.6) is the second one in terms of frequency f. It states, for example, that to have a time resolution of 1 ms, or more precisely to resolve two pulses that are 1 ms apart in time, we require a system with a bandwidth of at least 1000 Hz. Other numbers are easily substituted. The wings of the frequency distribution are higher for the first pulse in Fig. 15.3 because of its sharp onset.

Figure 15.3 (a) An exponentially decaying pulse, as in the stridulation of an insect, with (b) its Fourier transform, or amplitude distribution in frequency space. (c) A symmetric pulse, as in the song of a frog, with (d) its Fourier transform. In each case the frequency distribution is centered on the carrier frequency ω_0 of the song, and the bandwidth of the spectrum is approximately the reciprocal of the pulse length.

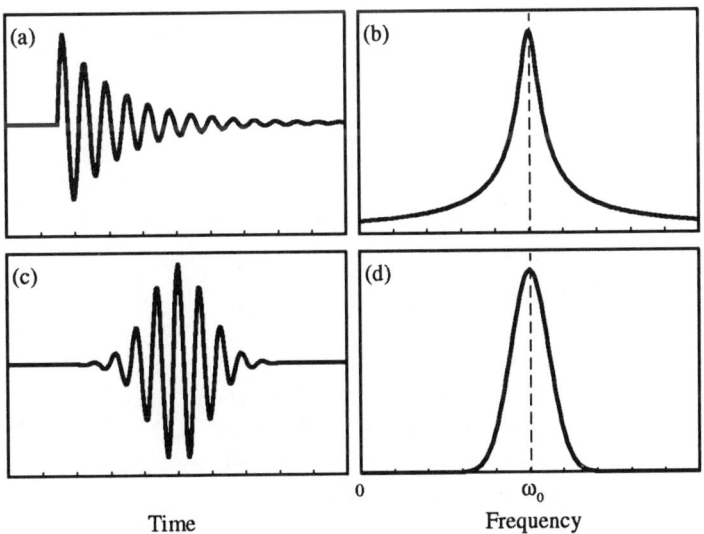

It is appropriate to examine the effect if the pulses of Fig. 15.3 are not isolated, but repeated by the animal in a regular series with time interval τ. The fact that the whole signal is now periodic with fundamental angular frequency $\omega_1 = 2\pi/\tau$ means that all the upper partials in the spectrum must be integer multiples of ω_1. This does not change the shape of the envelope of the power spectrum in frequency space, as shown in Fig. 15.3, but simply splits the previously continuous spectrum under the envelope into a set of closely spaced spectral components separated by a frequency difference ω_1.

Calculation of a Fourier transform for a complicated function using the formula (15.5) can be quite time consuming, even if numerical methods are used on a computer. It turns out, however, that there is a very efficient computational algorithm, known as the Fast Fourier Transform (FFT), which can be readily programmed for a small computer and which yields complex Fourier transforms in a very short time. The basis of the algorithm is outlined in the book *Numerical Recipes* listed in the Bibliography, together with a source-code listing that is easily copied. The FFT algorithm, either implemented in software on a general-purpose computer or in microchip hardware in a dedicated processor, is now so common that signals can be very readily converted from the time domain to the frequency domain and back again, after having been filtered or otherwise manipulated in either domain. Our main need here, however, is to understand the result of the transformation, as discussed above, and to be able to transform signals calculated or measured in the time domain into a power spectrum.

15.3 Signals and Noise

In the real world, signals containing wanted information are always in competition with other signals containing irrelevant information that can be described as noise. Clearly, what is signal and what is noise depends upon the listener! Noise can consist of other conversations at a party, of the sounds produced by machinery in a factory, or of the sounds of wind and rain. Each has its own spectral distribution and time evolution, and detailed discussion needs to take this into account. It turns out, however, that we can construct quite general descriptions of noise and its effect on communication that are useful guides in nearly all situations.

Two idealized types of noise are important in such a general discussion. The first is called white noise, in loose analogy with white light, and is defined to be a noise with constant power per unit frequency (per hertz) across the whole spectrum. Such a noise cannot be realized in practice, since its integrated power goes to infinity at the high-frequency end of the spectrum, but we can have noises with constant power distribution across a wide spectral band. An example of approximately white noise is the hiss emitted by compressed air or steam released from a nozzle, or the sound from a high-quality radio tuned to the space between stations. More exactly, these distributions should be referred to as band-limited white noise.

SIGNALS, NOISE, AND INFORMATION

More important among natural noises is the second type, called pink noise or $1/f$ noise, which has constant power per unit frequency ratio (per octave or per decade) across the whole spectrum. Again such an idealized noise cannot exist in nature, because its integrated power diverges at both the low-frequency and high-frequency limits, but natural noise sources have a nearly $1/f$ spectrum over very wide frequency ranges. An example of approximately $1/f$ noise is the sound of waves breaking on a beach. In fact, however, if we simply measure sound pressure level, we find that natural background noise has a nearly $1/f$ spectrum down to frequencies of less than a microhertz, corresponding to periods of weeks, for which the atmospheric pressure has a typical variation amplitude of about 10 millibars, corresponding to 1000 Pa or 153 dB relative to the standard level of 20 µPa. This increase of noise level at subsonic frequencies is one reason that our ears are vented by Eustachian tubes to decrease response at frequencies below about 20 Hz, as illustrated in Figs. 9.4 and 9.5. The same sort of venting is applied to microphones for the same reason.

These two noise distributions are special cases of a distribution of the form $P(f) \propto 1/f^n$ where $P(f)$ is the power per unit bandwidth (e.g. per hertz) at frequency f. For the general noise background in the ocean the behavior is of approximately this form, with $n \approx 1.7$, at least over the range 100 Hz to 40 kHz. The spectral level at 1 kHz is between -30 and -60 dB rel 0.1 Pa, depending on surface wind speed. Because none of these distributions can be extended over the whole frequency range from zero to infinity, we generally deal with band-limited noise, the bandwidth being set by the frequency limits of the detecting system, or by the characteristics of the noise source itself. If the bandwidth under consideration is small, say less than 1 octave, then there is not a great deal of difference over this small range between any of the distributions.

When considering the performance of an auditory system, we need to know how much energy it absorbs from the signal and how much from the ambient noise, for we shall later see that the ratio of signal power to noise power in a communication channel is most important. From an elementary point of view, we can note that some specific fraction of the absorbed energy, from either source, is ultimately transferred to the neural transduction apparatus, and it is desirable that as much of its response as possible is due to signal rather than noise. We can get some insight into this problem by considering the behavior of a simple vibrator with mass m, resonance frequency ω_0, and damping resistance $\gamma = 2m\alpha$, exposed to a signal force $F_s e^{j\omega t}$ and a random noise force of amplitude $F_N(\omega)$. This vibrator was analyzed in detail in Chapter 2, where we found that the velocity response to the signal force is

$$v_s = \frac{j\omega F_s/m}{(\omega_0^2 - \omega^2) + 2j\omega\alpha}. \tag{15.7}$$

The power absorbed is just the resistance $2m\alpha$ multiplied by half the square of the absolute value of v_s. If we suppose the system to be tuned so that its resonance frequency ω_0 corresponds to the signal frequency ω, then the signal power absorbed is

$$P_S = \frac{F_S^2}{4m\alpha} = \frac{F_S^2 Q}{2m\omega_0} \tag{15.8}$$

where $Q = 2\alpha/\omega_0$ is the quality factor of the vibrator. The absorbed power thus increases as the damping is reduced and the resonance made sharper.

The power P_N absorbed from the noise signal can be evaluated similarly. The algebra is simplest if we assume that we are dealing with $1/f$ noise, so that we can write

$$F_N(\omega) = (\omega/\omega_0)^{1/2} F_N \tag{15.9}$$

so that F_N, without any ω argument, is the noise amplitude at the resonance frequency ω_0 of the system. We substitute $F_N(\omega)$ into an equation similar to (15.7) to find the magnitude $v_N(\omega)$ of the noise response velocity, and then integrate to calculate the total noise power absorbed, according to the equation

$$P_N = \int_0^\infty \frac{1}{2} m\alpha [v_N(\omega)]^2 df \tag{15.10}$$

the integration being with respect to f rather than ω, in accord with (15.4). This integral is not particularly easy to evaluate, but we can find its value very efficiently by consulting a table of integrals, such as that in the book by Gradshteyn and Ryzhik listed in the Bibliography. Provided the Q of the vibrator is significantly greater than 1, so that $\alpha \ll \omega_0$, we find that the total noise power absorbed is

$$P_N \approx \frac{F_N^2}{8m}. \tag{15.11}$$

F_N is the noise amplitude in the vicinity of the system resonance, and this equation tells us that the noise power absorbed is independent of the system Q or bandwidth. This result is not exact for low-Q systems, but the discrepancy is not important.

The reason behind this result is simple. By (15.8), for a single-frequency disturbing force close to the resonance, the energy absorbed is proportional to the system Q. For a broad-band noise, however, the energy absorption is also proportional to the width of the resonance curve, which varies as $1/Q$. When these two factors are multiplied together, the Q-dependence is eliminated.

Actually the discussion in both these cases has been oversimplified as far as auditory systems are concerned, since we have assumed a constant force, implying zero source impedance, whereas the source impedance for an auditory system is actually equal to the radiation resistance R_R at its input, as discussed in Chapter 9. By the maximum power transfer theorem, discussed in Section 9.6, the neural transducer, with acoustic resistance R_N, will receive the greatest possible power from the external signal if $R_N = R_R + R_T$, where R_T is the contribution of the tympanum to the losses in the system. If the system is optimized in this way, then, since the total acoustic resistance is $R_R + R_T + R_N = 2R_N$, the damping coefficient is $\alpha = R_N/m$ and the optimum Q value is

$$Q_{opt} = \frac{m\omega_0}{2R_N}. \qquad (15.12)$$

Even if internal losses in the tympanum could be reduced to zero, the matched value of R_N cannot be made less than R_R, so that there is a limit to the amount by which the Q can be increased before performance begins to decline.

The optimization strategy for designing, or evolving, an auditory system with good noise rejection is thus intimately involved with the choice of system bandwidth. The system should have a bandwidth, determined by its Q value or damping, just sufficient to encompass the bandwidth of the signal. If the purpose of the auditory system is to locate prey or give warning of predators, and if the ecological system in which the animal lives is narrowly defined, then good narrow-band system optimization is possible. If the ecological niche is wider, then auditory efficiency must be sacrificed in the interests of versatility.

Narrow-bandwidth communication systems have good noise immunity and are used by many relatively simple animals such as insects. The signal consists of a series of tone bursts with a carrier frequency characteristic of the species, and the signal bandwidth is effectively that given by the pulse-length spreading, as illustrated in Fig. 15.3. The auditory system need only have enough bandwidth to cover the frequency range of the signal, which may imply a Q value in excess of 10. If the channel bandwidth is made too narrow, then the auditory system will be unable to follow the time variation of the signal.

Since all auditory and vocal systems have limited bandwidth, it is interesting to understand the physical constraints that have led to the evolution of particular frequency ranges for communication within particular species. Broadly speaking, three determining factors enter. The first is the efficiency with which sound can be produced and radiated at a particular frequency, the second is the sensitivity with which it can be detected, and the third involves the transmission characteristics of the medium, including the influence of noise. We have given a good deal of attention to the first two matters in earlier chapters, though from a slightly different viewpoint, so that we summarize the conclusions rather briefly before discussing the propagation channel.

All sound-production systems rely finally upon coupling to a radiating wave-field, and we have seen in Chapter 7, and particularly in Figs 7.1, 7.3, and 7.5, that radiation efficiency, as measured by radiation resistance, decreases steadily once the diameter of the radiating structure is less than about 0.6 times the wavelength. This argues for concentrating the signal into frequencies above this limit. The losses in biological tissue, however, generally increase with increasing frequency, which suggests that optimization might result in a rather lower operating frequency. In fact, the communication energy of most animals is

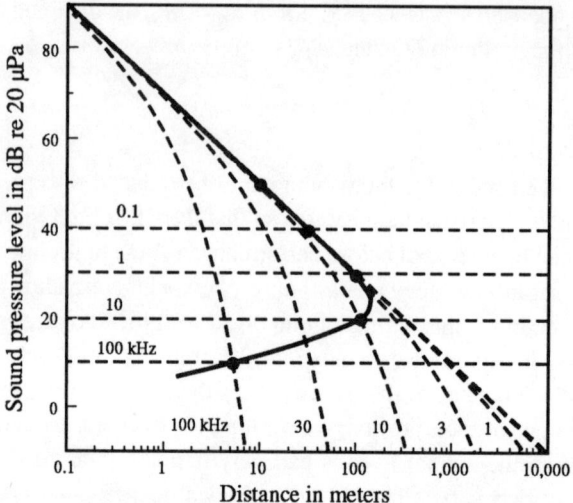

Figure 15.4 Propagation characteristics of sound of various frequencies, shown in kHz as a parameter, from a source of power 0.1 mW (broken curves) together with typical outdoor octave-band noise levels (about 30 dB(A) overall) at the same frequencies (broken lines). The intersections of these two sets of curves (full curve) define the communication distance for which the signal-to-noise ratio is unity (0 dB), as a function of frequency.

concentrated into a spectral region lower than this criterion would suggest by a factor of 3 to 5, as we see later. The auditory system presents similar optimization criteria, though they are generally more easily met.

With regard to the propagation medium, there are two things to consider. The first, which we discussed in some detail in Chapter 7, relates to the decrease in intensity caused by spherical spreading and by wave attenuation when the source and receiver are separated. We recall that the intensity decrease from spherical spreading amounts to 6 dB for a doubling of the propagation distance, while attenuation losses increase as the square of the frequency. The first of these phenomena affects all frequencies equally, while the second argues for the use of low frequencies if long propagation distances are involved. There is, however, a competing argument, for acoustic noise is generated moderately uniformly at the earth's surface by turbulence and other effects, and its intensity increases at low frequencies roughly as $1/f$. This in turn argues for the use of higher frequencies for long-distance communication so that the signals are not obscured by locally generated noise.

These competing effects are illustrated in Fig. 15.4, where two sets of curves have been superimposed. The horizontal broken lines show the average background acoustic noise level in an octave band at several frequencies in a quiet outdoor environment in which the A-weighted sound level is about 30 dB(A).

SIGNALS, NOISE, AND INFORMATION

These levels do not depend on position. The broken curves show the relative sound pressure level as a function of distance for sources of the same power operating at different frequencies, as discussed in Section 6.6 and plotted in Fig. 6.5. These curves should be slid up or down across the noise-level lines as the power of the source is changed. Equally, the noise-level lines can be slid up and down as the background noise level changes. We take an acceptable limit to the range of communication to be that for which the signal power and the noise power in an octave band around the signal frequency are equal, so that the intersection of the signal curve and the noise line for a given frequency then gives the maximum communication range. These intersections are shown circled and joined by a full curve in the figure. For high frequencies the communication range is always small; it increases as the frequency is lowered, reaches a maximum, and then decreases again. The frequency for which the range is a maximum decreases as the power is increased.

The performance deduced from Fig. 15.4 for a range of source powers is summarized in Fig. 15.5. Given the available acoustic power, assumed concentrated in a bandwidth of about an octave, an auditory system well matched to the signal, and assuming an environmental noise background with level as shown by the parameter, this diagram gives, in semiquantitative terms, the optimal frequency and the range at this frequency for a signal-to-noise ratio of unity (0 dB). The figures given are, in all cases, optimistic in relation to communication distance because they assume free-air propagation and neglect attenuation by vegetation

Figure 15.5 Optimum frequency and communication distance for a signal-to-noise ratio of 0 dB for narrow-band sources of specified power in ambient noise of the level shown as a parameter. The figure is approximate only.

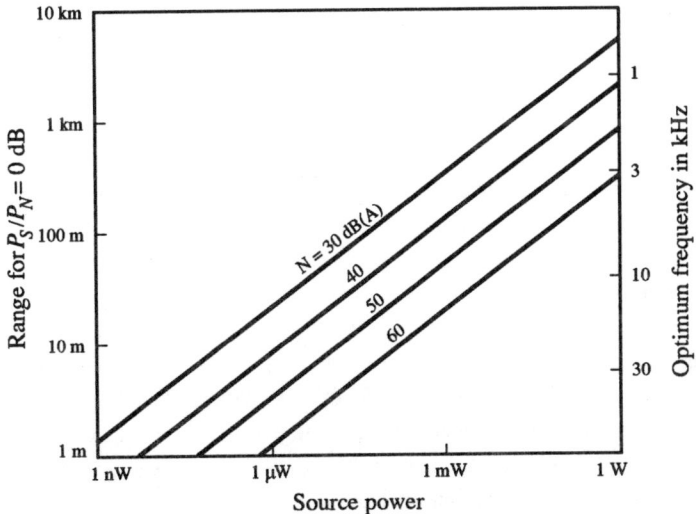

and other obstacles. At large distances they also ignore the effects of refraction and of turbulent scattering in the atmosphere. We should therefore regard Figs 15.4 and 15.5 as only roughly approximate, and rather optimistic. They show an effective range of about 100 m for a source power of 0.1 mW and frequencies in the band 300 Hz to 3 kHz, corresponding to our common experience with human speech or bird song. At a much greater power level, the wide-band noise from a jet aircraft or the sharp sizzling snap of a lightning strike, both with acoustic power in the kilowatt range, can be heard to a distance of 10 km or so, where the audible sound is almost entirely at very low frequencies.

When we combine this range data with the radiation efficiency criterion discussed before, we see the sort of strategy that can be adopted in optimizing acoustic communication systems. Insects are very small and generally have high population density, so that they need communicate with low power over distances of only a few meters. This argues for a best communication frequency in the 5 to 10 kHz range. Because the information content of insect signals is not large, a relatively narrow frequency range and simple coding system suffices, as we discuss later, and reduces the effects of noise interference. Birds and mammals are larger, so they can more easily produce lower frequencies, their population density is much lower, and they are mobile, which suggests the need for a communication range up to perhaps one or two hundred meters. A communication frequency band centered around 1 kHz is therefore appropriate. Because of the greater complexity of information communicated, the bandwidth required is larger. In all cases, of course, the auditory system has evolved to match the vocal system in center frequency and bandwidth.

15.4 Information and Coding

The purpose of acoustic communication, or indeed of any communication, is to pass information from one individual to another. This statement may appear self-evident, but on closer examination we see that it requires a more precise understanding of the concept of information, particularly if we wish to develop any quantitative conclusions. The development of a formal understanding of information and its communication rests largely on the work of Nyquist and Shannon at the Bell Telephone Laboratories some fifty years ago, and is now an essential tool for communications engineers. We do not need such a detailed understanding for our present purposes, but an outline of what is involved throws considerable light on many aspects of animal acoustics.

Communication can be defined as the passing of a message between two individuals, and the simplest message that can be passed is the answer "yes" or "no". We can call such an answer a "bit" of information. The word bit is actually a contraction of the phrase "binary digit", which arises since we could represent a "yes" answer by a 1 and a "no" answer by a 0. The digits 1 and 0 are all that are required to represent a number in base 2 arithmetic. Ordinary arithmetic is represented in base 10 using the digits 0 to 9. In base 2 arithmetic, 1 means 2^0 or 1, 10 means 2^1 or 2, 100 means 2^2 or 4, and so on. The binary number 111 is 7 and 1001 is 9. Any sequence of "yes" and "no" answers can therefore be represented by an appropriately large binary number.

Now suppose we had a list of all possible messages that could be passed between two individuals. The individual messages might be very complicated, but as long as their number is finite this makes no difference. Suppose we write out all the possible messages as an ordered list in a book. Then the simplest way to send one of the messages is for the recipient to effectively ask the question "Is the message in the first half of the list?" The answer can be sent as a simple "yes" or "no", or equivalently as a binary digit 1 or 0, and contains one bit of information. On receiving this elementary communication, the recipient then asks "Is the message in the first half of this part of the list?" to which the reply is again 1 or 0, and so on. Once we have reduced the section of the list to one containing only two messages, then the answer to the final question uniquely specifies the message. Since the questions asked are always the same, it is unnecessary to ask them, and the message can be sent simply by transmitting a string of bits such as 110100..., giving the location of the message in the list. If the number of messages on the original list is N, then we require just $\log_2 N$ questions to locate the particular message being sent, or in other words just this number of binary digits to specify the message. We say that the message contains $\log_2 N$ bits of information. Notice that this is completely independent of the content of the message and depends only on the number of possible messages. This somewhat surprising definition will be shown in a moment to be quite sensible!

When we consider the actual sending of the message, we need to use some sort of code. The simplest is to agree to send the elements of the signal one after the other at regular intervals, with, for example, one particular tone representing a 1 and a different tone representing a 0. The rate at which we send the tones is then the information content of the message in bits per second. This is very much like the communication system used between simple computers. We can see, however, that there is a problem if the communication channel is not perfect—if any single bit symbol of the message is incorrectly received, or incorrectly heard by the recipient, then we no longer know exactly which message was sent, but have a choice of two. The ambiguity doubles for each extra missing bit. This problem is generally the result of noise on the channel, for noise is a statistically random signal and its intensity can rise momentarily to a high level to obscure part of the signal. The problem will clearly be reduced if the two tones are far apart in frequency, and increased if they are nearly the same. The solution is to use more signal power so as to dominate the noise, to send the message more slowly, so that each tone will last long enough that it can be heard through any noise bursts, or to code the signal in such a way that missing parts of the message can be reconstructed. In one possible coding, known as redundancy coding, the message is simply sent more than once. Comparison of two received messages allows identification of the errors, and if three or more copies of the message are sent then the errors can be corrected with reasonable reliability by accepting the majority version at each bit location. This strategy, which is not very efficient, is used by many animals, but partly in order to maintain a continuous signal coverage of their territory.

One of the important contributions of Shannon was to show that there is a simple relationship between the rate C at which information can be sent along a noisy channel, the frequency bandwidth W of the channel in hertz, the peak signal power S, and the average white noise power N. Shannon's relation is

$$C = W \log_2 \left(1 + \frac{S}{N}\right). \tag{15.13}$$

C, given in bits per second, is called the capacity of the communication channel. This result, or something like it, looks entirely reasonable. If we let the signal strength decrease, for example by moving away from the source, until the signal to noise ration S/N is unity, then we find that $C = W$, so that the capacity of the channel in bits per second is equal to its bandwidth in hertz. Increasing the signal power increases the channel capacity, but only logarithmically rather than linearly. The important outcome of Shannon's work, however, is that he also showed that, by appropriately coding the signal, one could actually transmit messages at rates up to the capacity of the channel without any errors at all due to the noise. The crucial thing is use of an appropriate coding scheme and, not surprisingly, the communications industry has devoted a very large amount of attention to developing such error-correcting or noise-immune codes. Nature has evolved a rather similar strategy.

As we pointed out above, an elementary code simply sends two different tones to represent the 1s and 0s of the binary message. In an even simpler version, the 0s can be represented by silent spaces, so that only a single tone is involved. This is rather like the coding used in insect stridulation, which generally consists of a simple amplitude-modulated carrier. The amplitude modulation may be just a set of repetitive pulses, or may have some sort of fine structure. The information content, estimated from the modulations regarded as binary digits, is typically around 100 bits per second (100 baud). The coding is, however, generally simply repetitive, with the same message being sent over and over again. An individual message sequence may last only for a fraction of a second, so that its information content appears to be only a few tens of bits. There is a further variable, and this is the carrier frequency. It is certainly characteristic of the species, which makes its approximate frequency important, and it is possible, though not very likely, that small variations in frequency convey additional information. A total information content of a few tens of bits is, however, adequate for insect communication purposes. This allows the transfer of species identification (6 bits will identify a single species out of $2^6 = 64$ possibilities), sex (although often only males sing), and a few other pieces of relevant information. The most important additional information, which is given by intensity and wave propagation direction, is the location of the animal, since this serves both as a spacing signal for other males and as a beacon for females.

Many more sophisticated physical coding systems exist in higher animals, as clearly indicated in the songs of birds. As a single example, however, let us consider the physical coding involved in human speech. The range of human hearing extends from about 50 Hz to 20 kHz, and the range of frequencies generated in speech is similar. The important part of this range is that from about 300 Hz to 3 kHz, for this is where hearing sensitivity is maximal, and the more extreme parts of the range can be regarded as incidental. If we keep again the criterion that signals should be intelligible with a signal-to-noise ratio of unity, then this suggests a channel capacity, and so a potential information rate, of about 3000 bits per second for human speech. We shall see that the actual information content is much smaller than this, reflecting not a physical limitation on the channel or even the auditory system, but rather a limitation at a neural level.

English speech can be written down using an alphabet of 27 characters—26 letters and one space (we can include punctuation marks without making a significant difference). Other languages use rather different systems, but most can be rendered phonetically with about this number of characters. Let us consider a message of 100 characters and estimate its information content. If all arrangements of characters represented possible messages, then the number of such possible messages would be 27^{100} or about 10^{140}. The rules of language, however, restrict the possibilities greatly. For one thing, the message must be composed of English words, of which there are only about 10,000 in common use. The rules of grammar and syntax further restrict the possible messages within the coding system of the English language. In a sentence of 100 characters we expect to find about 17 words, since the average length of English words is 6 letters and a terminating space. Of these 17 words, no more than about 10 can be nouns, verbs, adjectives, or adverbs, and the rest must be grammatical linking words such as articles, conjunctions, and prepositions. Most of the words in the dictionary belong to the first class, suggesting about 10^4 possibilities for each, or 10^{40} altogether. The second class of words is much smaller, with perhaps 100 choices for each, making about 10^{14} possibilities. In all, this gives about 10^{54} possible sentences of 100 characters that comply with the rules of English grammar and syntax, quite apart from their meaning. Since $10^{54} \approx 2^{170}$ this implies an average information content of rather less than 2 bits per character in English. Actually this is an overestimate for speech, since spoken English has a vocabulary closer to 1,000 than to 10,000 words, which would reduce the information content to a little more than 1 bit per character. In speech, of course, we are not actually concerned with the written language based on alphabetic symbols, but rather with a set of elementary speech sounds known as phonemes. The discussion can be equally carried out in terms of these units.

Ordinary rapid speech is not much more than about 6 words or 40 characters per second, which indicates an information rate of only about 50 to 100 bits per second. Even if we include other information such as speaker identification, this changes the result very little. For example, it might take us about 3 seconds to identify a known speaker on the telephone, in the absence of spoken identification. Our circle of possible speakers is unlikely to exceed 100 people, and to identify one of these is only $\log_2 100 \approx 7$ extra bits of information, an addition to the information rate of only 2 extra bits per second.

The total coding system of human speech, which involves vocabulary and syntax as well as simply speech sounds, is quite efficient from an error-correction point of view. If speech were simply coded then, even at 1 bit per character, our chance of guessing correctly a missing character should be only 1 in 2, while at 1.7 bits per character the chance would be less than 1 in 3. In fact we can guess a missing character with nearly 100 percent certainty, and the same is nearly always true for two or even three consecutive missing characters. The coding of human speech clearly extends over most of the length of each individual sentence, for it is clues at the syntactic level of sentence structure as well as those at the word-recognition level that enable us to perform this feat, quite apart from the actual meaning of the message.

We see therefore that the human vocal communication system uses a channel of potential capacity up to 3,000 bits per second to transmit information at a rate that probably does not exceed 200 bits per second, but that it uses a coding scheme that is quite efficient as far as error correction is concerned. This sort of performance seems typical of very many biological systems—physical limits are not approached as far as speed is concerned, but there is a great deal of subtlety about message coding and later neural processing. We shall discuss this in more detail in Section 15.6.

It is appropriate to say just a little about the physical nature of the message coding in biological systems, because there is a good deal that is common between different animals. The most particular feature is that the calls of most animals consist predominantly of modulated quasi-continuous sounds produced by non-linear systems. As we have discussed in Chapters 13 and 14, the acoustic outputs produced by such systems consist, apart from transients, of waves with many exact harmonics locked in phase and frequency to the fundamental. In this they are quite distinct from noise, which has a random character. A particular feature of such harmonic signals is that their waveforms repeat exactly in time, so that they can be identified by correlating the signal at one instant with its form at later times. This sort of pattern recognition task is something at which animal brains are particularly adept, and provides a way of separating signals from competing random noise, which has no such correlation. The information in the message can then be coded in terms of the frequency of the fundamental and the relative strengths of the harmonics (i.e. formant bands) as well, of course, as in terms of temporal variations in these quantities and in the signal intensity. Interpolation of nonharmonic noise-like sounds serves both to break up the message into units and to add additional coding. The fact that different individuals use slightly different fundamental frequencies and may have particular sharp formant resonances aids in separating messages when the background noise is not random but consists of competing messages from others.

15.5 Animal Sonar

Most animals make some use of sounds that they generate to deduce information about their surroundings. This is true even of humans, who are able to detect in the dark the presence of a large enclosed space, or of a nearby wall, by evaluating the quality of sound reflections from small noises that they make when moving. Other animals, particularly bats and dolphins, have evolved sophisticated systems for producing appropriate probing sounds and for analyzing the information contained in the returning echoes. The subject is so extensive that we can do no more than outline some of the essential physics here.

Suppose a wave of frequency ω is emitted by an animal and reflected by an object located at a distance large compared with the sound wavelength. What is the strength of the scattered wave back at the position of the animal? Suppose the range to the target object is R and that the intensity of the probing sound measured at a distance of 1 meter from the source is I_0. At the position of the target, the intensity is reduced by spherical spreading and by attenuation in the medium, and has the value

$$I(R) = \frac{I_0}{R^2} e^{-2\alpha R} \tag{15.14}$$

where α is the attenuation coefficient as discussed in Section 6.5. Suppose the target is approximated as a disc of radius a. Then the echo reflected back towards the source will be essentially the same in strength as if the disc were vibrating in still fluid with an amplitude equal to that of the sound wave falling upon it. This velocity amplitude is $v = [2I(R)/\rho c]^{1/2}$ and, from the discussion in Section 7.5, the total power scattered from the target is found to be about $\rho c k^4 a^6 v^2$ where $k = \omega/c$ as usual. The scattering is proportional to the sixth power of the radius of the target and to the fourth power of frequency. This result holds provided $ka < 2$. For targets larger than the wavelength λ, the simple disc approximation is no longer valid, and the scattering strength depends on target geometry, but is roughly proportional to the cross-section area. It also no longer increases with increasing frequency. If the target is smaller than the probing wavelength, then the scattering is nearly isotropic.

The returning echo from the target also spreads as a spherical wave and undergoes further attenuation. Combining these results, we can write the strength of the echo wave returned to the animal as

$$I_R = AI_0 R^{-4} e^{-4\alpha R} a^6 \omega^4 \qquad \lambda > \pi a \tag{15.15}$$

where A is a numerical constant equal to about 10^{-11} if everything is in SI units. This result holds only if the target dimensions are significantly less than the wavelength. For larger targets or shorter wavelengths, the echo intensity is of order

$$I_R \sim I_0 R^{-4} e^{-4\alpha R} a^2 \tag{15.16}$$

but can be either larger or smaller than this depending on the geometry of the target, since reflection begins to be specular (mirror-like) in this case.

The form of these results suggests that, for echoes of the highest possible signal strength, the animal should generally use a wavelength comparable with the size of the object being sought. Signals of higher frequency are unduly attenuated in propagation, since α increases as ω^2 and there is no further increase in echo strength. For bats seeking small insects as prey, the optimal signal frequency is therefore in the range 30 to 100 kHz, depending on the strategy used. For marine animals the fact that the velocity, and hence the wavelength, of sound in water is about 4 times that in air implies that marine animals such as dolphins should use a much lower frequency than bats, and this is made further appropriate by the generally larger size of their prey.

Bats and dolphins also use sonar techniques for general orientation during motion, and in this case the obstacles and targets are generally much larger than a wavelength in size. In such cases a relatively short wavelength is still an advantage, for it allows identification of features of near-wavelength size in the echo. Processing of the echo information is, however, by no means a readily formulated procedure.

An animal using an active sonar technique is, of course, interested in determining the location of its prey, and there are several aspects to this. The first is the determination of direction, and this can be accomplished by a combination of using a relatively narrow transmitted beam, achievable with a high operating frequency, and a binaural auditory system. In the animals concerned, the ears are generally acoustically uncoupled, so that directional determination must be made largely by analysis at the neural level, though the horn directionality effects discussed in Chapters 10 and 11 also come into play.

High frequency sound is generally limited in angular spread because the source size is comparable with the wavelength, and this gives some degree of directionality to the sonar probe. Some bats use a more subtle strategy by radiating sound through their nostrils, rather than through an open mouth, and, since the nostrils are separated by rather more than a wavelength, this leads to several fan-shaped beams in the transmitted cry. Suppose that a signal of frequency ω is radiated from two coupled sources separated by a distance d, as shown in Fig. 15.6. The radiated intensity in a direction making an angle θ with the symmetry plane is a combination of two waves having path difference $d \sin \theta$, and the resulting intensity is

$$I(\theta) = I_0 \left| 1 + e^{-kd \sin \theta} \right|^2 = 2I_0 [1 + \cos(kd \sin \theta)] . \tag{15.17}$$

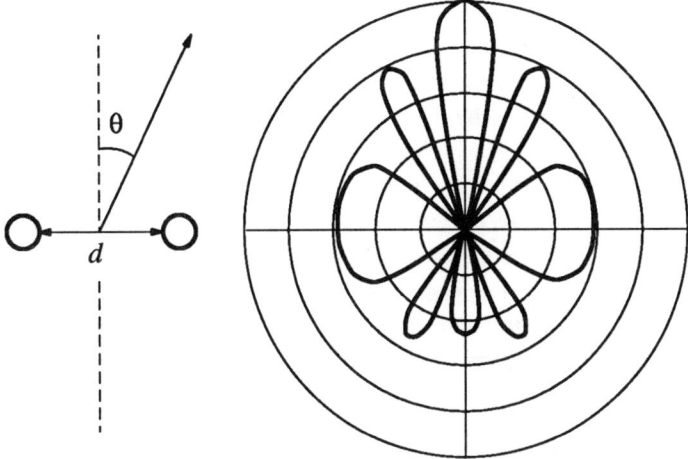

Figure 15.6 Fan beams produced by two sources, such as the nostrils of a bat, separated by a distance d equal to twice the sound wavelength. The intensity pattern is shown on a linear scale, and has been rather arbitrarily corrected for the shielding effect of the animal's body.

This has maxima when $\sin\theta = 2n\pi/kd$, for $n = 0, 1, 2, \ldots$, and nulls for $\sin\theta = (2n+1)\pi/kd$. These fan beams could well assist materially in prey location. The animal could steer the beams by moving its head or, rather more subtly, by introducing a phase shift between the nostrils by varying in some way the shape of its nasal passages.

The second necessary clue for location is that of range. The simple strategy is to use a very short cry duration and to note the delay time τ before the echo returns. The range is then $R = c/2\tau$. Clearly the cry duration must be less than 2τ at the minimum range, or else the echo will be obscured by the cry itself. For a minimum range of 10 cm in air, this requires a cry duration of less than about 0.5 ms. For a cry frequency of 60 kHz this gives 30 cycles of the oscillation within the cry and a frequency spread, by (15.6), of about 2 kHz. Of course the bat can use longer cries when the prey is more distant.

There is a problem when an animal uses a short cry and is hunting a small prey, for the total reflected energy is, of course, proportional to cry length, and short weak signals are difficult to detect. There is also a limit to the intensity that the animal can produce, so that it cannot simply increase the instantaneous power radiated. Some animals use an ingenious strategy called chirping, which is also employed for the same reason in sophisticated radar and sonar systems. The idea is as follows. The spectrum of a short tone burst, as shown in Fig. 15.3, is spread over a range centered on the carrier frequency. We can express this by writing the tone burst in the simple form

$$p(t) = \int g(\omega - \omega_0) \cos \omega t \, df \qquad (15.18)$$

where $g(\omega - \omega_0)$ is a bell-shaped function centered on the carrier frequency ω_0. At time $t = 0$ all the cosine functions add in-phase and there is a large pressure. At times progressively before or after $t = 0$ the cosine functions become increasingly out of phase and cancel each other out. Now suppose we pass this pulse through a dispersive delay line that delays low frequencies more than high frequencies—transverse waves in stiff plates and surface waves on liquids have this property. This will spread the pulse out in time so that it takes on the character of a descending whistle, beginning at the top of the frequency band of the pulse and ending at the bottom of this band, as shown in Fig. 15.7. The total transmitted energy is the same, but the average power is low. This gliding whistle is called a chirp. If we transmit a chirp and receive its echo, then we can effectively reverse the frequency dispersive chirping process, by using a delay line with opposite characteristic, and restore the shape of the pulse. The transit time through the chirping and de-chirping lines is constant and known, so that we can derive the echo delay time τ.

The purpose of using a chirped pulse is, of course, to keep the peak power comfortably low while increasing the total energy in the pulse, without increasing its effective pulse length after de-chirping. In a sonar system, whether animal or electronic, we therefore do not go through this whole chirping and de-chirping process. The animal simply emits a cry that is already a gliding tone of chirp type.

Figure 15.7 (a) A short pulse having a bandwidth of about an octave (2:1). (b) A similar pulse chirped to a descending sweep, maintaining the peak power constant. Clearly the chirped pulse contains much greater total energy for a given maximum amplitude. In a real system the chirp duration could be much longer than shown here.

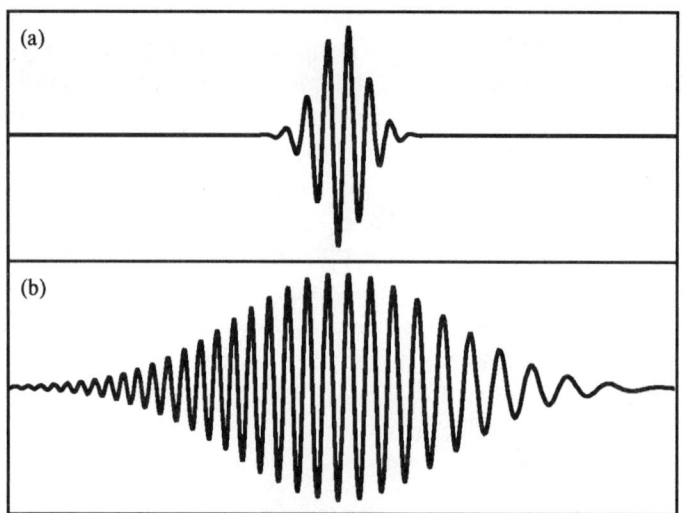

SIGNALS, NOISE, AND INFORMATION

The intensity can be at the limit possible for the animal during most of the glide, so that the energy in the reconstructed pulse is effectively higher than this by the ratio of the chirp duration to the reconstructed pulse duration. When the echo chirp is received, the animal could conceivably have a physical dispersive delay line in its auditory system to de-chirp the signal, but in fact this is not necessary, and the computation required can be done neurally. We return to this in the next section.

With a sufficiently sophisticated processor for the return echo, the animal can extract even more information than this. The most important extra facts concern the motion of the target relative to the predator. This is based on an wave phenomenon called the Doppler effect. We need only a particular version of this effect appropriate to the predator-prey problem. Suppose a source, the predator, emits a sound pulse of frequency ω while moving towards a target, the prey, with velocity v_S. If the target is moving away from the source with velocity v_T relative to the still air, then the distance between source and target is increasing at a rate $v_T - v_S$. To a sufficient approximation for our present purpose, this means that each successive pressure maximum in the wave emitted by the source must travel an additional distance to reach the prey and be reflected back to the source, so that each is delayed relative to the one immediately before it, and the frequency of the echo is lower than the transmitted frequency ω. The approximate result for the echo frequency ω' is

$$\omega' \approx \omega \left[1 - \frac{2(v_T - v_S)}{c} \right] \tag{15.19}$$

where c is the speed of sound in still air. If $\omega' > \omega$ then the predator is catching up on the prey, and the amount of the frequency shift, together with the range information from the pulse delay, allows the predator to calculate the time until interception occurs. Clearly this sort of information is valuable in carrying out an interception strategy. It is difficult to apply this technique to chirped pulses, in which the frequency range is already spread.

More detailed examination of the structure of the echo can give the predator even more information about the target. If the prey has flapping wings, then these will periodically increase and decrease the apparent diameter of the target and make a quite large difference to the target strength and the intensity of the received echo. Flapping wings will also cause fine-structure on the Doppler shift of the echo, though this is likely to be a much smaller effect. Armed with information on the distance of the prey and strength of its echo, the predator knows its approximate size. Addition of information about wing-beat frequency and flight speed generally gives an adequate identification of target species.

15.6 Neural Processing

In this book we have given very little attention to the transduction of mechanical vibration into neural impulses, and as yet no consideration to the neural processing of those impulses to derive useful information. Each of these topics could occupy several volumes in its own right, and we can do little here except outline a few of the most important aspects of these subjects.

In our acoustic models for auditory systems we treated the neural transduction apparatus simply as a mechanical load connected to the tympanum or other vibrating structure. This is, in fact, an adequate description for most of the calculations we need to carry out. Even at this mechanical level, however, we must recognize that the neural transduction apparatus in a living animal consumes metabolic energy and may act as a source of mechanical energy. To a first approximation, certainly, the main difference between a living and a non-living transduction apparatus appears in terms of mechanical tensions and fluid pressures, and these are important in determining mechanical impedance, itself a rather complex function of frequency when considered in detail. The active transduction cells are themselves, however, ultimately linked mechanically to the rest of the ear, and any mechanical disturbance produced by their excitation can affect the system. Such disturbance may be simply linear, as when distortion of a transducer cell produces an electric potential that affects the mechanical properties of neighboring cells, perhaps causing them to contract in sympathy with the original excitation. Such interaction, if appropriately phased so that the mechanical feedback is positive, can effectively raise the Q value of that part of the system to which the transducer cells are connected. If the positive feedback is marginally too great, then the system may break into spontaneous oscillation that, being mechanically coupled to the tympanum, leads to oto-acoustic emission. At this stage, however, the system may still be essentially linear throughout, except for some saturation phenomenon limiting the amplitude of the spontaneous oscillation. Because neural transducer cells generate impulses, however, they are nonlinear, probably mechanically as well as electrically, when this excitation threshold is reached. The neural aspects of such excitation have been extensively studied, but investigation of any mechanical consequences is still a very recent topic.

As discussed in Chapter 12, the neural transduction apparatus in virtually all animals has significant mechanical functions. The prime one of these appears to be to perform an initial frequency dispersion of the vibrational signal. This is performed in the simplest cases by the attachment of a number of transducer cells, each with different mechanical properties and consequently different frequency response. In more complex systems such as the human cochlea, extensively studied by Georg von Békésy, there is a fluid-loaded basilar membrane that has dispersive properties and localizes vibrations of different frequencies to different parts of its length, where they excite appropriate sensor cells. The basilar membrane and its associated apparatus is generally treated as passive, but there is no intrinsic reason why it and similar structures in other animals might not be under some sort of active muscular control to change their properties. A comparable adjustment certainly occurs in focusing the eyes.

Many nerve fibers carry information to the brain from the auditory transduction apparatus. These fibers may in fact interact with their neighbors in the transduction apparatus to do some form of preliminary processing, but this certainly happens in detail in the brain. The neurophysiology of this interaction is also beyond the scope of this book, but it is important that we say just a little about the nature of neural processing and its relation to the operations implied by the mathematical analysis in this chapter.

The sort of computation carried out in this book, as in all conventional scientific work, is based upon simple deterministic algorithms—addition, multiplication, differentiation, integration, and so on. Because they are deterministic we can write a computer program to carry them out, and a generalpurpose computer can then be programmed to do the work, whether it is the solution of a differential equation, the calculation of a Fourier transform, or the de-chirping of a pulse signal. Going one step further, if we want to do a particular calculation in the most efficient manner possible, then we can build an integrated-circuit chip with the logic of the algorithm embedded in the design. Such a chip, for example, as well as containing significant memory, might be designed to perform a Fast Fourier Transform in very nearly real time, so that any signal can be examined in the time domain and the frequency domain simultaneously. All these computation mechanisms are called von Neumann machines, after the Hungarian-American mathematician John von Neumann who did much early work on electronic computers. Von Neumann computers take a defined algorithm and work through it sequentially.

While it would be possible for neural systems to work in this way, and indeed to some extent they do when we perform mathematical operations in our heads, most neural computation appears to operate on entirely different principles. A form of computer, called a neural network, throws some light on the possible computational strategy of animal brains. Basically a neural network computer consists of cells each with a number of inputs and a single output. The inputs are all connected to the outputs of a previous layer of cells, and the output to one or more of the inputs to the next layer of cells. Such a network, with just three layers, is illustrated in Fig. 15.8.

Each input can either be active, if it is sent a signal from a cell in the previous layer, or inactive. If the input to cell n from the output of cell i is active, then it contributes a stimulus α_{ni} to the cell, where α_{ni} is an adjustable parameter for that particular transmission link, otherwise it contributes nothing. If the parameter α_{ni} is positive, then that input signal is called excitatory; if α_{ni} is negative then the input signal is inhibitory. All the cell is required to do is to add up all the stimuli it receives, and then to send out a signal if this sum exceeds a certain threshold value. The analogy to excitatory and inhibitory synapses and to the firing of a nerve cell is obvious. The whole system may be dynamic in time, so as to respond to sequential inputs, or may simply settle into a static configuration of on and off cells in response to a steady input.

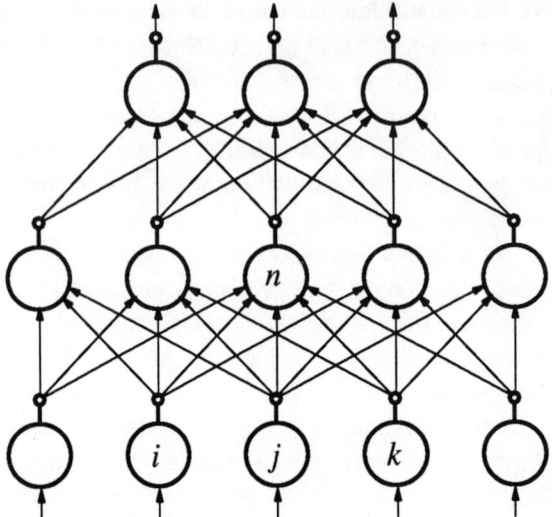

Figure 15.8 A simple neural network of three layers. Each cell receives inputs from a number of cells in the next lower layer and sends its own output to a number of cells in the next higher layer. The cells in the lowest layer receive external inputs, and the cells in the top layer give outputs. The network "learns" through adjustment of the weights assigned to all links such as ni, nj, and nk.

In such a neural network computer, the first layer of cells is connected to some sort of input sensor. In an acoustic case, for example, each cell in the input layer might be fed the output level of one of a bank of octave or one-third-octave band filters, each corresponding to one class of auditory neural transducers. The final layer of cells produces the output and might, for example, drive a set of lights indicating a coded result.

The neural network is programmed by adjusting the weights α_{ni} associated with all its interconnections. This adjustment could be carried out by an operator, but in practice it is very time-consuming and is done by a supervising program in an ordinary computer. The objective is that the neural computer should learn to recognize certain patterns presented to its input, for example spoken vowels, and give appropriate output, for example by lighting just one of a set of lights marked A, E, I, O, and U. Initially the computer is hopeless at this task so some of the interconnection weights are varied and there is another trial. Whenever the computer gets a nearly right answer, for example by lighting two lights, one of which is correct, the interconnection weights are made a little more extreme in their current direction and, when the answer is completely wrong, they are moved a little in the reverse direction. In this way the network effectively "learns" by examining repeated examples, being "rewarded" for correct answers and "punished" for wrong answers. Once it has learned a task, the supervising computer has no further function and can be removed.

Neural computers built, trained, and operated in this fashion can perform very impressively on quite complex recognition tasks, even if the number of cell units involved is as small as 30 or so, each with about 10 inputs, arranged in three layers. As the number of cell units, the number of layers, and the complexity of interconnection increase, the potential of the neural computer expands greatly, but the time involved in the necessary "training" task also increases very rapidly.

The important thing about a neural network computer is that its operation does not depend on its precise mode of construction, again a good model for a real neural system, and its operation does not follow an algorithmic procedure. It learns a task by recognizing patterns, either spatial or temporal, and gives an answer based on experience and analogy, rather than on precise logic. Applied, for example, to the processing of a chirped pulse and the extraction from it of time information, the network would certainly not perform a de-chirping transform followed by a time adjustment and a simple division to find range, rather it would recognize the pattern relationship between the transmitted chirp and the received chirp and interpret this in terms of range and other parameters. The extent to which this processing step is genetically determined, for example by the initial weights given to the various interconnections in our simple model, and the extent to which it is learned by experience is not yet clear.

Whether or not the current generation of neural networks represents a good model for the behavior of real animal systems, the analogy is most persuasive and explains many of the features of those systems that we know from experimentation or from introspection. Its most important contribution in our present context is perhaps to warn us against concentrating on simple algorithmic methods as explanations of perceptual abilities in even very primitive animals.

15.7 Conclusion

At the beginning of this book we stated our purpose to be the application of physics to the understanding of acoustic systems in biology. In the course of developing this theme we emphasized that the essence of understanding, at a physical level, is to be able to build a simple conceptual model of the system under discussion, to calculate its behavior, and to see that this gives an adequate approximation to the behavior of the real system. We have seen that this is quite reasonably possible for the peripheral auditory system and for the vocal system of most animals, and gives good insight into their operation. If a model cannot be made to fit the experimental facts to an appropriate approximation, then it is either incorrect in its assumptions, or at least omits something important. Unfortunately we cannot assume with certainty that a model that fits the experimental facts is correct, but at least it has a good chance of being correct, and the fact that it is quantitative allows us to make further quantitative predictions that can be tested by measurement. Conversely, a hypothesis about the way in which a system works must remain simply informed speculation until it has been embodied in a testable quantitative model.

It is possible to proceed in this way to a discussion of the mechanics of the neural transducer itself, and indeed there is a large and significant body of literature and much current research on this subject. To keep the book within a reasonable compass, we have been able to include only a brief discussion of this field, but our more detailed treatment of the peripheral auditory system provides necessary information about the mechanical input to the transduction organ, and therefore sets the background against which discussion of the transduction mechanism must occur.

Finally, since the evolutionary purpose of acoustical systems is to allow an animal to receive information from its environment and to communicate with members of its own species, we have given the briefest possible outline of this subject. Here theories are, for the present, rather abstract, since we have little knowledge of the detailed functioning of the brain. Once again, however, application of physical principles provides a background against which possible theories can be tested. The literature is again very extensive, particularly in relation to the human auditory system, since it extends through neurophysiology to psychophysics. Only a specialist text could do it justice.

References

Fourier transforms: [53]
Communication theory: [54], [55]
Bats and sonar: [34] Ch 3
Sound propagation and noise: [28]; [29]; [30]
Human hearing: [41]–[45]
Neural networks: [56]

Discussion Examples

1. A signal consists of a regular series of rectangular pulses, each of unit amplitude and duration δ, repeating N times per second. Calculate the power spectrum of the signal.
2. Calculate the Fourier transform of a sinusoidal tone burst of N cycles at frequency ω_0. Examine how the spread of the spectrum varies with N.
3. Examine the increase in channel capacity that can be achieved under noisy conditions by doubling the transmitter power, (a) when the received power is small compared with the noise power, and (b) when the received power is large compared with the noise power.

Solutions

1. Fundamental frequency is $\omega_1 = 1/N$, and the signal in each period is 1 for $0 < t < \delta$ and 0 otherwise. From (15.13) its power spectrum is concentrated at frequencies $n\omega_1$, with power (amplitude squared) given by

$$C_n^2 = (2/n^2\pi^2)(1 - \cos n\omega_1 \delta).$$

SIGNALS, NOISE, AND INFORMATION   299

Note that, for very small δ, $\cos n\omega_1\delta \approx 1-(n\omega_1\delta)^2/2$, so that in this case there is a large run of harmonics of nearly equal power.

2. Let the tone burst extend from $-T/2$ to $T/2$, where $T = 2N\pi/\omega_0$. Using (15.5), the transform can be shown to be

$$\frac{2j\omega}{\omega_0^2-\omega^2}\sin\left(\frac{N\pi\omega}{\omega_0}\right).$$

Writing $\omega = \omega_0 + \delta$ to examine behavior near the burst frequency, we find that the magnitude of this transform is $N\pi$ at $\omega = \omega_0$ and the width of this central peak out to the first zero is about $2\omega_0/N$. There are sidebands of much smaller amplitude. This agrees generally with (15.6), the factor 2 discrepancy arising from the effect of the sharp switching of the tone burst, together with an overestimate of the peak width.

3. The results follow from (15.13). (a) If $x \ll 1$ then $\ln(1+x) \approx x$ and so $\log_2(1+x) = \ln(1+x)/\ln 2 \approx 1.44x$. Here $x \equiv S/N$, so that if $S \ll N$ then doubling S doubles the channel capacity C. (b) If $S \gg N$ however, then $\log_2(1+S/N) \approx \log_2(S/N)$, and doubling S/N increases C by only an additional W, or equivalently by a factor $[1+\log_2(S/N)]/\log_2(S/N)$ which much less than 2.

A. MATHEMATICAL APPENDIX

Most of the necessary mathematics has been introduced at an appropriate place in the text. The purpose of this appendix is to provide a little more detail on some topics and to serve as a convenient reference.

A.1 Complex Numbers

Ordinary real-number mathematics can be generalized by defining a quantity j with the property

$$j^2 = -1. \tag{A.1}$$

We refer to j as an imaginary number, but it is simpler to consider it as just a mathematical symbol with this property. Some texts use the symbol i for this purpose, and the relationship is that $i = -j$. Results in one convention can be converted to the other simply by making this substitution. A complex number is one with both real and imaginary parts, such as $1.3 + 2.5j$. In the text we have used **_bold Italic type_** for algebraic quantities that represent complex numbers and *ordinary Italic type* for real numbers. For example, we write $z = x + jy$.

Complex numbers can be added or subtracted simply by remembering that their real and imaginary parts are different, like apples and oranges, so that we must keep them apart.

$$(a + jb) \pm (c + jd) = (a \pm c) + j(b \pm d). \tag{A.2}$$

Similarly, complex numbers can be multiplied if we remember the relation (A.1), so that

$$(a + jb)(c + jd) = ac + j^2 bd + j(ad + bc) = (ac - bd) + j(ad + bc). \tag{A.3}$$

Division is a little more complicated, and we need the following result

$$\frac{1}{(c + jd)} = \frac{1}{(c + jd)} \times \frac{(c - jd)}{(c - jd)} = \frac{c - jd}{c^2 + d^2}. \tag{A.4}$$

Thus

$$\frac{(a + jb)}{(c + jd)} = \frac{(a + jb)(c - jd)}{c^2 + d^2} = \frac{(ac + bd) + j(bc - ad)}{c^2 + d^2} \tag{A.5}$$

which expresses the result as a standard complex number with real and imaginary parts. One should not try to remember these results, but simply the way to get to them.

For many purposes it is useful to think of real numbers as plotted along a line

301

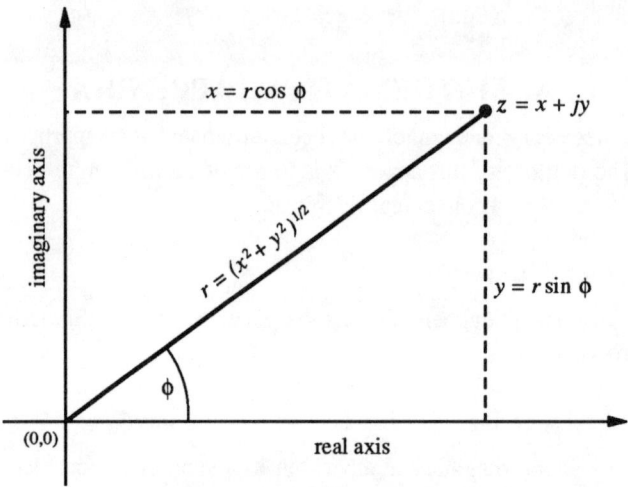

Figure A.1 The complex plane is defined by the real and imaginary axes. Any complex number $z = x + jy$ can be represented by a point on this plane with rectangular coordinates (x, y). If polar coordinates are used, then the point is represented as (r, ϕ) where r is the magnitude and ϕ the phase angle associated with z.

drawn as the x axis. We can then do something similar for imaginary numbers by plotting them along another line in the direction of the y axis. The difference in direction of these two axes illustrates that there is an essential difference between real and imaginary numbers. A complex number $z = x + jy$ is then represented as a point in the (x, y) plane, which we usually refer to as the complex plane, as shown in Fig. A.1. The origin where the axes cross is the point $(0,0)$. This diagram suggests an alternative way of specifying a complex number by using polar coordinates (r, ϕ) for its representative point in the complex plane. From elementary trigonometry $x = \cos \phi$ and $y = \sin \phi$ so that

$$z = x + jy = r(\cos \phi + j \sin \phi). \tag{A.6}$$

We call r the modulus or absolute value of the complex quantity z and ϕ the phase angle. ϕ must always be in radians when we use it in a mathematical procedure, but can be referred to in either radians or degrees (π radians = 180°). Usually we use the same symbol for the modulus of a complex quantity as for the quantity itself, but in a different typeface, so that z is the absolute value of the complex quantity z. Sometimes, to be even more explicit, we write the absolute value of z as $|z|$. From the right-angled triangle in the figure,

$$z \equiv |z| \equiv r = (x^2 + y^2)^{1/2}. \tag{A.7}$$

APPENDIX A

When evaluating the modulus of an expression of the form z_1/z_2 it is simplest to calculate this as $|z_1|/|z_2|$.

A.2 Exponential and Logarithmic Functions

The exponential function $\exp x$ is most conveniently defined to be the solution of the differential equation

$$\frac{dz}{dx} = \alpha z \quad \Rightarrow \quad z = A \exp \alpha x \tag{A.8}$$

where A is an arbitrary constant. The exponential function turns up in applications such as the description of simple population growth or radioactive decay. Consideration of the equation

$$\frac{dz}{dx} = (\alpha + \beta) z \tag{A.9}$$

shows that

$$\exp[(\alpha + \beta) x] = \exp \alpha x \exp \beta x \tag{A.10}$$

from which it follows that we can write

$$\exp \alpha x = e^{\alpha x} \tag{A.11}$$

where e is a simple real number. Further investigation shows that

$$e = 2.71828\ldots. \tag{A.12}$$

If α is real and positive, then $e^{\alpha x}$ grows smoothly and without limit as x increases, the rate of growth being always proportional to the value of the function itself, as expressed by (A.8). Similarly, the value of $e^{-\alpha x}$ declines steadily towards zero. We discuss the behavior of the exponential function when its argument αx is imaginary or complex in Section A.3.

In much the same way, the logarithmic function $\ln x$ can be defined to be the solution of the differential equation

$$\frac{dz}{dx} = \frac{1}{x} \quad \Rightarrow \quad z = A + \ln x. \tag{A.13}$$

Logarithms defined in this way are natural logarithms to the base e. To convert them to common logarithms to the base 10 we need to multiply by $\log_{10} e = 0.43429\ldots$ so that

$$\log_{10} z \approx 0.4343 \ln z. \tag{A.14}$$

Common logarithms to the base 10 are used for expressing quantities such as decibels, but natural logarithms to the base e are more convenient for algebraic manipulations.

A.3 Trigonometric and Hyperbolic Functions

We are used to sine, cosine, and tangent functions defined as the ratios of the sides of a right-angled triangle, but another equivalent definition arises when we deal with complex numbers. Suppose we look at the differential equation

$$\frac{d^2z}{dx} = k^2 z \quad \Rightarrow \quad z = A e^{kx} + B e^{-kx} \tag{A.15}$$

From (A.8) the general solution is simply the sum of exponential functions shown to the right of (A.15), with A and B being arbitrary constants. The occurrence of two arbitrary constants is associated with the fact that the differential equation is of second order, meaning that it involves the second differential d^2z/dx^2. If we now insert a minus sign in (A.15), or equivalently replace k by jk, then the solutions become complex exponentials

$$\frac{d^2z}{dx^2} = -k^2 z \quad \Rightarrow \quad z = e^{\pm jkx}. \tag{A.16}$$

But we know that $\cos kx$ and $\sin kx$ are also solutions of this equation, which means that we must be able to express them in terms of the complex exponentials $e^{\pm jkx}$. A little bit of algebra, using the result $\cos^2 x + \sin^2 x = 1$, readily convinces us that the necessary relation is

$$e^{\pm jkx} = \cos kx \pm j \sin kx \quad \text{or} \quad e^{\pm j\phi} = \cos \phi \pm j \sin \phi \tag{A.17}$$

where ϕ is any real angle in radians. We can write, equivalently,

$$\cos \phi = \frac{(e^{j\phi} + e^{-j\phi})}{2} \qquad \sin \phi = \frac{(e^{j\phi} - e^{-j\phi})}{2j}. \tag{A.18}$$

We can now return to our representation of complex numbers in (A.6) and see that a complex number z can be represented in terms of its magnitude r and its phase angle ϕ by the expression

$$z = r e^{j\phi}. \tag{A.19}$$

The phase ϕ must always be expressed in radians, not degrees, for this relation to hold. The elementary imaginary number j is clearly just $j = e^{j\pi/2}$. The notation (A.19) is used very frequently in the text. The expression (A.19) also allows us to see that multiplying a complex number by j is equivalent to increasing its phase by $\pi/2$ or $90°$. Indeed an alternative strategy for multiplying together two complex

APPENDIX A

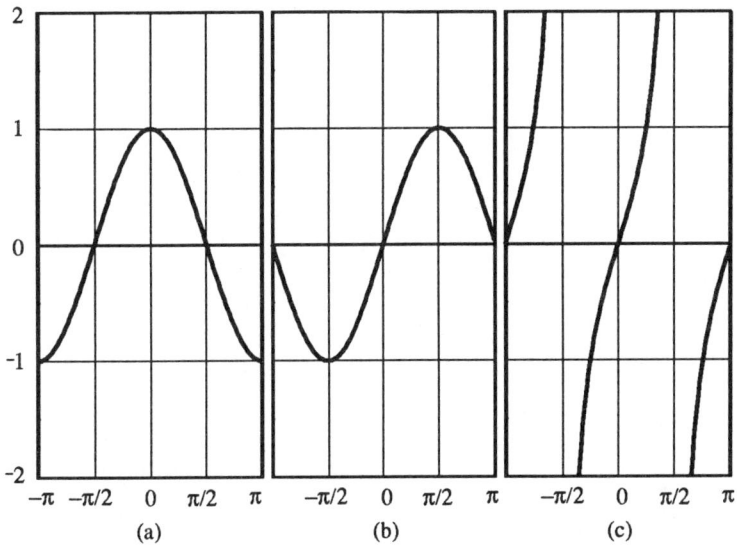

Figure A.2 The trigonometric functions (a) $\cos\phi$, (b) $\sin\phi$, and (c) $\tan\phi$ over the range $-\pi \le \phi \le \pi$. These functions are all cyclic and repeat over successive intervals of 2π.

numbers is to convert each to (r, ϕ) form, multiply their magnitudes r, which are just real numbers, and add their phases ϕ.

For reference, the cosine, sine, and tangent functions are shown over one period of their behavior in Fig. A.2. All three functions repeat in a cyclic manner so that their value is unchanged if their argument is increased by 2π.

Just as the equation (A.16) leads to the trigonometric cosine and sine functions defined in terms of complex exponentials, so equation (A.15) leads to new functions called hyperbolic cosines and sines, defined in terms of real exponentials. The definitions are

$$\cosh\phi = \frac{e^{\phi}+e^{-\phi}}{2} \qquad \sinh\phi = \frac{e^{\phi}-e^{-\phi}}{2}. \qquad (A.20)$$

As for the trigonometric functions, we define a hyperbolic tangent function by $\tanh\phi = \sinh\phi/\cosh\phi$. There are simple relations between hyperbolic functions and ordinary trigonometric functions, namely

$$\cos j\phi = \cosh\phi \qquad \sin j\phi = j\sinh\phi. \qquad (A.21)$$

The hyperbolic functions are not periodic like the trigonometric functions. The cosh and sinh functions increase exponentially in magnitude for large values of the argument, while the tanh function saturates at the values ± 1, as shown in Fig. A.3.

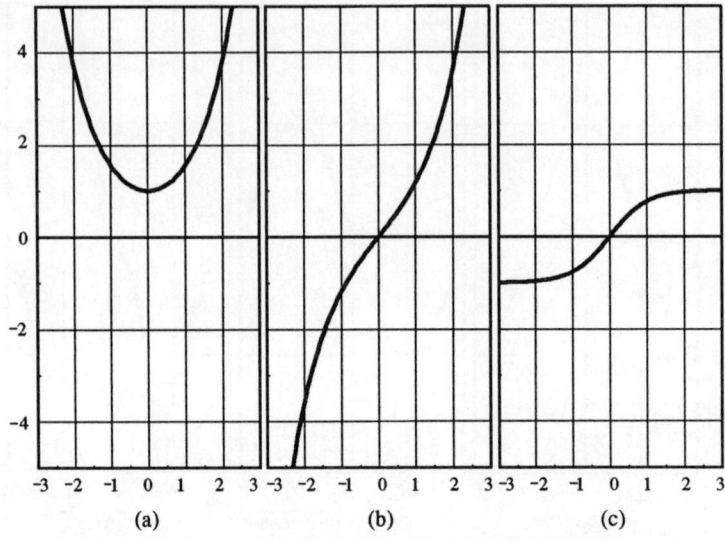

Figure A.3 The hyperbolic functions (a) $\cosh\phi$, (b) $\sinh\phi$, and (c) $\tanh\phi$ over the range $-3 \leq \phi \leq 3$. These functions are not cyclic, and the trends shown in the graphs extend to all lower and higher values of ϕ.

It is useful to remember the relations

$$\cos^2\theta + \sin^2\theta = 1 \qquad \cosh^2\theta - \sinh^2\theta = 1 \qquad (A.22)$$

which are easily proved from (A.18) and (A.20).

Finally we should note some useful formulae for expanding both trigonometric and hyperbolic functions. These are necessary in the context of the present book when the argument is kx and the wave number k is affected by attenuation and has the complex form $k = \omega/c - j\alpha$. All of the results below can be converted to those with a minus sign by noting that $\cos(-B) = \cos B$, $\cosh(-B) = \cosh B$, $\sin(-B) = -\sin B$, and $\sinh(-B) = -\sinh B$.

$$\cos(A + B) = \cos A \, \cos B - \sin A \, \sin B$$

$$\sin(A + B) = \sin A \, \cos B + \cos A \, \sin B$$

$$\cosh(A + B) = \cosh A \, \cosh B + \sinh A \, \sinh B$$

$$\sinh(A + B) = \sinh A \, \cosh B + \cosh A \, \sinh B$$

$$\cos(A + jB) = \cos A \, \cosh B - j \sin A \, \sinh B$$

$$\sin(A + jB) = \sin A \, \cosh B + j \cos A \, \sinh B. \qquad (A.23)$$

A.4 Bessel Functions

The exponential function and its close relatives, the trigonometric functions, arise as solutions of the differential equation (A.16), which in our present application represents the propagation of sinusoidal waves in one dimension, or rather the propagation of waves that have no spatial dependence in the other two dimensions and are thus plane waves. They are familiar elementary functions because of their simple occurrence also as the ratios of the sides of a right-angled triangle.

Every differential equation has similar characteristic mathematical functions associated with it. The one-dimensional equation for propagation of spherical waves away from a point source has characteristic functions that, fortunately, can also be expressed in terms of trigonometric functions, though matters get rather more complicated if we want to include a description of the angular variation. The equation for propagation of waves in two dimensions is simpler in angular terms, since the angular behavior is simply $\cos n\phi$ or $\sin n\phi$, where ϕ is the coordinate angle and n is determined by the angle ϕ_0 between the boundaries, if any, of the propagation region. For cylindrical waves in free space, n is zero or an integer. The resulting equation for the radial part of the propagating wave has the general form

$$\frac{1}{r}\frac{d}{dr}\left(r\frac{dz}{dr}\right) - \frac{n^2}{r^2} = -k^2 z \qquad (A.24)$$

or, multiplying by r^2 and writing $x \equiv kr$,

$$x^2 \frac{d^2 z}{dx^2} + x \frac{dz}{dx} + (x^2 - n^2) z = 0. \qquad (A.25)$$

This equation is known as Bessel's equation, and its solutions as Bessel functions. There are two independent Bessel functions $J_n(x)$ and $N_n(x)$, the second kind also being known as Neumann functions. Both have an oscillatory behavior that decays in amplitude with increasing x. The first few J_n for integral n are illustrated in Fig. 5.3 of the main text. The Neumann functions N_n are also oscillatory and decaying, but they are infinite for $x = 0$.

Just as trigonometric functions were combined in (A.17) to give a complex exponential function, which we identified in the text with a propagating plane wave, so the two kinds of Bessel functions can be combined as $H_n = J_n + jN_n$ to give new functions H_n, called Hankel functions, which represent propagating cylindrical waves. We can also define Bessel functions for imaginary arguments, known as modified Bessel functions, which are the analogs of the hyperbolic functions. These additional functions need not concern us here, but are mentioned to emphasize the fact that Bessel functions are really just as straightforward as ordinary trigonometric functions, but happen to be rather more complicated mathematically because of the special nature of the point $x = 0$ in their domain of definition. Full details, formulae, and tables for Bessel functions are given in the *Handbook of Mathematical Functions* listed in the Bibliography.

A.5 Orthogonal Functions

Most of the mathematical functions with which we are concerned here arise as the solutions of differential equations describing the vibrational behavior of simple extended systems. If these systems are unbounded, then the solutions describe propagating waves. Often, however, we are concerned with the vibrations of systems within closed boundaries upon which they are either clamped or completely free. The solutions then form standing waves or normal modes, and we have denoted these quite generally by the symbol ψ_n where the subscript n, which may be multiple, is used to label the different normal modes. The ψ_n are known mathematically as the eigenfunctions of the problem.

It is a quite general property of normal modes for a linear system that they are physically independent. Mathematically this is expressed by saying that they are mutually orthogonal, so that

$$\iint \psi_n(x,y)\psi_m(x,y)\,dx\,dy = 0 \quad \text{if} \quad n \neq m \tag{A.26}$$

where the integral is over the whole vibrating surface. This is true, for example, of trigonometric functions and Bessel functions satisfying simple boundary conditions, though not in the absence of those conditions. It is convenient to choose the magnitude of the functions ψ_n so that

$$\iint \psi_n^2(x,y)\,dx\,dy = 1. \tag{A.27}$$

Functions satisfying this condition are said to be normalized. Functions satisfying both (A.26) and (A.27) are said to be orthonormal.

B. PIPE AND HORN IMPEDANCE COEFFICIENTS

This appendix gives explicit expressions for the impedance coefficients Z_{ij} for a pipe, including allowance for wall losses, and for several types of horn. Note that in Chapter 11 we have sometimes written the coefficients Z_{ij} for a pipe as P_{ij} and those for a horn as H_{ij} to avoid confusion.

We can generalize the pipe impedance coefficients

$$Z_{11} = Z_{22} = -jZ_0 \cot kl$$

$$Z_{12} = Z_{21} = -jZ_0 \operatorname{cosec} kl \qquad (B.1)$$

to include a damping parameter α by substituting $k = \omega/c - j\alpha$ and using the methods of expansion of trigonometric functions given in equations (A.19). Writing $k' \equiv \omega/c$ for convenience, the results are

$$Z_{11} = Z_{22} = Z_0 \frac{\sinh\alpha l \cosh\alpha l - j\cos k'l \sin k'l}{(\sin k'l \cosh\alpha l)^2 + (\cos k'l \sinh\alpha l)^2}$$

$$Z_{12} = Z_{21} = Z_0 \frac{\cos k'l \sinh\alpha l - j \sin k'l \cosh\alpha l}{(\sin k'l \cosh\alpha l)^2 + (\cos k'l \sinh\alpha l)^2}. \qquad (B.2)$$

Clearly these reduce to (B.1) as $\alpha \to 0$.

Evaluation of the coefficients Z_{ij} for a finite horn follows just the same lines as the method used in Chapter 10 for a cylindrical pipe. For clarity we shall always take the radius of the large end or mouth (port 1) to be a, so that the mouth area is $S_1 = \pi a^2$, and the radius of the small end or throat (port 2) to be b, so that the throat area is $S_2 = \pi b^2$. The length of the horn between these ends is taken to be l.

For a conical horn, the Z_{ij} are given by

$$Z_{11} = -\frac{j\rho c}{S_1}\left[\frac{\sin(kl + \theta_2) \sin\theta_1}{\sin(kl + \theta_2 - \theta_1)}\right]$$

$$Z_{22} = \frac{j\rho c}{S_2}\left[\frac{\sin(kl - \theta_1) \sin\theta_2}{\sin(kl + \theta_2 - \theta_1)}\right]$$

$$Z_{12} = -\frac{j\rho c}{(S_1 S_2)^{1/2}}\left[\frac{\sin\theta_1 \sin\theta_2}{\sin(kl + \theta_2 - \theta_1)}\right] \qquad (B.3)$$

where the coordinates x are measured from the apex of the cone to ports 1 and 2, and

$$k = \omega/c \qquad \theta_1 = \tan^{-1} kx_1 \qquad \theta_2 = \tan^{-1} kx_2. \qquad (B.4)$$

The numerical behavior of these impedance coefficients is complicated, and we clearly have zeros when one or other of the terms in the numerator vanishes, with infinities when the denominator vanishes. They are all pure imaginary numbers if wall losses are neglected.

The expressions for an exponential horn are rather more involved. If we define the flare constant m of the horn, and several other quantities, by

$$k = \omega/c \qquad m = \frac{1}{2l}\ln\left(\frac{S_1}{S_2}\right) \qquad \kappa = (k^2 - m^2)^{1/2} \qquad \theta = \tan^{-1}\left(\frac{m}{\kappa}\right) \qquad (B.5)$$

then for $k > m$ the impedance coefficients are given by

$$Z_{11} = -\frac{j\rho c}{S_1}\left[\frac{\cos(\kappa l - \theta)}{\sin \kappa l}\right]$$

$$Z_{22} = -\frac{j\rho c}{S_2}\left[\frac{\cos(\kappa l + \theta)}{\sin \kappa l}\right]$$

$$Z_{12} = -\frac{j\rho c}{(S_1 S_2)^{1/2}}\left[\frac{\cos \theta}{\sin \kappa l}\right]. \qquad (B.6)$$

For lower frequencies at which $k < m$, however, we must re-express these by defining

$$\kappa' = (m^2 - k^2)^{1/2} \qquad \theta' = -\frac{1}{2}\ln\left(\frac{1 + m/\kappa'}{m/\kappa' - 1}\right) \qquad (B.7)$$

this last being a re-expression of $\theta = j\theta' + \pi/2 = \tanh^{-1}(m/b')$. In terms of these quantities, the impedance coefficients are

$$Z_{11} = \frac{j\rho c}{S_1}\left[\frac{\sinh(\kappa' l - \theta')}{\sinh \kappa' l}\right]$$

$$Z_{22} = -\frac{j\rho c}{S_2}\left[\frac{\sinh(\kappa' l + \theta')}{\sinh \kappa' l}\right]$$

$$Z_{12} = -\frac{j\rho c}{(S_1 S_2)^{1/2}}\left[\frac{\sinh \theta'}{\sinh \kappa' l}\right]. \qquad (B.8)$$

In an infinite horn, waves do not propagate if their frequency is below the cutoff for which $k = m$, but are exponentially attenuated.

The corresponding expressions for the impedance coefficients of a parabolic horn involve Bessel functions.

$$Z_{11} = -\frac{j\rho c}{S_1}\left[\frac{J_0(kx_1) N_1(kx_2) - J_1(kx_2) N_0(kx_1)}{J_1(kx_1) N_1(kx_2) - J_1(kx_2) N_1(kx_1)}\right]$$

APPENDIX B

$$Z_{22} = -\frac{j\rho c}{S_2}\left[\frac{J_0(kx_2)N_1(kx_1) - J_1(kx_1)N_0(kx_2)}{J_1(kx_1)N_1(kx_2) - J_1(kx_2)N_1(kx_1)}\right]$$

$$Z_{12} = -\frac{j\rho c}{(S_1 S_2)^{1/2}}\left[\frac{2}{\pi k(x_1 x_2)^{1/2}}\right]\left[\frac{1}{J_1(kx_2)N_1(kx_1) - J_1(kx_1)N_1(kx_2)}\right] \quad (B.9)$$

where

$$x_1 = lS_1/(S_1 - S_2) \qquad x_2 = lS_2/(S_1 - S_2). \quad (B.10)$$

J_0 and J_1 are Bessel functions, and N_0 and N_1 are Neumann functions, or Bessel functions of the second kind, as discussed in Appendix A.

C. PHYSICAL QUANTITIES AND UNITS

In developing physical theories, and particularly in calculating numerical results, it is important that we use a consistent set of units for the physical quantities involved. Several such consistent sets have been developed over the past century or so, but it is now almost universal practice in science to use the Système Internationale (SI) convention, previously known as the MKS system after its three basic units, the meter, the kilogram, and the second. Tables C.1 and C.2 below give the names of the SI units for quantities of interest in the present book. Also listed are approximate numerical values for some of the more common quantities encountered in the text. It must be emphasized that some of these values, particularly those relating to the physical properties of biological materials, are only very roughly approximate. They are adequate for estimating the magnitude of an effect or a response, but in careful work should be replaced by more accurate values for the system concerned.

Table C.1 SI units for physical quantities

Quantity	Unit	Unit symbols
acoustic impedance	acoustic ohm	$Pa\ m^{-3}\ s$
angle	radian	—
angular frequency	radian per second	s^{-1}
area	square meter	m^2
density	kilogram per cubic meter	$kg\ m^{-3}$
elastic modulus	pascal (or newton per sq. meter)	Pa or $N\ m^{-2}$
force	newton	N
frequency	hertz	Hz
kinematic viscosity	meter sq. per second	$m^2\ s^{-1}$
length	meter	m
mass	kilogram	kg
power	watt	W
pressure	pascal (or newton per sq. meter)	Pa or $N\ m^{-2}$
time	second	s
velocity	meter per second	$m\ s^{-1}$
viscosity	pascal second	$Pa\ s$
volume	cubic meter	m^3
wave impedance	rayl	$kg\ m^{-2}\ s^{-1}$
Young's modulus	pascal (or newton per sq. meter)	Pa or $N\ m^{-2}$

Table C.2 Approximate values for some common quantities

Quantity	Value
atmospheric pressure	10^5 Pa
density of air	1.2 kg m^{-3}
density of biological tissue	1000 kg m^{-3}
density of water	1000 kg m^{-3}
gravitational acceleration	9.8 m s^{-2}
Poisson's ratio (typical)	0.3
reference acoustic intensity	1×10^{-12} W m^{-2}
reference acoustic pressure (air)	20 µPa r.m.s.
reference acoustic pressure (water)	0.1 Pa r.m.s.
velocity of sound in air	340 m s^{-1}
velocity of sound in water	1480 m s^{-1}
viscosity of air	1.8×10^{-5} Pa s
viscosity of water	1.0×10^{-3} Pa s
wave impedance of air	415 rayl
wave impedance of water	1.5×10^6 rayl
Young's modulus of biological tissue	$10^8 - 10^{10}$ Pa

GLOSSARY

acoustic admittance [Y or $Y^{(A)}$]   the complex ratio of acoustic volume flow to acoustic pressure at a given frequency; the reciprocal of acoustic impedance

acoustic compliance   spring-like behavior of an acoustic load; the reciprocal of acoustic stiffness

acoustic impedance [Z or $Z^{(A)}$]   the complex ratio of acoustic pressure to acoustic volume flow at a given frequency; the reciprocal of acoustic admittance

acoustic inertance   mass-like behavior of an acoustic load

acoustic ohm   unit for acoustic impedance; equal to 1 pascal per (cubic meter per second)

angular frequency [ω]   frequency measured in radians per second, $\omega = 2\pi f$

basilar membrane   flexible partition along the middle of the cochlea to which are attached the neural-transducer hair cells

bit   a binary digit; the unit of measure for information, consisting of one yes/no answer, conventionally 1 or 0

bronchus   used generally for the airways between the lungs and the vocal mechanism

characteristic impedance [Z_0]   the acoustic input impedance of an infinitely long cylindrical pipe

chirp   transformation in which the spectral components of a pulse are dispersed in time

cochlea   the neural transduction organ in vertebrates

complex number   a general number having both real and imaginary parts

compliance   spring-like behavior of a load; the reciprocal of elastic stiffness

contralateral   on the side of the animal away from the sound source

decibel [dB]   logarithmic unit of power ratio; 10 dB is a factor 10 in power, 3 dB a factor of about 2

dispersion   propagation of waves with velocity depending upon frequency

eigenfunction   one of the characteristic solutions of a differential equation with simple boundary conditions; the mathematical expression of a normal mode

Fourier transform   representation of a time-varying quantity by a sum (or integral) of sinusoidal waves

fundamental   lowest vibration frequency of a system; lowest frequency component of a wave

harmonic spectral component with frequency that is an integer multiple of the frequency of the fundamental; the fundamental is the first harmonic

Helmholtz resonator enclosed volume with single aperture, generally with a tubular neck

imaginary number the product of a real number with the operator j; the square of an imaginary number is negative, since $j^2 = -1$

inertance mass-like behavior of a load

intensity acoustic power flow per unit area

intensity level $[L(I)]$ the acoustic intensity expressed in decibels relative to 10^{-12} watts per square meter

ipsilateral on the same side of the animal as the sound source

isotropic the same in all directions

kinematic viscosity $[\nu]$ coefficient of viscosity divided by density, η/ρ

larynx the vocal mechanism of a mammal or similar animal

linear response amplitude directly proportional to stimulus amplitude; mathematically, an equation involving only the first power of the dependent variable

mechanical admittance $[Y^{(M)}]$ the complex ratio of velocity to force; the reciprocal of mechanical impedance

mechanical compliance spring-like behavior; the reciprocal of stiffness

mechanical impedance $[Z^{(M)}]$ the complex ratio of force to velocity at a given frequency; the reciprocal of mechanical admittance

mechanical ohm unit for mechanical impedance, equal to 1 newton per (meter per second)

nonlinearity response amplitude not directly proportional to stimulus amplitude, usually expressed with quadratic and cubic response terms in addition to a linear term; mathematically an equation with powers of the dependent variable higher than one

normal mode one of a set of simple vibrations of a body in which all parts move in the same phase; the shape of the displacement remains constant but its amplitude varies sinusoidally in time

normalization an adjustment of the amplitude of a function so that the integral of its square over the domain (length, area, or volume) of the problem is equal to unity

octave a factor 2 in frequency

ohm unit of electrical resistance; see mechanical ohm, acoustic ohm

orthogonality mathematical independence; two functions are orthogonal if the integral of their product over the domain (length, area, or volume) of the problem vanishes

ossicle small bone, particularly the bones of the middle ear

otolith a mineral body supported on sensory cells in the auditory system of some animals; generally sensitive to acceleration

overtone one of the spectral components of a vibration with frequency higher than that of the fundamental; not necessarily a harmonic of the fundamental

pinna the external ear

Poisson's ratio [γ] ratio of the fractional decrease in diameter to the fractional increase in length for a thin rod under tension; typically $\gamma \approx 0.3$

rayl unit of wave impedance ρc, equivalent to 1 kg m^{-2} s^{-1}

sound pressure level [$L(p)$] the acoustic pressure expressed in decibels relative to 20 µPa r.m.s.

syrinx the vocal mechanism of a bird

thermal conductivity heat flow per unit area for unit temperature gradient

trachea used generally for the airway between the vocal mechanism and the mouth

tymbal used generally for a thin buckling shell in an insect's sound-producing system

tympanum used generally for the tympanic membrane or eardrum

viscosity [η] dissipative resistance to change in shape of a fluid

wave impedance [z] the ratio of acoustic pressure to acoustic particle velocity for a medium

wave number [k] $2\pi/\lambda$ or ω/c

Young's modulus [E] ratio of elastic stress (force/area) to elastic strain (extension/length) for a thin rod

BIBLIOGRAPHY

Since this is a textbook rather than a research monograph, it is not appropriate to give extensive references to the current scientific literature. Such a list would be long indeed, and would contain many excellent papers devoted to experimental examination of the behavior of auditory systems, particularly at the behavioral and neurophysiological levels. The literature on the physical analysis of biological sensory systems is much more sparse. The bibliography below concentrates instead on giving a list of rather general monographs on relevant parts of acoustics, and a selection of similarly general books on biological matters, together with some useful reference books. The list is representative, rather than complete, and simply indicates some starting points for further reading in the field. Comments are given only when the title is not self-explanatory. Most of the books listed have extensive bibliographies, which should be consulted as a guide to the research literature.

General Acoustics Texts

[1] *Vibration and Sound,* P.M. Morse, New York: McGraw-Hill (1948), reprinted New York: American Institute of Physics (1976). This classic text takes a mathematical approach to acoustics and vibrations. It is the standard reference in the field.

[2] *Theoretical Acoustics,* P.M. Morse and K.U. Ingard, New York: McGraw-Hill (1968). An expanded version of *Vibration and Sound* with introduction of several modern topics. Rather heavily mathematical.

[3] *Vibrations and Waves in Physics,* I.G. Main, Cambridge: Cambridge University Press (1978). A more introductory account of the subject.

[4] *The Physics of Vibration,* A.B. Pippard, (Vol. 1) Cambridge: Cambridge University Press (1978). The first half of this book covers material relevant to our present purpose, before going on to more physics-related topics.

[5] *The Theory of Sound,* Lord Rayleigh, (2 vols) London: Macmillan (1896), reprinted New York: Dover (1945). The classic book in the field of acoustics. Obviously superseded in many respects by later work, but contains a wealth of mathematical and physical discussion, all of which is still relevant.

[6] *Acoustics,* L.L. Beranek, New York: McGraw-Hill (1954), reprinted New York: American Institute of Physics (1986). A more practically oriented text.

[7] *Simple and Complex Vibratory Systems,* E. Skudrzyk, University Park: Pennsylvania State University Press (1968). An advanced text, giving particular attention to extended systems such as beams and plates.

[8] *Fundamentals of Acoustics,* L.E. Kinsler, A.R. Frey, A.B. Coppens and J.V. Sanders, New York: Wiley (1982). An good general text on acoustics.

Includes a treatment of underwater acoustics. The second edition by Kinsler and Frey (1962) is equally good for our purposes.

[9] *Acoustical Engineering,* H.F. Olson, Princeton: Van Nostrand (1957). A practically oriented book that makes extensive use of electrical analogs.

[10] *The Physics of Musical Instruments,* N.H. Fletcher and T.D. Rossing, New York: Springer-Verlag (1991). There are many resemblances between the physics of musical instruments and the physics of auditory and vocal systems, and the analysis methods are similar. This is a fairly quantitative and mathematical book, going back to fundamentals.

Special Topics in Acoustics

[11] "On the propagation of sound waves in a cylindrical conduit," A.H. Benade, *J. Acoust. Soc. Amer.* **44,** 616–623 (1968). Gives a good treatment of sound propagation in narrow tubes.

[12] "On the effect of the internal friction of fluids on the motion of pendulums," G.G. Stokes, in *Mathematical and Physical Papers,* Vol. 3, pp.1–140 Cambridge: Cambridge University Press (1922). The original treatment of the damping of vibrating cylinders and spheres.

[13] "On vibrations of shallow spherical shells," E. Reissner, *J. Appl. Phys.* **17,** 1038–1042 (1946). A discussion of the lower modes, with approximate formulae.

[14] "Mode locking in nonlinearly excited inharmonic musical oscillators," N.H. Fletcher, *J. Acoust. Soc. Amer.* **64,** 1566–1569 (1978). A more extended discussion of mode-locked harmonicity.

[15] *Response and Stability,* A.B. Pippard, Cambridge: Cambridge University Press (1985). An extended discussion of nonlinearity, with examples of chaos in simple systems.

[16] "Calculation of the steady-state oscillations of a clarinet using the harmonic balance technique," J. Gilbert, J. Kergomard, and E. Ngoya, *J. Acoust. Soc. Amer.* **86,** 35–41 (1989). The method of harmonic balance explained and applied.

[17] "Ab initio calculations of the oscillations of a clarinet," R.T. Schumacher, *Acustica* **48,** 71–85 (1981). The method of Chapter 13 explained in detail and applied to a reed instrument.

[18] "On the oscillation of musical instruments," M.E. McIntyre, R.T. Schumacher, and J. Woodhouse, *J. Acoust. Soc. Amer.* **74,** 1325–1345 (1983). Methods of calculating the behavior of self-excited oscillators in the time domain, as used in Chapter 13. Specialized here to stringed instruments.

[19] "Aerodynamic whistles," R.C. Chanaud, *Scientific American* **222** (1), 40–46 (1970). A simple qualitative discussion of the subject.

[20] "Experiments on the fluid mechanics of whistling," T.A. Wilson, G.S. Beavers, M.A. DeCoster, D.K. Holger, and M.D. Regenfuss, *J. Acoust. Soc.*

Amer. **50,** 366–372 (1970). Experimental results with some theoretical discussion.

Special Topics in Animal Acoustics

[21] "Physical models for the analysis of acoustical systems in biology," N.H. Fletcher and S.Thwaites, *Quart. Rev. Biophys.* **12,** 25–65 (1979). A condensed account of the methods treated in detail in the present book.
[22] "Biophysical aspects of sound communication in insects," A. Michelsen and H. Nocke, *Adv. Insect Physiol.* **10,** 247–296 (1974). A very good general survey.
[23] "External-ear acoustic models with simple geometry," R. Teranishi and E.A.G. Shaw, *J. Acoust. Soc. Amer.* **44,** 257–263 (1967). The model discussed in Section 10.9.
[24] "The external ear," E.A.G. Shaw in *Handbook of Sensory Physiology,* Vol. V/1, pp.455–490, edited by H. Autrum et al., New York: Springer-Verlag (1974). Discusses the acoustics of the human external ear.
[25] "Obliquely truncated simple horns: Idealized models for vertebrate pinnae," N.H. Fletcher and S. Thwaites, *Acustica* **65,** 195–204 (1988). Discussion of short normal horns and approximate calculations for obliquely truncated conical horns. Our discussion uses slightly different notation and is improved in other ways.
[26] "Bird song—a quantitative acoustic model," N.H. Fletcher, *J. Theor. Biol.* **135,** 455–481 (1988). A detailed discussion of the methods of Chapter 13 applied to bird song.
[27] "The structural basis of voice production and its relationship to sound characteristics," J.H. Brackenbury in *Acoustic Communication in Birds,* Vol. 1, pp.53–73, edited by D.E. Kroodsma and E.H. Miller, New York: Academic Press (1982). A review of anatomy and physiology, with many references.
[28] "Sound reception in different environments," A. Michelsen in *Sensory Ecology,* pp.345–373, edited by M.A. Ali, New York: Plenum (1978).
[29] "Environmental aspects of sound communication in insects," A. Michelsen in *Acoustic and Vibrational Communication in Insects,* pp.1–9, edited by K. Kalmring and N. Elsner, New York: Paul Parey (1985).
[30] "Adaptations for acoustic communication in birds: Sound transmission and signal detection," R.H. Wiley and D.G. Richards in *Acoustic Communication in Birds,* Vol. 1, pp.131–181, edited by D.E. Kroodsma and E.H. Miller, New York: Academic Press (1982).

General Books on Animal Acoustics

[31] *The Acoustic Sense of Animals,* W.C. Stebbins, Cambridge, Mass.: Harvard University Press (1983).

[32] *Comparative Studies of Hearing in Vertebrates,* edited by A.N. Popper and R.R. Fay, New York: Springer-Verlag (1980).

[33] *Bioacoustics: A Comparative Approach,* edited by B. Lewis, London and New York: Academic Press (1983).

[34] *Ultrasonic Communication in Animals,* G. Sales and D Pye, London: Chapman and Hall (1974).

[35] *Acoustic Communication in Birds,* 2 Vols, edited by D.E. Kroodsma and E.H. Miller, New York: Academic Press (1982).

[36] *Bird Song: Acoustics and Physiology,* C.H. Greenewalt, Washington: Smithsonian Institution Press (1968).

[37] *Acoustic and Vibrational Communication in Insects,* edited by K. Kalmring and N. Elsner, Berlin and Hamburg: Verlag Paul Parey (1985).

[38] *The Vertebrate Inner Ear,* E.R. Lewis, E.L. Leverenz, and W.S. Bialek, Boca Raton, Florida: CRC Press (1985).

[39] *The Reptile Ear: Its Structure and Function,* E.G. Wever, Princeton: Princeton University Press (1978).

[40] *The Amphibian Ear,* E.G. Wever, Princeton: Princeton University Press (1985).

Human Speech and Hearing

[41] *Experiments in Hearing,* G. von Békésy, translated by E.G. Wever, New York: McGraw-Hill (1960), reprinted New York: American Institute of Physics (1989).

[42] *The Auditory Periphery: Biophysics and Physiology,* P. Dallos, New York: Academic Press (1973).

[43] *Speech and Hearing in Communication,* H. Fletcher, New York: Van Nostrand (1953).

[44] *Hearing: Physiology and Psychophysics,* W.L. Gulick, New York: Oxford University Press (1971).

[45] *Speech Analysis, Synthesis and Perception,* J.L. Flanagan, New York: Springer-Verlag (1972).

Mathematical and Physical Reference Books

[46] *Handbook of Mathematical Functions,* edited by M. Abramowitz and I.A. Stegun, Washington: National Bureau of Standards (1972), reprinted New York: Dover. The standard source for information on all mathematical functions, as formulae, graphs and extensive tables.

[47] *Tables of Functions,* E. Jahnke and F. Emde, Leipzig: Teubner (1938), reprinted New York: Dover (1945). A classic forerunner of Abramowitz and Stegun, particularly notable for its 3-dimensional plots of functions in the complex plane.

[48] *A Survey of Applicable Mathematics,* K. Rektorys, London: Iliffe (1969).

A source-book for nearly all the mathematical techniques used in science.

[49] *Tables of Integrals, Series and Products,* I.S. Gradshteyn and I.M. Ryzhik, New York: Academic Press (1965). Almost all the known indefinite and definite integrals listed. A great help if analytic solutions are sought.

[50] *Numerical Recipes: The Art of Scientific Computing,* W.H. Press, B.P. Flannery, S.A. Teukolsky and W.T. Vetterling, Cambridge: Cambridge University Press (1986). A superb collection of algorithms for solving most numerical problems met in scientific work. The mathematical background for each method is carefully explained, and a well-documented computer routine given, both as a listing and as a machine-readable disk file. The original edition uses Fortran, but transcriptions to Pascal, C, and Basic are available.

[51] *A Physicist's Desk Reference,* edited by H.L. Anderson, New York: American Institute of Physics (1989). A good reference on the basic facts and formulae of physics, with tables of many useful quantities.

[52] *Linear and Nonlinear Waves,* G.B. Whitham, New York: Wiley (1974). An advanced comprehensive treatment of waves, with emphasis on nonlinear effects. The sections on surface waves on water are, however, easily readable.

Information and Communication Theory

[53] *The Fourier Transform and its Applications,* R.N. Bracewell, New York: McGraw-Hill (1986). Written for engineers, but the introductory chapters are useful in the present context.

[54] *The Mathematical Theory of Communication,* C.E. Shannon and W. Weaver, Urbana: University of Illinois Press (1962). The classic book on the subject.

[55] *On Human Communication: A Review, a Survey, and a Criticism,* C. Cherry, New York: Wiley (1957). An easily readable classic.

[56] *An Introduction to Neural and Electronic Networks,* edited by S.F. Zornetzer, J.L. Davis and C. Lau, New York: Academic Press (1990). A large collection of papers with a biological emphasis.

INDEX

Acoustic source. *See* Radiation
A-weighting, 95–96
Acceleration sensor, 63–67
Acoustic admittance. *See* Acoustic impedance
Acoustic impedance
  of active neural transducer, 164
  of cavity, 140
  of cochlea, 226
  definition, 113
  and mechanical impedance, of neural transducer, 162–64
  of pipe, 138–39, 179–83
  of tympanum, 143–47
Admittance, acoustic. *See* Acoustic impedance
Admittance, mechanical
  definition, 20
  of oscillator, 20–22
  point, 39–42
  transfer, 39–42
Aeolian tones, 266
Aerodynamic sound, 265–70
Air
  attenuation of sound, 99–100
  sound speed, 93
  viscosity, 54
Air pressure, vocal, 249, 262
Amplitude
  definition, 10
  complex, 14, 35–36
  r.m.s., 94–95
Analogs, electric network
  for cavity, 140
  electro-acoustic, 137–47
  electro-mechanical, 130–37
  for Helmholtz resonator, 142–43, 184
  high-frequency, 181, 184, 190–91
  for lever, 134–37, 147–49
  low-frequency, 129–50
  for simple oscillator, 133–34
  for tympanum, 143–47
Angular frequency, 10
Angular response. *See* Response
Anti-resonance, 42

Aperture, end correction at, 141
Aperture whistle, 269–70
Attenuation
  of spherical waves, 101–2
  of surface waves, 104
Attenuation coefficient, 99
Attenuation of sound
  in air, 99–100
  in environment, 282
  in pipes, 182–82
  in water, 100
Auditory canal, 207–9
Auditory system. *See* Ear, Ears
Auditory threshold. *See* Sensory threshold

Bandwidth
  of communication system, 277, 280, 284–88
  of horn, 188–93
Bar
  angular compliance, 47
  boundary conditions, 44, 46, 48
  driven, 46–47
  eigenfunctions, 44, 47
  end-loaded, 47–50
  equation of motion, 42–43
  imperfectly clamped, 46
  mode frequencies, 44–45, 47, 49
  normal modes, 44–45, 47
  stiffness, 42
Basilar membrane, 76–78, 219–25
Bats, 202, 225, 265, 289–93
Beam. *See* Bar
Bernoulli equation, 249
Bernoulli force, 251
Bessel function, 74–75, 77, 187, 307–8
Biology, 3–4, 7
Bird song
  formants, 261–62
  power, 97, 125, 259
  voiced, 258–64
  waveform, 260
  whistled, 270
Birds
  auditory system, 182–83

325

Birds (*cont.*)
  muscular power, 231
  vocal system, 258–64
Boundary layer, 54–56, 120
Bubble
  Q value, 176
  resonance frequency, 108, 175
  scattering by, 107–8
  in sound field, 107–8, 175–76
Buckling of shell, 86
Bulk modulus, 91–93
  air, 93
  water, 92

Capacitance, analog, 130–31
Capacity, channel, 286
Cavity
  acoustic impedance, 140
  modes, 186
Channel capacity, 286
Chaotic vibration, 83, 240
Characteristic impedance of pipe, 180
Characteristic values. *See* Eigenvalues
Chirp, 291–92
Cicada, 164, 242–43
Cochlea, 76–78, 219–28
  active response, 228, 294
  impedance, 226
  matching, 225–28
  mechanics, 219–25
  microphonic, 228
Cockatoo, 240
Coding, 285–89
Complex amplitude, 14, 35–36
Complex number, 11–14, 301–3
Complex plane, 302
Compliance, angular
  of bar, 47
  of clamp, 46
  definition, 22
Computers
  neural-network, 295–96
  use of, 7, 149–50
  von Neumann, 295
Cricket, 212–13, 238
Critical damping, 15, 67
Cut-off, of horn, 188, 193, 198

Damping
  coefficient, 15
  critical, 15, 67

Damping (*cont.*)
  of oscillator, 14–15
  of sensory hair, 56–58, 66–67
  of string vibrations, 37–38
  of tympanum, 161, 163–64
  viscous, 14, 54–57, 182
De-chirp, 292
Decibel, 62, 95
Delta function, 256
Diffraction. *See* Scattering
Directionality. *See also* Response
  of horn, 194–95
  of open pipe, 122
  of piston in baffle, 121–22
  of source near reflector, 114–15
  of vibrating disc, 117
  of vibrating panel, 118
Disc, vibrating, 116–17
Dispersion, frequency
  in cochlea, 221–25
  by second filter, 217–19
  in tapered membrane, 76–78, 219
Dispersion, velocity
  of surface waves on water, 104
  of waves on bars and plates, 43, 105
Distortion, harmonic, 27–28
Doppler effect, 293
Duffing equation, 27

Ear. *See also* Ears
  aquatic, 174–76
  baffled diaphragm, 157–59
  cavity backed, 159–62
  horn-loaded, 204–8
  human, 198–200
  inner, 215–28
  middle, 225–28
Eardrum. *See* Tympanum
Ears. *See also* Ear
  cavity-coupled, 164–68, 173–74
  insect-like, 212–13
  pipe-coupled, 208–12
Echo, sonar, 289–90
Edge tones, 267–68
Efficiency
  of aerodynamic source, 269
  of frictional excitation, 237–38
  of loudspeakers, 125
  of pick and file excitation, 242
  of pneumatic vocal system, 262
  of tymbal buckling system, 242–43

INDEX

Eigenfunction expansion
 for membrane, 78
 for string, 35–37
Eigenfunctions. *See also* Normal modes
 bar, 44
 membrane, annular, 81
 membrane, circular, 74
 membrane, rectangular, 72
 membrane, tapered, 77
 orthonormality, 36
 string, 35–37
Eigenvalues. *See* Mode frequencies
Elastic moduli, 32, 42, 91–93, 105, 250
Emission, oto-acoustic, 164, 228
End correction, 141
Equation
 Duffing, 27
 horn, 186
 linear, 23
 nonlinear, 27
 Stokes, 56–57
 Van der Pol, 235
 Wave. *See* Wave equation
 Webster, 186
Eustachian tube, 160–62, 168, 208, 212
Excitation
 aerodynamic, 265–70
 frictional, 234–38
 pick and file, 240–42
 pneumatic, 246–65
Experiment, 172–74
Exponential function, 12, 303

Far field, 122
Fast Fourier transform (FFT), 278
Field, acoustic
 far, 122
 free, 154–56
 near, 122–24
Filter, second, 217–19
Fish
 boundary layers on, 43
 line organ, 115, 176
 swim bladder, 108, 175–76
Flare cut-off, 188
Fletcher-Munson curves, 95–96
Formants, 261–62, 264
Fourier
 analysis, 23, 54, 233, 273–78
 integral, 275–76
 series, 275

Fourier (*cont.*)
 transform, 259, 275–76
Free field, 154–56
Frequencies, mode. *See* Mode frequencies
Frequency
 of aeolian tones, 266
 angular, 10
 of edge tones, 268
 optimal for communication, 281–84
 optimal for sonar, 290–93
 of simple vibrator, 9–10
Frequency dispersion. *See* Dispersion
Frequency response. *See* Response
Friction, 14, 234–38
Frog
 hearing, 164, 168–71
 sound production, 264–65
Function
 Bessel, 74–75, 77, 187, 307–8
 delta, 256
 exponential, 12, 303
 hyperbolic, 43, 304–6
 logarithmic, 303
 Neumann, 81, 307–8
 orthogonal/orthonormal, 36, 308
 trigonometric, 13, 304–6

Gravity sensor, 65, 67
Gravity waves on water, 103
Group velocity, 198
Gyration, radius of, 42–43

Hair, sensory. *See* Sensory hair
Hair cells
 in cochlea, 219–20, 228
 in insects, 57–63
 and otoliths, 63–67
Harmonic
 distortion, 27–28
 generation, 27–28, 236, 238–40, 259
 series, 35
Harmonics, 35, 273–74
Head, diffraction round, 106
Hearing, human, 96
Heisenberg uncertainty relation, 277
Helmholtz frequency, 142
Helmholtz resonator
 advanced discussion, 183–86
 mouth as, 232, 257, 267–68
 simple discussion, 142–43
Hertz, 10

Higher modes
  in horn or pipe, 196–98
  of tympanum, 144–45
Horn
  asymmetric, 195–96
  Bessel, 187
  conical, 187, 192–94, 310
  cut-off frequency, 188, 192–93
  cylindrical, *See* Pipes
  directionality, 194–95, 196
  exponential, 187, 192–93, 205–6, 310
  gain, 192–94, 196, 202
  higher modes in, 196–200
  hybrid reflector, 201–3
  impedance coefficients, 187–88, 310–11
  matched, 191, 194
  obliquely truncated, 195–96
  optical model, 201–2
  parabolic, 187, 192–93, 201–2, 310–11
  pass-band, 192–93
  stopped, 191, 194
Horn equation, 186
Hyperbolic function, 43, 304–6

Imaginary numbers, 11–12, 301–3
Impedance. *See also* Acoustic impedance, Input impedance, Mechanical impedance
  characteristic, of pipe, 180
  point, 41
  transfer, 41
  wave, 94
Impedance coefficients
  for horn, 187–88, 310–11
  for pipe, 179–83, 309–10
  for lever, 136–37
  for transformer, 135–37
Impedance transformation, by lever, 135, 226–27
Inductance, analog, 130
Inertance, 22
Information, 284–89
Inner ear, 215–28
Input impedance
  of cochlea, 226
  of infinite pipe, 180
  of open pipe, 181
  of short open pipe, 138, 141
  of stopped pipe, 181
Insect
  acoustic radiation near ground, 114–15

Insect (*cont.*)
  auditory system model, 164–68, 173–74, 212–13
  boundary layers on, 55
  muscular power, 231
  song structure, 232, 243, 286
  sound production, 86–87, 125, 232–43
Intensity, 96–97
Intensity level, 97
Interface reflection, refraction, transmission, 97–99

Jet, air, 265–70

Kinematic viscosity, 54
Kirchhoff's laws, 148

Laboratory studies, 171–74
Larynx, 247–54, 264–65
Lateral-line organ, 115, 176
Level
  intensity, 97
  sensory threshold, 62–63
  sound pressure, 95
  stimulus, 62
  velocity, 62
Lever
  arm resonance, 136
  frequency response, 147–50, 227
  impedance coefficients, 136
  impedance transformation, 135, 226–27
  transformer analog, 134–37
Linearity, 23–24
Load, radiation, 112–14. *See also* Radiation impedance
Logarithmic function, 303
Logarithmic scales, 18, 62, 95, 97
Lungs, 254–55, 259

Matching
  of cochlea, 162–64, 225–28
  of horn, 191, 193–94
  of neural transducer, 162–64
Mathematical models, 4–7
Maximum power transfer theorem, 163
Meatus, 207–9
Mechanical admittance. *See* Admittance, mechanical
Mechanical impedance. *See also* Impedance
  definition, 20

INDEX 329

Mechanical impedance (*cont.*)
  of oscillator, 20–23
  relation to acoustic impedance, 114
Membranes
  annular, 81
  basilar, 76–78, 219–25
  circular, 73–75
  driven, 78–79
  eigenfunctions, 72, 74, 77, 81
  elastically braced, 79–81
  frequency-dispersive, 77–78, 217–19, 220–25
  loaded, 81–83
  mode frequencies, 72–74, 77, 82
  rectangular, 71–73
  shape effects, 72–73
  slack, 83
  tapered, 76–78
  wave equation, 71, 73
  wave speed, 71
Middle ear, 137, 217
Mode frequencies.
  bar, 44
  bar, compliantly clamped, 47
  bar, end-loaded, 49
  Helmholtz resonator, 142, 185
  membrane, circular, 73, 74
  membrane, effect of shape, 72–73
  membrane, loaded, 82
  membrane, rectangular, 72–73
  membrane, tapered, 77
  plate, 84
  shell, 86
  string, 35
Mode functions. *See* Eigenfunctions
Mode locking, 236, 238–40
Model
  avian vocal system, 258–64
  baffled-diaphragm ear, 157–59
  cavity-backed ear, 159–62
  cavity-coupled ears, 164–68
  cochlea, 219–25
  frog-like system, 168–71
  horn-loaded ear, 204–8
  insect auditory system, 212–13
  insect sound production, 232–43
  pinna, 198–200
  pipe-coupled ears, 208–13
  second filter, 217–19
Models
  appropriately complex, 5

Models (*cont.*)
  construction of, 153
  mathematical, 4–7
  simplified physical, 4–7, 297–98
Modes. *See* Normal modes, Higher modes
Mouth
  directionality, 122
  as Helmholtz resonator, 232, 257, 267–68
  as horn, 254
Musical instruments, 233, 236, 247, 255, 268

Nares, 168–71, 212
Near field
  of dipole, 117, 123–24
  of monopole, 113, 123
  in spherical wave, 101–2
Network, neural, 295–96
Networks, analog
  computer solution, 149–50
  Kirchhoff's laws, 148
  solution procedure, 147–49
Neumann function, 81, 307–8
Neural network, 295–96
Neural processing, 293–97
Neural transducer
  at base of sensory hair, 57, 62–63
  cochlear, 215–28
  frequency dispersion in, 76–78, 217–19, 221–25
  matching to tympanum, 146, 161–64, 225–28
Noise
  band-limited, 279
  environmental, 23, 279, 282–84
  pink, 279
  white, 278
  $1/f$ 279, 282
Noise power
  absorbed by oscillator, 279
  and signal power, 278–84
Nonlinearity
  of cochlea, 220, 228
  elastic, 27, 250–51
  and harmonic generation, 27–28, 236, 238–40, 259
  in multi-mode systems, 238–39
  of sensory hair, 67–68
  of slack membrane, 83
  in sound-generating systems, 240, 250–53, 259

Normal modes
  bar, 44–45
  bar, compliantly clamped, 47
  bar, end-loaded, 50
  membrane, circular, 73–74
  membrane, loaded, 82
  membrane, rectangular, 72–73
  membrane, tapered, 77
  plate, 84
  shell, 85–86
  string, 33–37
Normalization, 36, 44, 50
Nostrils, 168–71, 212
Numbers, complex, imaginary, real, 11–14, 301–3

Orthogonality/Orthonormality, 36, 307–8
Oscillator, simple harmonic
  analog circuit, 133–34
  damped, 14–15
  driven, 16–19
  nonlinear, 26–28
Oto-acoustic emission, 164, 228, 294
Otolith
  end-loaded bar model, 47–50
  frequency response, 65–66
  as gravity sensor, 67
  Q value, 66–67
  sensory threshold, 66
  shear, 64, 67

Parameters, experimental determination, 172–74
Pass-band, of horn, 192
Period of vibration, 10
Phase
  definition, 10
  of driven oscillator, 16–19, 22
Phon, 95–96
Physics, 3–4, 297–98
Pick and file excitation, 240–42
Pinna
  diffraction by, 106–7
  human, 198–200
  mammalian, 186, 195–96, 198–203
  as oblique horn, 195–96
  as shallow horn, 198–200
Pipes
  attenuation in, 182
  end correction, 141
  impedance coefficients, 179–83, 309

Pipes (*cont.*)
  open, 138, 141, 181, 185
  porous, 182–83, 209
  radiation from, 120–22
  sound speed in, 183
  stopped, 181, 185
  wall losses in, 182–83
Place theory of audition, 225
Plates, vibration of, 83–84
Point admittance, 39–42
Poisson's ratio, 84, 86, 313
Power, acoustic. *See also* Radiation
  aircraft engines, 126
  birds, 97, 125, 259
  human voice, 125, 264
  insects, 125, 237–38, 242–43
  sonar reflection, 289–90
  sound systems, 125
  whistles, 266, 269
Power, available muscular, 231
Power spectrum, 276–77
Pressure. *See* Air pressure, Sound pressure
Pressure gain of horns, 191–94
Pressure generator, equivalent, 189–90
Propagation number. *See* Wave number
Pulse, spectrum of, 277–78

Q value
  definition, 17–18
  of biological systems, 18–19
  of horn resonance, 193
  of matched system, 163
  of otolith, 66
  of second filter, 218
  of sensory hair, 58–59
  and signal-to-noise, 279–81
  of simple oscillator, 17–19
  and time resolution, 25–26, 277
  of tympanum, 161, 163–64
Quadrature, 22
Quality factor. *See* Q value
Quantum Mechanics, 3–4

Radiation, acoustic
  from dipole, 118, 123–24
  from line of sources, 115
  from open pipe, 119–22
  from pair of sources, 115
  from piston in baffle, 120
  from pulsating sphere, 111–12
  from source near reflector, 114–15

INDEX   331

Radiation, acoustic (*cont.*)
  from spherical source, 111–12, 122–23
  from vibrating disc, 116–17
  from vibrating panel, 117–19, 237
  from tymbals, 242–43
  under water, 125–26
  from wings, 117, 119
Radiation impedance
  acoustic, 112–14
  electric analog, 140
  mechanical, 112–13
  on open pipe end, 120–21
  on piston in baffle, 120
  on pulsating sphere, 113, 117
  on vibrating disc, 116
  under water, 125–26
Radius of gyration, 42–43
Rayl, 94
Reciprocity, 40, 99, 124–25, 181, 188
Reflection coefficient
  at interface, 98
  in horn, 256–57
Reflection of sound at interface, 97–99
Refraction of sound, 99
Resistance, analog, 131
Resonance
  in multimode systems, 38–42
  in simple oscillator, 16–19
  tracheal, 261–62, 264
Resonance frequencies. *See* Mode frequencies.
Resonance frequency
  of bubble, 107, 175
  Helmholtz, 142, 185
  large amplitude, 251
  of otolith, 66
  of sensory hair, 59, 63
  of simple oscillator, 9, 17
  of tympanum, 157–77, 204–12
  of vocal valve, 250–51, 259–60
Resonator, Helmholtz. *See* Helmholtz resonator
Resonators, coupled, 217–19
Response
  of baffled-diaphragm ear, 158
  behavioral, 172
  of cavity-backed ear, 162
  of cavity-coupled ears, 166–67
  displacement, 17, 19
  of driven oscillator, 16–19, 23–24
  of driven string, 38–39

Response (*cont.*)
  of frog-like system, 171
  of horn-loaded ear, 206
  of human ear, 95–96, 198–200
  neurophysiological, 172
  of otolith sensor, 65–66
  phase, 16–19, 23–24
  of pinna models, 195–96, 198–200
  of pipe-coupled ears, 210
  of sensory hair, 59–63
  transient, 24–26
  velocity, 18, 19
R.m.s. amplitude, 94–95
Rod. *See* Bar

Scattering
  by bubble, 107–8
  by cylinder, 106–7
  by disc, 105–6
  by sphere, 106
Second filter, 217–19
Sensor
  acceleration, 63–67
  gravity, 65, 67
  pressure. *See* Ear, Ears
  velocity, 58–63
Sensory cell
  in cochlea, 219–20, 228, 294–95
  firing rate, 61, 216
  in sensory hair root, 57, 62–63, 67
Sensory hair, 46–50, 53–68
  Q value, 58–59
  response bandwidth, 59, 63
  viscous force on, 56–57
Sensory threshold
  auditory, 95–96, 156, 227
  for sensory hair, 61–63
  for otolith detector, 66
Shells
  buckling, 86–87
  mode frequencies, 86
  ribbed, 87
  stiffness, 85
  vibrations, 233–34
Signal, in noise, 278–84
Simple harmonic oscillator. *See* Oscillator
Sonar, 289–93
Sound, aerodynamic, 265–70
Sound pressure
  in horn throat, 192–94, 196
  in plane wave, 94–97

Sound pressure (cont.)
  reference, 95
  in spherical wave, 101, 111–12
Sound pressure level
  in air, 95
  in water, 96
Sound production
  frictional, 234–38
  impulsive, 232–34
  mechanical, 231–43
  pick and file, 240–42
  pneumatic, 246–65
Sound speed
  in air, 93
  in pipes, 183
  in water, 92
Specific acoustic impedance, 94. See also Wave impedance
Speech, human
  coding, 264, 287–89
  frequency range, 264
  information content, 287–88
  acoustic power, 125, 264
Sphere, pulsating, 111–12, 122–23
Spherical waves, 100–102
Spiracles, 212
Standing waves. See Normal modes.
Stiffness
  nonlinear, 26–27
  of bar, 42
  of plate, 84
  of spring, 9, 22
Stimulus level, 62
Stokes equations for viscosity, 56–57
Stridulation. See Sound production
String, vibrating, 30–42, 45
Sub-genual organ, 134
Superposition, 23–24
Surface tension, 103–4
Surface waves
  on solids, 104–5
  spreading of, 104
  on water, 103–4
Swim bladder, 108, 175–76
Syrinx, 247–48, 254–64

Theorem
  maximum power transfer, 163
  Thévenin, 189
Threshold. See Sensory threshold
Time resolution, 26, 277

Tone burst, 24–26, 291–92
Transducer, neural. See Neural transducer
Transfer admittance, 39–40, 124
Transformer, analog, 134–37
Transient response, 24–26, 67
Transmission coefficient, 98
Transmission of sound through interface, 97–99
Trigonometric functions, 13, 304–6
Tubes. See Pipes
Turbulence, 265–66
Tymbal, 87, 242–43
Tympanic membrane. See Tympanum
Tympanum
  acoustic impedance, 144
  analog circuit, 143–47
  higher modes, 145
  matching to cochlea, 225–28
  neural transducer load, 146
  resonance frequency, 157–77, 204–12

Units, SI, 11, 312–13

Valve, vocal
  acoustic conductance, 253–54
  configurations, 247–48
Van der Pol equation, 235
Velocity, acoustic particle
  near dipole, 123–24
  in plane wave, 94
  near pulsating sphere, 111
  in spherical wave, 101, 111, 123
  near vibrating disc, 117
Velocity sensor, 58–63
Vibrator. See Oscillator
Viscosity
  and boundary layers, 54–56, 120
  and damping, 14, 54–57, 182
  kinematic, 54
Viscous drag
  on cylinder, 56–57
  on hair, 57
  on oscillator, 14, 23, 27
  on sphere, 56
Vocal folds, 248–49
Vocal system
  avian, 258–64
  human, 247–58, 264–65
Vortex street, 266

Wall losses

Wall losses (*cont.*)
  in horns, 189
  in pipes, 138–39, 182–83
Water
  attenuation of sound, 100
  sound speed, 92
  surface waves on, 103–4
  viscosity, 54
Water-gauge pressure,
Wave equation
  for acoustic waves, plane, 91–92
  for acoustic waves, spherical, 100–102
  for bar, 42
  for membrane, 71, 73
  for string, 31–33
Wave impedance
  of air, 94, 313
  definition, 94
  and reflection/transmission, 97–99
  in spherical wave, 102
  of water, 94, 313
Wave number, 93
Wave speed. *See also* Sound speed
  on bar, 43
  on basilar membrane, 223
  in fluid medium, 92
  on membrane, 71

Wave speed (*cont.*)
  on solid surface, 104–5
  on string, 33
  on water surface, 103
Wavefront curvature, 188
Wavelength, 93
Waves
  acoustic, plane, 90–100
  acoustic, spherical, 100–102
  attenuation, 99–100
  on basilar membrane, 219, 221–25
  capillary, on water, 103
  in cochlea, 221–25
  gravity, on water, 103
  in horns, 188
  Love, 105
  in pipes, 180
  Rayleigh, 105
  spreading of, 101–2, 104
  surface, on solids, 104–5
  surface, on water, 103–4
Webster equation, 186
Whistle, 265–70
Wing beats, sound generation by, 101, 119

Young's modulus, 32, 42, 250, 313